MECHANICS AND PHYSICS OF BUBBLES IN LIQUIDS

MECHANICS AND PHYSICS OF BUBBLES IN LIQUIDS

Proceedings IUTAM Symposium, held in Pasadena, California, 15–19 June 1981

edited by
L. VAN WIJNGAARDEN

Reprinted from *Applied Scientific Research*, vol. 38 (1982)

1982

MARTINUS NIJHOFF PUBLISHERS
THE HAGUE / BOSTON / LONDON

MECHANICS AND PHYSICS OF BUBBLES IN LIQUIDS

Proceedings IUTAM Symposium, held in Pasadena, California, 15–19 June 1981

Distributors:

for the United States and Canada

Kluwer Boston, Inc.
190 Old Derby Street
Hingham, MA 02043
USA

for all other countries

Kluwer Academic Publishers Group
Distribution Center
P.O. Box 322
3300 AH Dordrecht
The Netherlands

Library of Congress Cataloging in Publication Data CIP

Main entry under title:

Mechanics and physics of bubbles in liquids.

1. Bubbles--Congresses. 2. Liquids--Congresses.
3. Biological physics--Congresses. I. Wijngaarden,
L. van. II. International Union of Theoretical
and Applied Mechanics. III. IUTAM Symposium on
the Mechanics and Physics of Bubbles in Liquids
(1981 : Pasadena, Calif.)
QC183.M497 532'.6 81-22379
AACR2

ISBN-13: 978-94-009-7534-7 e-ISBN-13: 978-94-009-7532-3
DOI: 10.1007/ 978-94-009-7532-3

Contents

Part Four: Bubble dynamics

Part Five: Collective phenomena in bubbly liquids

Part Six: Physico-chemical effects

Preface

A IUTAM (International Union of Theoretical and Applied Mechanics) Symposium 'Mechanics and Physics of Bubbles in Liquids' was held at Pasadena, Calif., USA from 15 through 19 June 1981.

The present volume contains the printed version of nearly all papers read at the Symposium. The study of the behaviour of bubbles in liquids was originally stimulated by problems in cavitation and in boiling of liquids. Today research is initiated by problems in many other fields as well. In this respect a growing interest from the side of biomechanics may be mentioned.

Ordering of the papers could be done either according to the various mechanical and physical aspects of the subject or according to the fields of application. The presentaton at the Symposium contained a bit of both; there was a session on physico-chemical aspects for example and also a session on biological applications. The subdivision in this volume follows roughly the sessions in the Symposium. Most of them start with a paper of a survey nature, reporting progress made in recent years. Here, as in other fields of engineering science, one notes the important part played by experimental techniques and by numerical analysis.

As a whole the collection of papers takes stock of the present state of the subject. I hope that this volume, apart from serving the specialist, may also be of use to those wishing to obtain an overview of what is going on in this fascinating field. During the preparation of these Proceedings it was a pleasure to work with the staff of Martinus Nijhoff Publishers.

I wish to acknowledge the valuable help of my collaborator A. Biesheuvel.

L. van Wijngaarden

List of participants

IUTAM SYMPOSIUM on the Mechanics and Physics of Bubbles in Liquids, held in Pasadena, Calif., 15–19 June 1981

Acosta, Allan J.,
Mechanical Engineering, 104–44
California Institute of Technology
Pasadena, CA 91125 USA.

Akulichev, Victor A.,
Pacific Oceanological Institute
Far East Science Center
Academy of Sciences of the USSR
7 Radio Street
Vladivostok 690032, U.S.S.R.

Apfel, Robert E.,
Mason Laboratory
Yale University
P.O. Box 2159
New Haven, CT 06520 USA.

Bernier, Robert,
Mechanical Engineering, 104–44
California Institute of Technology
Pasadena, CA 91125 USA.

Bjørnø, Leif,
The Acoustic Laboratory
Technical University of Denmark
Building 352
DK-2800 Lyngby, Denmark.

Boubee, Philippee,
Commissariat A L'Energie Atomique
Centre D'Etudes de Limeil
Boîte Postale 27
94190 Villeneuve St. Georges, France.

Bouré, Jean A.,
Service des Transferts Thermiques
Centre d'Etudes Nucléaires de Grenoble
Boîte Postale 85X
F-38041 Grenoble Cedex, France.

Brennen, Christopher E.,
Mechanical Engineering, 104–44
California Institute of Technology
Pasadena, CA 91125 USA.

Catton, Ivan,
Department of Chemical, Nuclear and
Thermal Engineering
2567 Boelter Hall
Univesity of California, Los Angeles
Los Angeles, CA 90024 USA.

Chahine, L. Georges,
Hydronautics Incorporated
7210 Pindell School Road
Howard County
Laurel, MD 20810 USA.

Chen, Yi-Shung,
U.S. Nuclear Regulatory Commission
ACRS Fellow
California Institute of Technology,
104–44
Pasadena, CA 91125 USA.

Chesters, Allan K.
Laboratory for Aero- & Hydrodynamics
Delft University of Technology
Rotterdamseweg 145
Delft, The Netherlands.

Cole, Robert,
Department of Chemical Engineering
Clarkson College of Technology
Potsdam, NY 13676 USA.

Cooper, Michael G.,
Department of Engineering Science
Oxford University, Parks Road
Oxford OX1 3PJ, England.

Coutanceau, Madeleine,
Laboratoire de Mécanique des Fluides
Université de Poitiers
40, Avenue du Recteur Pineau
86022 Poitiers Cedex, France.

Crum, Lawrence A.,
Department of Physics and Astronomy
University of Mississippi
University, MS 38677 USA.

Dhir, Vijay K.,
Department of Chemical, Nuclear and Thermal Engineering
5405 Boelter Hall
University of California, Los Angeles
Los Angeles, CA 90024 USA.

Duffey, Romney, V.,
Electric Power Research Institute
3412 Hillview Avenue
Palo Alto, CA 94303 USA.

Ellis, Albert T.
University of California, San Diego
Department of Applied Mechanics and Engineering Sciences
Code B-010
La Jolla, CA 92093 USA.

Fanelli, Michele,
Ente Nazionale per l'Energia Elettrica DSR-CRIS
via Ornata nr. 90/14
20162 Milano, Italy.

Fujikawa, Shigeo,
Department of Mechanical Engineering
Faculty of Engineering
Kyoto University
Yoshida-hon-machi
Sakyo-ku, Kyoto City, 606 Japan.

Gibson, Donald C.,
Csiro, Division of Energy Technology
P.O. Box 26, Highett
Victoria 3190 Australia.

Harper, John F.,
Mathematics Department
Victoria University
Wellington, New Zealand.

Hentschel, Werner,
Drittes Physikalisches Institut
Universität Göttingen
Bürgerstraße 42–44
D-3400 Göttingen 1
FRG.

Hsieh, Din-Yu,
Division of Applied Mathematics
Brown University
Providence, RI 02912 USA.

Inge, Claes,
Royal Institute of Technology
Institute for Mechanics
10044 Stockholm, Sweden.

Jacobi, Nathan,
Jet Propulsion Laboratory, 169–327
California Institute of Technology
4800 Oak Grove Drive
Pasadena, CA 91109 USA.

Kezios, Peter,
Department of Chemical Engineering
Princeton University
Princeton, NJ 08544 USA.

Kieffer, Susan W.,
U.S. Department of the Interior
Geological Survey
2255 North Germini Drive
Flagstaff, AZ 86001 USA.

Koffman, Larry,
School of Mechanical Engineering
Georgia Institute of Technology
Atlanta, GA 30332 USA.

Kuiper, Gert,
Netherlands Ship Model Basin
Haagsteeg 2
6700 AA Wageningen, The Netherlands.

Labernede, Francis,
Commissariat A L'Energie Atomique
Centre d'Etudes de Limeil
Boîte Postale 27
94190 Villeneuve St. Georges, France.

Langley, Dean S.,
Physics Department
Washington State University
Pullman, WA 99164 USA.

Lauterborn, Werner,
Drittes Physikalisches Institut
Universität Göttingen
Bürgerstraße 42–44
D-3400 Göttingen 1, FRG.

Lewin, Peter A.,
The Acoustics Laboratory
Technical University of Denmark
Building 404
DK-2800 Lyngby, Denmark.

Marston, Philip L.,
Department of Physics
Washington State University
Pullman, WA 99164 USA.

Martin, Wallace W.,
Department of Mechanical Engineering
University of Toronto
5 King's College Road
Toronto M5S 1A4 Canada.

Mørch, Knud Aage,
Laboratory of Applied Physics I
Technical University of Denmark
Building 307, Lundtoftevej 100
DK-2800 Lyngby, Denmark.

Nigmatulin, R. I.,
M. V. Lomonosov Institute of Mechanics
Moscow State University
Moscow V-234 U.S.S.R.

Nyborg, Wesley L.,
Department of Physics
Cook Physical Science Building
University of Vermont
Burlington, VT 05405 USA.

Oldenziel, Daniel,
Delft Hydraulics Laboratory
Rotterdamseweg 185
Delft, The Netherlands.

Pham, D. T.,
Centre d'Etudes Nucléaires de Grenoble
Service des Transferts Thermiques
Boîte Postale 85X
F-38041 Grenoble Cedex, France.

Plesset, Milton S.,
Department of Engineering Science, 104–44
California Institute of Technology
Pasadena, CA 91125 USA.

Prosperetti, Andrea,
Istituto di Fisica
Via Celoria 16
20133 Milano, Italy.

Rath, Hans Josef,
Institute of Mechanics
University of Hannover
Appelstr. 11, 3000 Hannover, FRG.

Sadhal, Satwindar-Singh,
Department of Mechanical Engineering and Applied Mechanics
University of California
Los Angeles, CA 90007 USA.

Shepherd, J. E.,
Sandia Laboratories
Division 2513
Albuquerque, NM 87185 USA.

Shima, Akira,
Institute of High Speed Mechanics
Tohoku University
Sendai, Japan.

Starrett, J.,
Department of Applied Mechanics and Engineering Sciences
Code B-010
University of California, San Diego
La Jolla, CA 92093 USA.

Sturtevant, Bradford,
Graduate Aeronautical Laboratories, 301–46
California Institute of Technology
Pasadena, CA 91125 USA.

Trinh, Eugene,
Jet Propulsion Laboratory, 183–901
California Institute of Technology
4800 Oak Grove Drive
Pasadena, CA 91109 USA.

Tulin, Marshall P.,
Hydronautics Incorporated
7210 Pindell School Road
Laurel, MD 20810 USA.

Van Beek, P.,
Department of Mathematics
Delft University of Technology
Julianalaan 132
Delft, The Netherlands.

Van Wijngaarden, Leen,
Technische Hogeschool Twente
P.O. Box 217
Enschede, The Netherlands.

Vaughan, Patrick W.,
Department of Medical Physics
Royal Postgraduate Medical School
Hammersmith Hospital
Ducane Road, London W12 OHS
England.

Wu, Theodore Y.,
Dept. of Engineering Science, 104–44
California Institute of Technology
Pasadena, CA 91125 USA.

Yates, George T.,
Dept. of Engineering Science, 104–44
California Institute of Technology
Pasadena, CA 91125 USA.

Yount, David E.,
Department of Physics and Astronomy
University of Hawaii
Honolulu, HI 96822 USA.

Zuber, Novak,
U.S. Nuclear Regulatory Commission
Room 1222, Mail Code 1130SS
Washington, DC 20555 USA.

PART ONE

Biological effects

Biophysical implications of bubble dynamics

WESLEY L. NYBORG and DOUGLAS L. MILLER

Physics Department, University of Vermont, Burlington, VT 05405, USA

Abstract. Evidence is reviewed, from theory and experiment, that biological systems can be affected by ultrasound at low levels, if resonant gas bodies are present. In a suspension of cells or other particles a pulsating gas bubble causes the particles to migrate toward its surface via radiation force. This motion, in addition to acoustic microstreaming, transports particles into the bubble near-field where they are subjected to highly localized stress fields. In plant leaves containing gas-filled channels, the ultrasonic intensity required to produce cell death varies with frequency, showing minima in ranges corresponding roughly to calculated frequencies for resonance of the channels.

Introduction

In an earlier paper [7] examples of bubble behaviour were discussed, which are observed when sound is applied to suspensions of biological cells, or other biosystems; the frequencies were mainly in the upper audible and lower ultrasonic range. In this lecture we review findings in the last few years which extend the observations to frequencies in the 'megahertz range', from about 0.5 MHz to 5 MHz, a range typical of medical ultrasound. A motivation for research at megahertz frequencies is to obtain information relating to questions of safety and effectiveness of medical diagnostic and therapeutic ultrasound.

Of special interest are consequences of bubble activity which occurs under conditions characteristic of diagnostic ultrasound, since this modality is now used very widely in medical practice. For continuous mode operation the pressure amplitude is often less than 0.1 bar but sometimes it is as high as 1.0 bar, or higher.

In pulse-echo operation the ultrasound is generated in a continuous series of pulses, the pulse-duration and repetition period being roughly one microsecond and one millisecond, respectively. During each short pulse the pressure amplitude may be less than 1.0 bar but frequently reaches 10 bar, and sometimes exceeds this.

In applications of ultrasound to physical therapy continuous sound (sometimes modulated) is applied, with spatial average intensity about 1–3 W/cm^3, the pressure amplitude reaching up to 5 bars at spatial maxima.

In therapeutic applications the ultrasound produces observable changes in the patient (as is, of course, intended) whereas there have been no reports of effects resulting from examinations with diagnostic ultrasound. Nevertheless diagnostic ultrasound seems to be viewed with the greatest concern, very likely because it is used so widely in examining pregnancies.

Applied Scientific Research 38: 17–24 (1982) 0003–6994/82/0381–0017 $01.20.

It cannot be excluded theoretically that diagnostic ultrasound might have biological consequences. Available evidence indicates that these consequences will depend critically on the extent to which small gaseous bodies of appropriate size are present in the human body. This latter topic is discussed elsewhere in this volume by Lewin and Bjørnø, as is also the likelihood that these will grow by rectified diffusion. When the pressure amplitude is no more than a few tenths of a bar, only microbubbles of near-resonant size (a few microns in diameter at low megahertz frequencies) are likely to perturb the biological system very much.

Porous membranes

In one method for obtaining biophysical evidence on the significance of small gaseous bodies, use is made of hydrophobic membranes perforated with cylindrical holes a few microns in diameter. (Prepared by Nuclepore Corporation, Pleasonton, CA.) When these are immersed in water the air remains in many of the holes. In a sound field, each air-filled hole becomes the site of acitvity which can be studied by various means.

Especially useful is an arrangement involving a modified optical microscope through which specimens can be viewed and photographed during exposure to ultrasound. With this microscope many studies have been made in which small strips of the perforated hydrophobic membrane are immersed in a suspension of biological cells or other particles and viewed during sonation at a frequency of 1 MHz or higher. When this is done one usually finds that many of the holes are sites of particle movements of several kinds; these include migration toward the hole as well as circulatory and rotational movements. A consequence of the movement toward the hole is that a collection of particles forms there while part of the surrounding region is cleared of particles. Typical results are seen in Figure 1. Here a quantity of human blood diluted in physiological saline was subjected to low amplitude ultrasound at a frequency of 2.1 MHz in the presence of perforated membrane (about 720 holes per mm^2 with diameter 3.4 μm). Before the sonation, many of the cells had settled onto the membrane, forming a continuous layer in which they were fairly uniformly distributed. During sonation the cells moved, evidently in response to vibration of gas bodies contained in the holes; in a short time they were redistributed as shown in Figure 1a. Each active hole is surrounded by a small cluster of cells (see example in Figure 1b) beyond which is an area cleared of cells.

For explanation of movements which lead to cellular distributions such as those seen in Figure 1, we turn first to theory for acoustic radiation forces. Near an active hole, if the contained bubble is resonant or nearly so, the acoustic field is dominated by a secondary field (approximately a spherical wave) radiated by the bubble. At points near the hole, whose distance r from the center is small compared to the compressional wavelength, the oscillatory

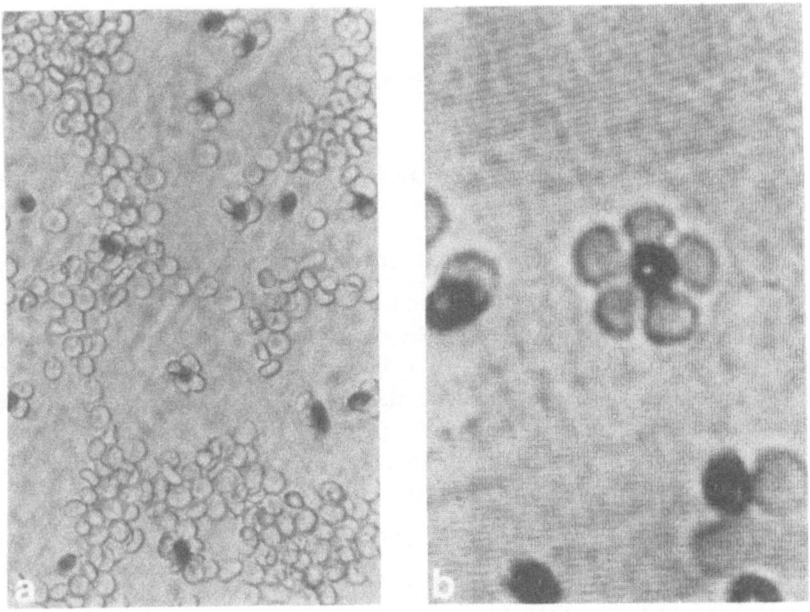

Figure 1(a). Distribution of red cells on a membrane with gas-filled holes after ultrasound is applied. Viewing is with transmitted light and holes appear black. Not all holes are 'active'.
Figure 1(b). Cluster of five red cells around an active hole.

motion of the liquid is nearly the same as if it were incompressible. Thus the velocity in this 'near field' region is nearly radial and, at any instant, falls off rapidly with distance. For a small particle of volume V in suspension the radiation force is given, in the near field of a hole, by the vector [2]

$$\mathbf{F} = DV\nabla\langle KE\rangle; \qquad (1)$$

here $\langle KE\rangle$ is the time-averaged kinetic-energy density in the acoustic field, '∇' is the gradient operator, and D is a nondimensional factor given by $3(\rho-\rho_0)/(2\rho+\rho_0)$, where ρ is the density of the particle and ρ_0 that of the surrounding liquid. (More generally, the expression for $\mathbf{F}$ in equation 1 should involve the potential-energy density but this term is negligible in the near field of a bubble.) Letting the oscillatory velocity of the liquid be u the quantity $\langle KE\rangle$ is equal to $\langle \rho u^2/2\rangle$.

To obtain a convenient expression for u, and thus for the radiation force, we avoid consideration of the detailed velocity distribution near a pore and, instead treat a simplified model which possesses the main features of the actual situation. Specifically, we consider the gaseous oscillator in the pore to be equivalent, for this purpose, to a spherical bubble of radius R_0. In the near field we then assume u given by UR_0^2/r^2, where U is the velocity (varying sinusoidally with time) at $r = R_0$. Then $\mathbf{F}$ is found to be directed toward the origin with magnitude varying inversely with r^5. This radiation force causes the particle to migrate towards the center with speed v which is proportional to $\mathbf{F}$. A resulting expression for the centripetal speed v can be written

$$v = v_0(R_0/r)^5. \tag{2a}$$

Here v_0 is the migration speed at $r = R_0$ (in the liquid just outside the bubble surface) and is given by

$$v_0 = sd\omega^2\epsilon_0^2/R_0, \tag{2b}$$

where ϵ_0 is the displacement amplitude of the bubble surface, d is equal to $3\rho_0/(2\rho + \rho_0)$ and s is the sedimentation constant used in biophysics to characterize particles. Specifically s is defined, for a particle falling due to gravity through a viscous liquid, as the ratio of its terminal speed to the gravitational constant g. We find from equation (2b) that the speed v_0 exceeds the speed of fall in a gravitational field (sg) by a factor $G.F.$, the 'g-factor', which can be written in terms of the ratio (ϵ_0/R_0):

$$G.F. = v_0/sg = (\omega^2 R_0 d/g)(\epsilon_0/R_0)^2. \tag{3}$$

To obtain a representative value for $G.F.$, we set (ϵ_0/R_0) equal to 0.1, a condition which can be achieved by using a driving sound field of very modest pressure amplitude. Suppose also that the frequency is 2 MHz, R_0 is 2 μm and $d \approx 1$. Then we find for $G.F.$ the approximate value 300,000, a g-factor characteristic of high speed centrifuges.

For proteins in the blood s is probably 10^{-13} sec or greater; hence it follows from the above results that a protein molecule near the vibrating bubble in blood will migrate toward it at a speed $(sg)(G.F.)$ of, at least, 300 nm/sec. This motion will be superposed on diffusion flow if concentration gradients exist. (So far as we know such motion of macromolecules toward vibrating bubbles has not been observed.)

Sedimentation coefficients can be estimated from a well-known expression derived for a sphere of radius a and density ρ falling through a liquid of density ρ_0 and shear viscosity coefficient η (assuming Stokes drag):

$$s = 2a^2(\rho - \rho_0)/9\eta. \tag{4}$$

From this one finds that for a red cell in blood s is of the order of 10^{-7} sec. However for particles of this size it is necessary to consider larger distances r and hence recall, from equation 2, that v decreases rapidly with increasing r. Letting $v = \mathrm{d}r/\mathrm{d}t$ in equation 2 and integrating, one finds that the time required for a particle to migrate from $r = R$ to a general point r is given by

$$bt = R^6 - r^6, \tag{5}$$

where $b = 6v_0R_0^5$. Letting $r = R_0$ and $R = R_l$ the equation gives the time for all particles within a sphere of radius R_l to collect at the bubble surface. If R_l is appreciably greater than R_0, equation 5 becomes $bt = R_l^6$.

This theory can be used to estimate how fast particles accumulate at a bubble. The number N of particles originally in the space between R_0 and R_l is approximately $4\pi\rho^* R^3/3$, where ρ^* is the (original) number of particles per unit volume. It is then found that the number N of particles collected at the bubble in time t is

$$N = Bt^{1/2}, \tag{6}$$

where $B = (4\pi\rho^*/3)b^{1/2}$. Another result is obtained by dividing through by R_l^6 in equation 5 to obtain

$$\left(\frac{r}{R_l}\right)^6 = \left(\frac{R}{R_l}\right)^6 + \left(\frac{R_0}{R_l}\right)^6 - 1; \tag{7}$$

here t was eliminated by requiring that a particle originally at R_l arrives at R_0 in time t. Letting R_0/R_l be 0.2 values of r/R_l were calculated for a few values of R/R_l with results shown in Table 1. These reflect

Table 1. Plot of equation 7. $R_0/R_l = 0.2$

R/R_l	1	1.10	1.12	1.20	1.30	1.40
r/R_l	0.2	0.96	1.00	1.12	1.25	1.37
$(R-r)/R_l$	0.8	0.14	0.12	0.08	0.05	0.03

the rapid fall-off of migration speed with distance. While a particle at R_l moves a distance $(R-r) = 0.8\,R_l$ in time t, a particle at $1.2R_l$ moves only one-tenth this distance. A consequence of this is a 'clearing' or 'thinning out' of the space between R_0 and R_l (see Figure 1a). As a result of the bubble vibration during the time t, none of the original particles in this space remain there, and only a few (those originally between R_l and $1.12R_l$) have moved in to take their place.

It is found from theory and experiment that radiation forces produce particle migration near vibrating bubbles even at low pressure amplitudes such as are characteristic of diagnostic ultrasound. With the aid of this migration, cells and other biological particles are brought into the immediate neighbourhood of vibrating bubbles and hence are subject to fields of mechanical stress which exist there.

Another important contribution to the motion of particles near a vibrating bubble is of a different character, where the particles are carried along in a steady circulatory flow of the liquid. The liquid flow is called 'acoustic streaming' or, because of its small scale, 'microstreaming'. Approximate theory has been given for the flow [2] but further development of the theory is required for application to microstreaming near micron-sized bubbles vibrating at frequencies in the megahertz range.

When the cells are brought into the vicinity of vibrating bubbles they are subjected to highly localized and inhomogeneous stress fields which exist there. These stresses are, at least, partly in the form of viscous shear stresses associated with the microstreaming, such as are invoked in explaining results of experiments at lower frequencies. Theoretical and experimental work with shearing flow generated hydro-dynamically has shown that cells and large molecules are deformed, sometimes irreversibly, by the viscous stress. In addition, biological cells near a vibrating bubble are acted on by nonuniform fields of radiation torque and radiation pressure which also probably tend to bring about deformation and rupture of delicate structures [8].

Whatever the detailed mechanisms may be, it has been shown that cells of the human blood (platelets, red cells and white cells) are affected by vibrating bubbles under conditions typical of diagnostic ultrasound. Thus platelets undergo a release reaction accompanied by clumping [4] while all of the blood cells show release of adenosine triphosphate [9] when exposed to megahertz ultrasound in the presence of perforated membrane (containing micron-sized gas bodies) at pressure amplitudes of the order of 0.3 bar or lower.

Plant tissues and insects

Another source of information on the biological significance of small gas bodies is from studies involving plant tissues. Many of these incorporate stable gas-filled channels in spaces between the cells. Leaves of the water plant Elodea have proved especially convenient for biophysical studies. Theory for the resonance frequencies of the channels has been derived and applied to two groups of channels found in the leaf, taking into account their dimensions, the elasticity of the bounding cell walls and the tension in the walls produced by turgor pressure. As shown by the horizontal bars near the frequency axis in Figure 2, the calculated resonance frequency ranges for the two groups of channels wire 0.55–1.0 MHz and 2.0–3.7 MHz, respectively [6].

On the same leaves experimental studies were carried out in which they were exposed to ultrasound of various intensities and frequencies and threshold intensities determined (such that intensities above threshold values usually lead to cell death). The solid curve shows the threshold vs frequency and exhibits minima approximately in the predicted frequency ranges. At these minima the threshold intensity (spatial peak) is only 75 mw/cm^2 and 300 mw/cm^2, corresponding to pressure amplitudes (in a travelling wave) of about 0.5 bar and 1 bar, respectively. In other experiments the intercellular gas was removed (by centrifugation) and the thresholds were found to be much higher. These experiments show that even in a tissue vibration of gas bodies can be biologically significant.

Other investigations were done with Elodea using pulsed 1 MHz ultrasound. [4] For single bursts the intensity thresholds were proportional to the exposure time raised to the power -0.29 for times greater than 1 msec. For

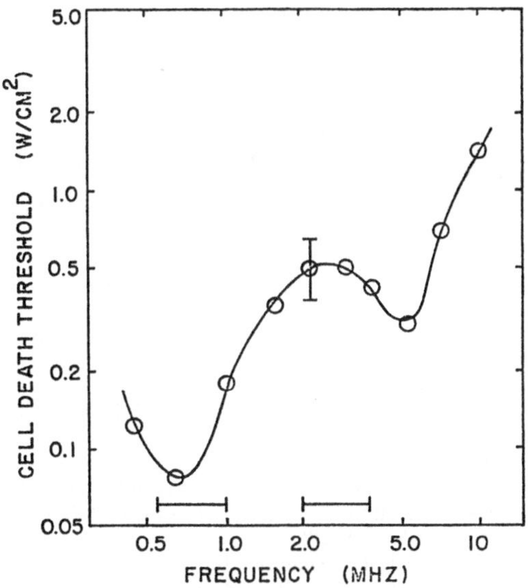

Figure 2. Cell death thresholds for 100 sec exposures of cells near the midrib of Elodea leaves. Horizontal bars indicate the theoretical ranges of resonance frequencies for gas-filled channels which adjoined these cells (from [6]).

10 μs pulses repeated every 10 ms the cell death threshold was about 6 bar for the pressure amplitude during the pulse.

Other investigators [1] have recently found insect larvae, with their microscopic gas-containing respiratory channels, to be valuable models for studying the significance of gas contained in tissues. They report that larvae of *Drosophila melanogaster* are affected by 1 μs pulses of 2 MHz ultrasound repeated every millisecond when the pressure amplitude exceeds about 6 bar. These are conditions typical of pulse-echo diagnostic ultrasound. That gas bubbles should respond to such short pulses is consistent with analysis by Lewin [3].

Conclusion

Evidence now exists that ultrasound with characteristics like those used in medical practice can produce significant biological effects when resonant gas bodies are present. This adds considerable interest to research now proceeding to determine the distribution of such gas bodies in animal (especially, human) tissue.

Acknowledgment

Research supported by the National Institutes of Health via GM-08209.

References

1. Child SZ, Carstensen EL and Lam SK (1981) Effects of ultrasound on Drosophila: III. Exposure of larvae to low-temporal-average-intensity, pulsed irradiation. Ultrasound in Med & Biol 7: 167–173.
2. Coakley WT and Nyborg WL (1978) Cavitation; dynamics of gas bubbles; applications. In Fry FJ (ed) Ultrasound: Its Application in Medicine and Biology, New York: Elsevier. Also, pp 1–75 in the same book: Nyborg WL, Physical principles of ultrasound.
3. Lewin PA (1978) Ultrasound induced damage of biological tissue. Report AFM 78–16 (Ph.D. thesis). Department of Fluid Mechanics, Technical University of Denmark.
4. Miller DL (1977) The effects of ultrasonic activation of gas bodies in Elodea leaves during continuous and pulsed irradiation at 1 MHz. Ultrasound in Med & Biol 3: 221–240.
5. Miller DL, Nyborg WL and Whitcomb CC (1979) In vitro clumping of platelets exposed to low intensity ultrasound.
6. Miller DL (1979) Cell death thresholds in Elodea for 0.45–10 MHz ultrasound compared to gas-body resonance theory. Ultrasound in Med & Biol 5: 351–357.
7. Nyborg WL (1974) Cavitation in biological systems. In Bjørnø (ed) Finite Amplitude Effects in Fluids, pp 245–251 (IPC Press).
8. Nyborg WL (1977) Physical mechanisms for biological effects of ultrasound. HEW Publication (FDA) 78-8062. Rockville, Md: Bureau of Radiological Health.
9. Williams AR and Miller DL (1980) Photometric detection of ATP release from human erythrocytes exposed to ultrasonically activated gas-filled pores. Ultrasound in Med & Biol 6: 251–256.

Thresholds for rectified diffusion and acoustic microstreaming by bubbles in biological tissue

PETER A. LEWIN and LEIF BJØRNØ

Technical University of Denmark, DK-2800 Lyngby, Denmark

Abstract. Results are presented of the calculation of the thresholds (in terms of peak acoustic pressure as a function of frequency of the incident ultrasonic wave), at which rectified diffusion may begin in environments represented by the liquid/cell structure of biological tissue, exposed to ultrasonic frequencies, 1–4 MHz, typical for clinical devices. Computations based on excitation by peak pressure amplitude values typical for continuous and pulse-echo diagnostic devices, suggests that rectified diffusion is unlikely to occur for the latter only. Acoustically induced shear stresses, caused by bubble pulsation produced microstreaming and affecting the integrity of cellular membranes, are evaluated and are found to lie above levels at which biological effects have been observed.

Introduction

The vigorously increasing medical uses of ultrasound lead to a continuing concern for the investigation of biological effects of the ultrasonic waves used. The physical mechanisms by which damage is believed to occur have been well reviewed and available literature data for the division between exposure levels that do or do not produce significant bio-effects have been compiled [12]. Recently, an in-depth review on biological effects of ultrasound including list of reports that have appeared in scientific journals up to the end of 1980, has been compiled [13].

One of the mechanisms often suggested for the biological action of ultrasonic beams irradiating human tissues is concerned with the presence in the tissue of minute gaseous bubbles which may, under the influence of the ultrasonic field, be stimulated to grow (due to rectified diffusion) to a size at which resonance or collapse occurs with associated severe shear stresses caused by bubble pulsation produced oscillatory and time independent flow (microstreaming).

A fairly comprehensive survey of the evidence for the existence of microbubbles in tissues has been given recently [7] from which it appears that the bubble sizes are most probably in the range from 0.1–5 μm. Very recently, ter Haar and her associates [15] have reported the existence of such microbubbles of some microns in diameter during ultrasonic irradiation in vivo.

It is possible that the bubbles under the influence of the ultrasonic field may be induced to grow to their resonant size by a process of rectified diffusion. The radii of resonant bubbles were calculated using [5] with insertion of characteristic values for surface tension, density, shear viscosity, etc., for bubbles in tissue [7]. Since it may be expected that an increased

Applied Scientific Research 38: 25–35 (1982) 0003–6994/82/0381–0025 $01.65.

number of pulsating resonant bubbles (stable cavitation) will result in reinforced biological effect, it is important to attempt to identify the acoustic conditions under which rectified diffusion may be induced.

1. Acoustic thresholds for the occurrence of rectified diffusion

The physical phenomena of rectified diffusion together with the mathematical theory describing the process has been treated in detail by several authors [2, 4, 14].

An expression for the acoustic peak threshold pressure amplitude required for rectified diffusion, taking account of dissipating effects may be written as [14]:

$$P_t = \rho_l \omega_0^2 R_0^2 \left\{ \left[1 - \left(\frac{\omega_A}{\omega_0} \right)^2 \right]^2 + b^2 \left(\frac{\omega_A}{\omega_0} \right)^2 \right\}^{1/2} \frac{\left(1 - \dfrac{c_i}{c_0} + \dfrac{2\sigma}{R_0 P_0} \right)^{1/2}}{\left[6 \left(1 + \dfrac{2\sigma}{R_0 P_0} \right) \right]^{1/2}}, \tag{1}$$

where P_t is the peak threshold pressure amplitude; and where ρ_l and P_0 are the liquid density and hydrostatic pressure, respectively; ω_0 and ω_A the linear resonance and the ultrasonic frequencies, respectively, while c_0 and c_i denote the saturation concentration and the ambient initial concentration of gas in the liquid; σ and b denote the surface tension and the damping constant.

Another expression for the acoustic peak threshold pressure amplitude for rectified diffusion given in [3] writes:

$$P_t = \frac{P_0}{\epsilon} \left| \frac{1 + \dfrac{2\sigma}{R_0 P_0} - \dfrac{c_i}{c_0}}{(3 + 4K)\dfrac{c_i}{c_0} - \left(1 + \dfrac{2\sigma}{R_0 P_0} \right) K} \right|^{1/2}, \tag{2}$$

where

$$\frac{1}{\epsilon} = \frac{\rho_l R_0^2}{P_0} \left[(\omega_0^2 - \omega_A^2)^2 + (\omega_A \cdot \omega_0 \cdot b)^2 \right]^{1/2}$$

and

$$K = \frac{1 - \frac{\rho_l \omega_A^2 R_0^2}{12P_0} + \frac{5\sigma}{3P_0R_0}}{1 + \frac{4\sigma}{3P_0R_0}}.$$

Equations (1) and (2) are both based upon the same physical model.

Both expressions are valid for small pressure amplitudes and have been derived for continuous wave pressure fields. The main difference between the two expressions is the correction factor K to account for the possible nonlinear character of the assumed radial pulsation of the bubble [3]. Equations (1) and (2) have been used to calculate the thresholds of acoustic pressure at which rectified diffusion may be induced for different relative concentration (c_i/c_0) of dissolved gas, and different relative frequencies f_A/f_0, see Figures 1A and 1B. The thresholds calculated are plotted in Figures 1A and 1B. The calculations were performed using the values of liquid properties appropriate for blood at atmospheric pressure ($P_0 = 10^5$ Pa) and for equilibrium bubble radii (R_0) of 1, 2 and 3.5 μm.

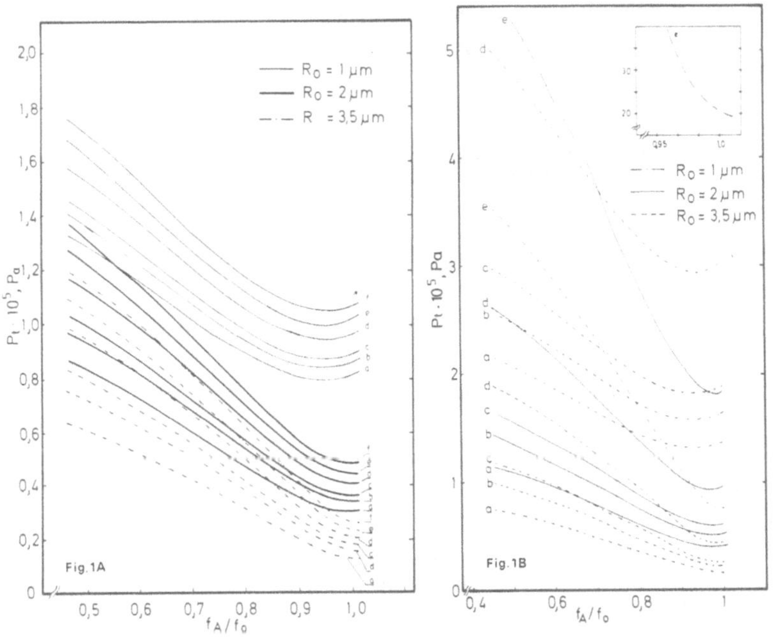

Figure 1. Peak pressure thresholds P_t for rectified diffusion calculated according to: (A) equation 1 and (B) equation 2 for different relative frequencies f_A/f_0 and concentrations c_i/c_0: (a) $c_i/c_0 = 1$, (b) $c_i/c_0 = 0.85$, (c) $c_i/c_0 = 0.75$, (d) $c_i/c_0 = 0.5$, (e) $c_i/c_0 = 0.3$ and (f) $c_i/c_0 = 0.1$.

Inspection of graphs of Figures 1A and 1B shows that P_t increases as the gas concentration in the liquid decreases. On the other hand P_t decreases as the frequency of the applied sound field approaches the resonant frequency of the bubble.

Further comparison of the figures reveals rather large discrepancies in the predicted thresholds, depending on the equation used for the calculation. While equation (1) leads to the lower (i.e. worst) resonant rectified diffusion thresholds (approximately $0.8 \cdot 10^5$ Pa, $0.3 \cdot 10^5$ Pa and $0.1 \cdot 10^5$ Pa for bubbles of 1, 2 and 3.5 microns respectively), equation (2) yields higher values.

The only experimentally based information on threshold for rectified diffusion in water for bubbles resonant in the low megahertz range seems to be given in [9] where an approximate agreement is obtained with both equations (1) and (2) for c_i/c_0 in the range 0.4–0.6 and for $\gamma = 1.2$. Although the experiment reported was performed under conditions which are not fully adequate for immediate comparison the conclusion in [9] is consistent with the results of the present work, that the threshold for rectified diffusion according to equation (1) is lower than the one obtained from expression (2) and that both thresholds decrease with increasing c_i/c_0.

It can be seen from Figures 1A and 1B that the expressions predict the lowest thresholds at frequencies which deviate slightly from the linear resonance frequencies f_0. The reason for this can be traced to the correction factor:

$$\left\{\left[1-\left(\frac{\omega_A}{\omega_0}\right)^2\right]^2+b^2\left(\frac{\omega_A}{\omega_0}\right)^2\right\}^{1/2}$$

in equations (1) and (2) which reaches a minimum value for ω_A/ω_0 slightly lower than unity.

It can also be seen that equation (2) is highly dependent on the parameter c_i/c_0 – the concentration of gas in the surrounding liquid – and ceases to be valid for $c_i/c_0 < 0.1$ as the expression under the square root sign becomes negative. For c_i/c_0 in the range 0.3 to 1.0, the difference between the thresholds given by equations (1) and (2) increases with decreasing c_i/c_0.

It should be remembered that the calculation is based on an analysis that assumes a free bubble and may not be fully applicable to the clinical situation. In the clinical situation, however, it is most likely that the bubbles are trapped in the cracks of the cell membranes. If this is the case, the surface area of the bubble which is directly exposed to the intercellular liquid is smaller, which in turn may have an influence on the rectified diffusion process and threshold. This influence, however, is difficult to estimate as no adequate theory describing the dynamic behaviour of bubbles in cracks so far exists.

It is relevant to compare the results of the calculations presented above with the levels of exposure used in medical applications. A recent survey [1] of the intensities output by different commercial diagnostic instruments suggests that the maximum output for pulse-echo equipments is of the order of 200 Watts cm^{-2} for durations of 10 μs, and for CW equipments is of the order of 375 $mW{\cdot}cm^{-2}$ (spatial and temporal average). The lowest thresholds for rectified diffusion from Figures 1A and 1B correspond to intensities of the order of 3 $mW{\cdot}cm^{-2}$ and 30 $mW{\cdot}cm^{-2}$, respectively. These are thus well below the intensities used by both pulsed and continuous wave devices, and are indeed within the exposure level quoted by AIUM [17].

It is probable that if continuous wave devices are used, rectified diffusion will take place leading to an increased number of resonant bubbles at frequencies close to 1 MHz ($R_0 = 3.5\,\mu m$) and 2 MHz ($R_0 = 2\,\mu m$) and more significant biological effects may result. For comparison, the lowest thresholds for rectified diffusion are shown in Figure 2 together with the maximum outputs from diagnostic devices and some summarising biological effects graph of [16].

As was mentioned above, the expressions (1) and (2) are valid only for steady state conditions and moderate pressures and thus may not be applicable to the pulsed mode of operation in which diagnostic scanners operate. As far as the authors are aware, no theory for rectified diffusion adequate for the case of pulse excitation of the bubble exists.

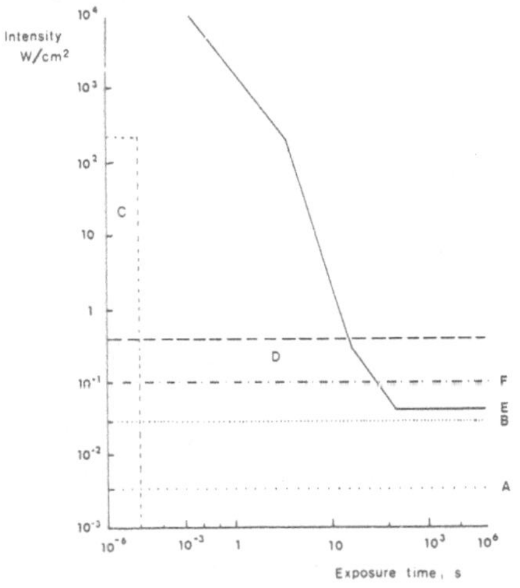

Figure 2. A comparison of the lowest thresholds for rectified diffusion from Figure 1A (line A) and Figure 1B (line B) corrected using plane wave relationships) with data on measured outputs from diagnostic ultrasonic devices [1, 6] (area C: pulse-echo, area D: continuous wave), the curve of the incidence of biological effects (curve E) [16], and the level specified in the AIUM statement (curve F) [17].

2. Calculation of the amplitude of displacement of the vibrating bubble wall

(a) Continuous excitation

If an oscillating pressure $p(t) \equiv P_A \sin \omega_A t$ is applied to a free bubble in an unbounded liquid the bubble will be set into forced radial oscillation and the motion of the bubble wall will be governed by the following second order non-linear differential equation in spherical coordinates [5]:

$$\rho_l R \frac{d^2R}{dt^2} + \frac{3}{2}\rho_l \left(\frac{dR}{dt}\right)^2 = P_i \left(\frac{R_0}{R}\right)^{3\gamma} - P_0 - \frac{2\sigma}{R}$$

$$- \frac{4\eta_l}{R}\frac{dR}{dt} - p(t), \qquad (3)$$

where $P_i = 2\sigma/R_0 + P_0$ is the pressure inside the bubble, R is the instantaneous radius and R_0 the equilibrium radius of the bubble.

Equation (3) was solved by means of numerical calculations performed on an IBM 370/165 computer. The programme permitted variation of parameters such as the intensity, the frequency, and the duty cycle of the driving sound pressure, as well as the bubble radius.

Representative examples of the steady state solution of equation (3) are shown in Figures 3(a) and (b). The curve labelled $P(t)$ is the driving sound pressure and the curve labelled $R(t)$ is the resulting resonant bubble oscillation radius-time curve. The examples depict quasi-linear (a) and highly non-linear (b) pulsations.

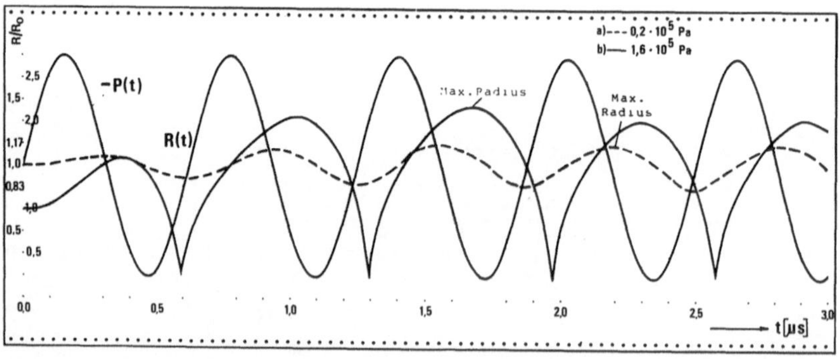

Figure 3. Examples of the steady state solution to the bubble equation of motion. $P(t)$ is the driving field pressure. The radius-time curve $R(t)$ was obtained for the following parameter set: $R_0 = 2\,\mu$m, $f_0 = 1.6$ MHz (natural frequency), $f_A = 1.6$ MHz, $\eta_l = 0.003$ Pa·s, $\sigma = 0.06$ N/m and $\rho_l = 1056$ kg/m^3. (a) Driving pressure amplitude $P_A = 0.2 \cdot 10^5$ Pa, (b) Driving pressure amplitude $P_A = 1.6 \cdot 10^5$ Pa.

The results show that for low pressure amplitudes in the steady state solution, the displacement amplitude of the bubble wall reaches its maximum value for a bubble driven at its resonance frequency. However in the case of non-linear oscillations the displacement amplitude at lower frequencies may be higher than the displacement amplitude at the resonance frequency due to the non-linear variation of the frequency response curve with excitation amplitude, which occurs at the higher excitation amplitudes.

(b) Transient excitation

A representative example of bubble response to an applied transient pressure typical of the pulses transmitted by ultrasonic diagnostic devices is shown in Figure 4. It appears that for relevant transient excitations, the bubble vibrations decay very rapidly. Thus on the basis of the solutions to the

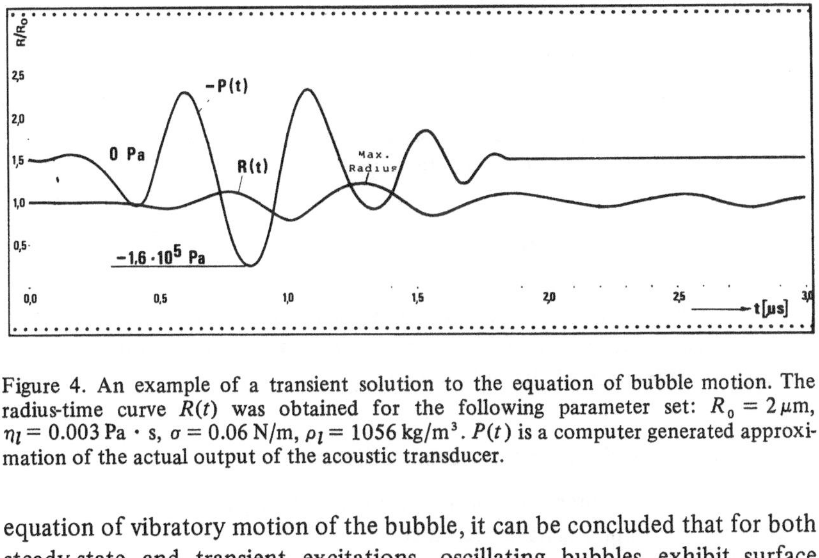

Figure 4. An example of a transient solution to the equation of bubble motion. The radius-time curve $R(t)$ was obtained for the following parameter set: $R_0 = 2\,\mu$m, $\eta_l = 0.003$ Pa · s, $\sigma = 0.06$ N/m, $\rho_l = 1056$ kg/m^3. $P(t)$ is a computer generated approximation of the actual output of the acoustic transducer.

equation of vibratory motion of the bubble, it can be concluded that for both steady-state and transient excitations, oscillating bubbles exhibit surface vibration of significant amplitude. It should be noted, that for the purposes of the calculations, the pressure amplitudes have deliberately been chosen to correspond with the levels most likely to be expected at the output of medical diagnostic equipment, i.e., in the continuous wave regime: $I = 10\,\text{mW/cm}^2$, or $P_A = 0.2 \cdot 10^5$ Pa to $I = 800\,\text{mW/cm}^2$ or $P_A = 1.6 \cdot 10^5$ Pa, and in the pulsed regime: up to $20 \cdot 10^5$ Pa [1, 6].

3. Estimation of the shear stresses

Determination of the shear stresses in the flow field around an oscillating bubble requires information on the velocity gradient in the vicinity of the bubble. Theoretical expressions for the acoustic streaming derived from the basic equations of hydrodynamics are given in [11].

For sinusoidal bubble motion the following approximate expression for the steady streaming velocity U in the boundary layer of thickness δ induced by a small hemispherical bubble of radius R_0 resting on a membrane is given by:

$$U = 2\pi f_A R_0 \left(\frac{\xi}{R_0}\right)^2 \left[1 - e^{-2(z/\delta)} - 3e^{-(z/\delta)} \sin\frac{z}{\delta}\right], \tag{4}$$

where: $2\pi f_A R_0 \left(\frac{\xi}{R_0}\right)^2 = U_m$ and $\delta = \left(\frac{\eta_l}{\pi \rho_l f_A}\right)^{1/2} = \left(\frac{2\nu}{\omega_A}\right)^{1/2}$.

ξ is the peak displacement amplitude of the bubble wall determined from the non-linear equation of bubble motion (3), U_m and δ denote the maximum streaming velocity and the acoustic boundary layer thickness, respectively, and z/δ is the dimensionless distance from the membrane.

The streaming velocity gradient $G = \frac{\partial U}{\partial z}$ changes rapidly in the vicinity of the membrane ($z = 0$). This means that a considerable shear stress $S = \eta_l G$ exists very close to the membrane and is almost insignificant for $z/\delta > 4$. A calculation of S for $z = 0$ leads to the expression:

$$S = \frac{\eta_l}{\delta} 2\pi f_A R_0 \left(\frac{\xi}{R_0}\right)^2. \tag{5}$$

This expression can be used for the evaluation of the steady state shear stresses generated in the vicinity of a vibrating resonant bubble for CW ultrasonic irradiation of the bubble field.

4. Limitations of the acoustic streaming theory

In deriving the expression for the acoustic streaming near a boundary a number of assumptions have been made:

– the pulse duration of the applied ultrasound must be much longer than the period of the applied frequency;

– the resulting expression for time-independent velocity applies only to minor amplitude variations of the acoustic pressure, as the derivation is based on perturbation theory;

– the radius of the pulsating bubble must be much greater than the thickness of the acoustic boundary layer (near boundary streaming); and

– it is assumed that the liquid is newtonian, i.e. its velocity is constant.

(a) Pulse duration

The derivation of U is based on time averaging over a large number of cycles and as far as the present authors are aware, no acoustic streaming theory fully adequate for the case of the very short, repetitive pulses exists. Therefore, for pulses which contain a small number of periods – say below 10 – the theory should be applied with caution, since it may not necessarily yield a correct value for the velocity gradient.

(b) Minor pressure amplitudes

All the fundamental expressions for the acoustic streaming theory have been derived using the general Rayleigh perturbation approach and thus require that the displacement amplitude of the bubble wall ξ is much smaller than the bubble equilibrium radius R_0 ($\xi/R_0 \ll 1$).

(c) Near boundary streaming approximations

The near boundary conditions require that the bubble radius R_0 is large compared to the acoustic boundary layer thickness δ, ($R_0/\delta \gg 1$). The R_0/δ ratios in blood have been calculated for frequencies of 1, 2 and 4 MHz, and bubble radii of 1, 2 and 3.5 μm. The results indicate that the condition $R_0/\delta \gg 1$ is hardly fulfilled, and the consequences of this are worth more detailed discussion.

(d) Viscosity

The acoustic streaming theory is developed using the assumption that the liquid considered is a newtonian liquid, such as water. For the purposes of calculating the displacement amplitude of the bubble wall, it was assumed that the blood viscosity was constant and the lowest measured value of 0.003 Pa·s was chosen [6]. However, blood can exhibit non-newtonian properties and thus its viscosity can change. Further calculations showed that a threefold increase in blood viscosity results in an approximate halving of the displacement amplitude of the bubble wall. This would lead to a decrease in the streaming velocity of a factor of four.

If all the above considerations are taken into account, it would appear that the velocity gradients calculated according to the theory [11] are of an order of magnitude too high. However, even those reduced velocity gradients lead to shear stresses which lie above levels at which biological effects have been observed.

5. Conclusion

The thresholds (in terms of peak acoustic pressure as a function of frequency of the incident ultrasonic wave) at which rectified diffusion may begin in environments represented by the liquid/cell structure of biological tissue have been calculated for both continuous wave and transient irradiations. It was shown that in the fields generated by currently available diagnostic devices operating in the continuous wave mode, the bubbles may grow into resonant bubbles in the size range 2–3.5 μm, while computation based on excitation by peak pressure amplitudes typical for pulse-echo diagnostic devices, suggests that rectified diffusion is unlikely to occur. Similarly, the steady state stresses in the vicinity of microbubbles irradiated with ultrasonic waves have been evaluated. It appeared that transient irradiation conditions are unlikely to cause biological effects, however the values for continuous wave irradiation do lie within the range in which bioeffects have been reported. The limitations of the calculation model have been carefully assessed.

The discussion presented, indicates the need for more accurate values for the physical constants of biological media forming the basis for the calculations. Moreover, it emphasizes the need for more extensive experiments on biological effects, not necessarily only those caused by the shear stresses. Investigation of the importance of another possible mechanism of interaction between ultrasound and biological tissue, among other things the possible generation of mechanical forces due to focusing effects in tissue is a topic for a separate work.

Acknowledgments

One of the authors, Peter A. Lewin, is indebted to Prof. Wesley L. Nyborg and Dr. Douglas L. Miller, Physics Department, University of Vermont, Burlington, for helpful comments on parts of the early draft of this paper.

References

1. Carson PL, Fischella PR and Oughton TL (1978) Ultrasonic power and intensities produced by diagnostic ultrasonic equipment. Ultrasound in Med and Biol 3: 341–350.
2. Eller AI and Flynn HG (1965) Rectified diffusion during nonlinear pulsations of cavitation. J Acoust Soc Am 37: 493–503.
3. Eller AI (1969) Growth of bubbles by rectified diffusion. J Acoust Soc Am 46: 1246–1250.
4. Hsieh DH and Plesset MS (1961) Theory of rectified diffusion of mass into gas bubbles. J Acoust Soc Am 33: 206–215.
5. Lauterborn W (1976) Numerical investigations of nonlinear oscillations of gas bubbles in liquids. J Acoust Soc Am 59: 283–292.
6. Lewin PA (1978) Ultrasound induced damage of biological tissue. Ph.D. Thesis, AFM 78-16. Copenhagen: Technical University of Denmark, Lyngby.
7. Lewin PA and Bjørnø L (1981) Acoustic pressure amplitude thresholds for rectified diffusion in gaseous microbubbles in biological tissue. J Acoust Soc Am 69: 846–852.
8. Miller DL (1977) The effects of ultrasonic cavitation of gas bodies in Elodea leaves during continuous and pulsed irradiation at 1 MHz. Ultrasound in Med and Biol 3: 221–240.
9. Miller DL (1977) Stable arrays of resonant bubbles in a 1 MHz standing wave acoustic field. J Acoust Soc Am 62: 12–19.
10. Nayfeh AS and Saric WS (1974) Nonlinear acoustic response of a spherical bubble. In Bjørnø (ed) Finite-Amplitude Wave Effects in Fluids, Proceedings of the 1973 Symposium, 272–276. Surrey, England: IPC Press.
11. Nyborg WL (1965) Acoustic streaming. Physical Acoustics Vol. II, Part B. (W.P. Mason ed.). New York: Academic Press.

12. Nyborg WL (1978) Physical mechanisms for biological effects of ultrasound. U.S.H.E.W. Publication (FDA) 78-8062.
13. Repacholi MH (1980) The effects of ultrasound on human lymphocytes. A search for dominant mechanism of ultrasound action. Ph.D. Thesis. Canada: University of Ottawa.
14. Safar MH (1968) Comment on papers concerning rectified diffusion of cavitation bubbles. J Acoust Soc Am 43: 1188–1189.
15. ter Haar GR, Daniels S, Morton K and Hill CR (1981) Evidence for ultrasonically induced bubbles 'in vivo'. In Ultrasonics International '81, p 426. Guildford, Surrey, UK: IPC Press.
16. Wells PNT (1977) Biomedical Ultrasonics. London, England: Academic Press Inc Ltd
17. Statement of the American Institute for Ultrasound in Medicine (1977) J Clinical Ultrasound 5: 2–4.

12. Nyborg WL (1978) Physical mechanisms for biological effects of ultrasound. US HEW Publication (FDA) 78-8062.
13. Repacholi MH (1968) The effects of ultrasound on human lymphocytes: A search for a mechanism of ultrasound action. Ph.D. Thesis, Canada: University of Ottawa.
14. Safar MH (1968) Comment on papers concerning rectified diffusion of cavitation bubbles. J Acoust Soc Am 43: 1188–1189.
15. ter Haar GR, Daniels S, Morton K and Hill CR (1981) Evidence for ultrasonically induced bubbles in vivo. In Ultrasonics International 81, p 428. Guildford, Surrey, UK: IPC Press.
16. Wells PNT (1977) Biomedical Ultrasonics. London, England: Academic Press Inc Ltd.
17. Statement of the American Institute for Ultrasound in Medicine (1977) J Clin Ultrasound 5: 2–4.

Bubble nucleation in aqueous media: implications for diving physiology

DAVID E. YOUNT

Department of Physics and Astronomy, University of Hawaii, Honolulu, Hawaii 96822

Abstract. Decompression sickness follows a reduction in ambient pressure and is a result of bubble formation in blood or tissue. Almost any body part, organ, or fluid can be affected, including bone. This generality suggests a common basis in the physical and chemical properties of water, particularly those relating to cavitation. In this paper, we review a cavitation model developed at the University of Hawaii in which spherical gas nuclei are stabilized by surface-active skins of varying gas permeability. The varying-permeability model provides a precise quantitative description of bubble counts made in supersaturated gelatin, and it accurately predicts levels of incidence for decompression sickness in several animal species, including salmon, rats, and humans.

1. Introduction

Ordinary samples of sea water, tap water, or even distilled water form visible bubbles when subjected to tensile, ultrasonic, or supersaturation pressures as small as 1 atm. This is several orders of magnitude below the theoretical tensile strength of pure water, and it implies that cavitation must be initiated by processes other than modest changes in pressure and the random motion of water and gas molecules.

Numerous experiments have demonstrated that cavitation thresholds can be significantly raised by degassing or by a preliminary application of static pressure [6, 16]. These are specific tests for gas nuclei. Furthermore, solid particles or container walls with smooth surfaces [9] are not expected to be effective in initiating bubble formation at tensile, ultrasonic, or supersaturation pressures less than about 1000 atm [7, 13].

The existence of stable gas nuclei is at first rather surprising. Gas phases larger than about 1 μm in radius should float to the surface of a standing liquid, while smaller ones should dissolve rapidly due to the surface tension. In Refs. [12, 14], the earlier proposals for coping with this dilemma are critically reviewed, and a new cavitation model, called the varying-permeability model, is introduced. In the remaining sections of this paper, we outline the new model and report on its current experimental status.

2. The varying-permeability model

According to the Laplace-Young equation,

$$p_{\text{in}} = p_{\text{amb}} + (2\gamma/r), \tag{1}$$

the pressure p_{in} inside a spherical gas phase of radius r is higher than the ambient hydrostatic pressure p_{amb} because of the surface tension γ. If the

Applied Scientific Research 38: 37–44 (1982) 0003–6994/82/0381–0037 $01.20.

liquid surrounding the cavity is in equilibrium with an external gas mixture at p_{amb}, a pressure gradient will exist across the boundary of the cavity, and gas will tend to diffuse outward until the radius diminishes to zero.

In the varying-permeability model [12, 14], collapse of a spherical gas nucleus is prevented by the compression strength of an elastic skin or membrane composed of surface-active molecules. Nuclear skins are ordinarily gas permeable and become impermeable only when subjected to large compressions. In the permeable region of the model, skins manifest themselves mainly through their skin pressure $(2\Gamma/r)$, which can be added to the left-hand side of equation (1) to give

$$p_{in} + (2\Gamma/r) = p_{amb} + (2\gamma/r). \tag{2}$$

Alternatively, one can think of the skin compression Γ as the reduction in surface tension γ that is brought about by the surfactant molecules. The new surface tension, $\gamma' = \gamma - \Gamma$, can be substituted for γ in equation (1).

The skin compression Γ is analogous to the 'surface pressure' Π that is measured when an 'insoluble monolayer' of surface-active molecules is spread across the liquid-gas interface in a Langmuir trough [5]. In a typical Π-A (surface pressure versus surface area) curve, the magnitude of Π increases as the monolayer area and the spacing between surfactant molecules are reduced. Eventually, the maximum value of Π is reached, and further reductions in surface area are accommodated by expelling surfactant molecules from the interface.

The Π-A curve assumed for Γ in the varying-permeability model is essentially a step function. 'Small-scale' changes in nuclear radius – those associated with variations in the spacing of a fixed number of skin molecules – are neglected, and only 'large-scale' changes – those associated with the addition or deletion of skin molecules – are actually calculated. The small-scale changes are important conceptually because they permit a stable mechanical equilibrium near the calculated large-scale radius with the fixed number of skin molecules appropriate to that radius. All large-scale processes take place at the maximum skin compression γ_C, which is referred to as the 'crumbling compression'. Whereas measured values of Π do not exceed the surface tension of the underlying substrate, γ_C must be larger than γ for a surfactant nucleus to survive. The greater compression strength of the nuclear membrane is attributed to the spherical geometry.

Two independent derivations of the varying-permeability model have been proposed [12]. The first, following the gas-impermeable organic skin model of Fox and Herzfeld [4], is an application of the equation given by Love [8] for an elastic shell. It is assumed that the shell is bounded by spherical concentric surfaces and that it is held strained by a difference between the internal pressure p_{in} and the external pressure

$$p_{\text{out}} = p_{\text{amb}} + (2\gamma/r). \tag{3}$$

The skin compression is identified as

$$\Gamma = \delta E/(1-\nu), \tag{4}$$

where δ is the skin thickness, E is Young's modulus, and ν is Poisson's constant. The assumption that Γ has a constant value γ_C for all large-scale changes in radius is equivalent to the assumption that the right-hand side of equation (4) is fixed.

The second derivation of the varying-permeability model is thermodynamic or chemical, rather than mechanical. The transport of surfactant molecules between the skin and the surrounding 'reservoir' is described by setting the electrochemical potentials equal in the two regions. A constant skin compression $\Gamma = \gamma_C$ corresponds essentially to a constant difference in the purely chemical potential between the skin and the reservoir or to a constant activation energy for the chemical reactions by which surfactant molecules are accreted or deleted by the skin.

An important by-product of the thermodynamic or chemical derivation is the prediction that γ_C, which is constant for a given nucleus throughout an arbitrary pressure schedule, should increase linearly with the initial value of the nuclear radius r_0:

$$\gamma_C = \gamma + (\beta_0/2)r_0. \tag{5}$$

The magnitude of the positive constant β_0 is determined mainly by the difference in the purely chemical potentials of the skin and the reservoir [12].

Application of the varying-permeability model to predict bubble counts in supersaturated gelatin is based on the 'ordering hypothesis' [12, 14]. Each gelatin sample is assumed to have the same initial distribution of nuclear radii, and the number of bubbles formed is equal to the number of nuclei larger than some minimum initial radius $r_0^{\min}$. The ordering hypothesis then states that nuclei are neither created nor extinguished when samples are subjected to a pressure schedule and that the initial ordering according to size is preserved. It follows that each bubble count is determined by the properties and behavior of a single critical nucleus and that a family of pressure schedules yielding the same bubble number N is characterized by the same critical radius $r_0^{\min}$ and (via equation (5)) by the same crumbling compression γ_C.

In the permeable region of the model, each isopleth of constant N is described by an equation of the form [12, 14]

$$p_{\text{ss}} = [2\gamma(\gamma_C - \gamma)/r_0^{\min}\gamma_C] + [p_{\text{crush}}(\gamma/\gamma_C)], \tag{6}$$

where p_{crush} is the maximum overpressure or crushing pressure to which the sample has been subjected and p_{ss} is the supersaturation pressure required to form the given bubble number.

When p_{crush} reaches a certain magnitude p^*_{crush}, the skin surrounding the critical nucleus becomes effectively gas-impermeable. Further compression will cause p_{in}, which was previously in diffusion equilibrium at the initial ambient pressure p_0, to increase. The increase in p_{in} strengthens the nucleus and makes p_{ss} less dependent upon p_{crush}. The model prediction for an isopleth of constant N in the impermeable region can be written [12, 14]

$$p_{ss} = [2\gamma(\gamma_C - \gamma)/r_0^{min}\gamma_C] + \{[p_m - p_0\,(r^*_{min}/r_m^{min})^3]\,(\gamma/\gamma_C)\}, \tag{7}$$

where p_m is the value of p_{amb} at the end of the compression, and r^*_{min} and r_m^{min} are the values of the nuclear radius at p^*_{crush} and p_{crush}, respectively. The magnitudes of r^*_{min} and r_m^{min} can be calculated from the model relations [12]:

$$2(\gamma_C - \gamma)\,[(1/r^*_{min}) - (1/r_0^{min})] = p^*_{crush}, \tag{8a}$$

$$2(\gamma_C - \gamma)\,[(1/r_m^{min}) - (1/r^*_{min})] = p_{crush} - p^*_{crush} + p_0\,[1 - (r^*_{min}/r_m^{min})^3]. \tag{8b}$$

3. Experimental status

The development of the varying-permeability model [12, 14] was originally motivated and guided by bubble counting experiments carried out in supersaturated gelatin [16]. Whereas ultrasonic cavitation experiments ordinarily determine only bubble formation thresholds and hence trace a single isopleth $N \sim 1$, gelatin data can be analyzed to yield a family of isopleths giving, for example, p_{ss} versus p_{crush} for $N = 1, 3, 10, 30, 100$, etc. In this sense, gelatin adds a new dimension to the study of heterogeneous cavitation and permits a much more incisive test of any nucleation model.

By adjusting the model parameters r_0^{min}, γ_C, and p^*_{crush} in equations (6–8b), a satisfactory description of nearly all of the original gelatin data [16] was obtained [12]. The combinations of (r_0^{min}, γ_C) required by the respective isopleths of constant N were consistent with equation (5), and the initial integral size distribution $N(r_0 \geqslant r_0^{min})$ was found to be exponential:

$$N(r_0 \geqslant r_0^{min}) = N_0 \exp\,(-r_0^{min}/b). \tag{9}$$

With the help of equation (5), equation (9) was rewritten as a Boltzmann distribution

$$N = N_0 \exp\,[-(\gamma_C - \gamma)S/kT], \tag{10}$$

and the parameter

$$S = 2kT/\beta_0 b \tag{11}$$

was interpreted as the area of an individual skin molecule in situ [12].

Granting that the varying-permeability model adequately summarizes the data in the original gelatin experiments [16], one may still doubt whether the model parameter $r_0^{\min}$ is the radius of an actual physical entity. To explore this point further, a new type of investigation was carried out [18] in which the initial size distribution of the objects that initiate bubble formation in gelatin was systematically altered by passing test samples through Nuclepore filters with pore radii of 0.18, 0.27, 0.36, 0.45, and 1.35 μm, accurate to better than ± 10%.

The results of the filter experiment are plotted in Figure 1. Within statistical errors of ± 2.6%, there is no difference between the no-filter points

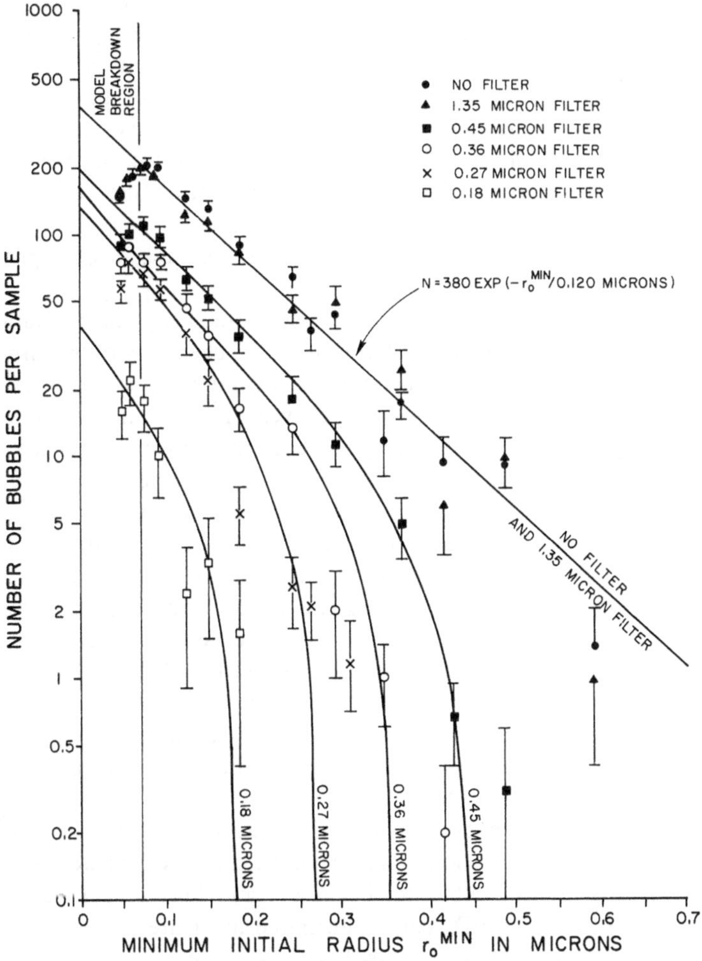

Figure 1. Number of bubbles per sample versus minimum initial radius for bubble formation in filtered and unfiltered gelatin.

and those found with filters having pore radii of 1.35 μm. It follows that the number of nuclei present in the original sample with physical radii larger than 1.35 μm was negligible. The distribution for the 0.18, 0.27, 0.36, and 0.45 μm filters are cut off sharply above the respective filter-pore radii, implying that r_0^{min} is, in fact, the radius of an actual physical structure capable of initiating bubble formation in gelatin. Finally, the primordial radial distribution is again exponential, and this result can now be inferred from the filter data alone, independently of any cavitation model.

As indicated in Figure 1, the varying-permeability model breaks down at the smallest values of r_0^{min}. To explore this phenomenon further, an experiment was recently completed [17] in which slow or stepped compressions were used to limit the effective overpressure p_{crush}, while still permitting large values of p_{ss}. Most of the new data points lay in the model-breakdown region. A satisfactory description of all of these results was obtained simply by assuming that the radius of the reservoir is slightly larger than the radius of the nuclear skin per se. The required difference in these radii, $\delta = 2.5$ Å, is too small to be the thickness of an insoluble monolayer; hence δ was interpreted as the 'active skin thickness', i.e., as the thickness of that portion of the nuclear membrane which is capable of supporting a pressure gradient or a gradient in chemical potential [17].

The values of β_0 determined for the above three experiments varied by a factor of 5, due presumably to differences in the chemistry of the surrounding media. Meanwhile, the area S determined for individual skin molecules, was nearly constant at 65, 48, 45 Å^2, respectively. This situation was interpreted [17] via equation (11) as providing evidence that any external change in the chemical potential or activation energy would be matched by a reciprocal internal change in the exponential slope parameter b. That is, any decrease in cavitation rate associated with a decrease in β_0 would be due mainly, and perhaps entirely to a steepening of the slope in the radial distribution $N(r_0 \geqslant r_0^{min})$. This exactly parallels the situation in which changes in the ambient pressure p_{amb} are also reflected in the slope parameter b, while the exponential form of equation (9) is again preserved [12]. With this association, the 'mechanical-chemical equilibrium duality', which allows two interpretations of the parameter δ and two derivations of the varying-permeability model, becomes a practical experimental tool for intercalibrating skin mechanics and membrane chemistry.

In the first applications of the varying-permeability model to decompression sickness [10, 11, 15], one additional assumption has been made, namely, that isopleths of constant bubble number N are also lines of constant stress or constant effective dose ED. This is illustrated in Figure 2, where a particular model calculation VP6 is compared with a compilation [15] of pressure reduction tolerances for rats subjected to various exposure pressures [1, 2, 3]. The exposure pressure is equal to $(p_{crush} + p_0)$, while the ED-50

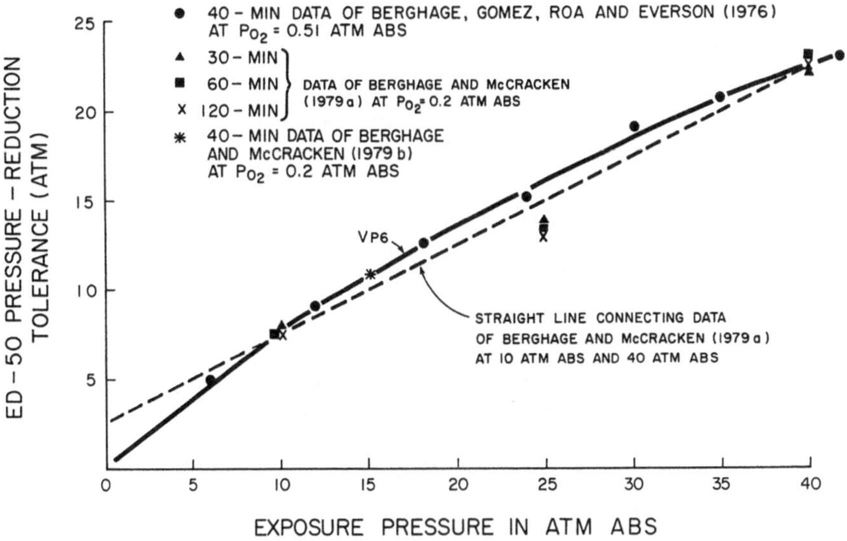

Figure 2. Compilation of pressure-reduction tolerances versus exposure pressure for albino rats. ED-50 is the pressure reduction needed to produce decompression sickness in 50% of the animals.

pressure reduction is just the critical supersaturation p_{ss} needed to induce signs of decompression sickness in 50% of the subjects.

Figure 2 demonstrates the importance of bubble nucleation as the primary and controlling even in decompression sickness. It also serves as an experimental test of the varying-permeability model that is independent of the original gelatin prototype. For example, there is evidence in Figure 2 for both a permeable and an impermeable region. The former, described by equation (6), is characterized by a linear increase of p_{ss} with p_{crush} and extends up to about 9 atm abs. The later, predicted by equation (7), exhibits a slower rise and a negative curvature from 9 to 42 atm abs and beyond. Whereas the onset of impermeability p^*_{crush} is a free parameter, the subsequent behavior is an intrinsic feature of the model determined mainly by Boyle's law. This behavior is tested and confirmed in Figure 2 over a range of exposures exceeding 33 atm abs.

Acknowledgements

It is a pleasure to thank my colleagues Ed Beckman, Tim Brown, Claude Harvey, Don Hoffman, and Jon Pegg for useful discussions and constructive comments during the preparation of this manuscript. This work is a result of research sponsored in part by the University of Hawaii Sea Grant College Program under Institutional Grant Number NA79AA-D-00085 from NOAA Office of Sea Grant, U.S. Department of Commerce.

References

1. Berghage TE, Gomez JA, Roa CE, and Everson TR (1976) Pressure-reduction limits for rats following steady-state exposures between 6 and 60 ata. Undersea Biomed Res 3:261–267.
2. Berghage TE and McCracken TM (1979) Equivalent air depth: Fact or fiction. Undersea Biomed Res 6:379–384.
3. Berghage TE and McCracken TM (1979) The use of oxygen for optimizing decompression. Undersea Biomed Res 6:231–239.
4. Fox FE and Herzfeld KF (1954) Gas bubbles with organic skin as cavitation nuclei. J Acoust Soc Amer 26:984–989.
5. Gaines GL Jr (1966) Insoluble monolayers at liquid-gas interfaces. New York: Interscience.
6. Harvey EN, Barnes DK, McElroy WD, Whiteley AH, Pease DC and Cooper KW (1944) Bubble formation in animals. I. Physical factors. J Cell Comp Physiol 24:1–22.
7. Kunkle TD and Yount DE (1975) Gas nucleation in gelatin. In Kent MB (ed). Sixth symposium on underwater physiology, pp 459–467.
8. Love AEH (1944) The mathematical theory of elasticity. New York: Dover. Reprint 4th ed, p 142.
9. Plesset MS (1969) The tensile strength of liquids. ASME fluids engineering and applied mechanics conference, Evanston, IL, pp 15–25.
10. Yount DE (1979) Application of a bubble formation model to decompression sickness in rats and humans. Aviat Space Environ Med 50:44–50.
11. Yount DE (1981) Application of a bubble nucleation model to decompression sickness in fingerling salmon. Undersea medical society annual scientific meeting. Undersea Biomed Res 8:34 (Abstract 42).
12. Yount DE (1979) Skins of varying permeability: A stabilization mechanism for gas cavitation nuclei. J Acoust Soc Am 65:1429–1439.
13. Yount DE and Kunkle TK (1975) Gas nucleation in the vicinity of solid hydrophobic spheres. J Appl Phys 46:4484–4486.
14. Yount DE, Kunkle TD, D'Arrigo JS, Ingle FW, Yeung CM and Beckman EL (1977) Stabilization of gas cavitation nuclei by surface-active compounds. Aviat Space Environ Med 48:185–191.
15. Yount DE and Lally DA (1980) On the use of oxygen to facilitate decompression. Aviat Space Environ Med 51:544–550.
16. Yount DE and Strauss RH (1976) Bubble formation in gelatin: A model for decompression sickness. J Appl Phys 47:5081–5089.
17. Yount DE and Yeung CM (1981) Bubble formation in supersaturated gelatin: A further investigation of gas cavitation nuclei. J Acoust Soc Am 69: 702–708.
18. Yount DE, Yeung CM and Ingle FW (1979) Determination of the radii of gas cavitation nuclei by filtering gelatin. J Acoust Soc Am 65:1440–1450.

The effect of dissolved gases on the dynamics of acoustic emission and sonoluminescence from cavitating liquids

P. W. VAUGHAN, E. GRAHAM and S. LEEMAN

Department of Medical Physics, Royal Postgraduate Medical School, Hammersmith Hospital, Du Cane Road, London, W12 0HS, UK

Introduction

In this experimental study of acoustic cavitation at 1.5 MHz, we investigate the effect of dissolved gas upon the ascendant and descendant acoustic cavitation thresholds of distilled water. An interesting hysteresis effect is noticed and discussed. The influence of aliphatic alcohol additions (in low concentration) upon sonoluminescence and upon oxidation of Fe^{++} is also studied, with simultaneous monitoring of both of these effects, as well as of subharmonic emission. Results are compared with those of Sehgal et al. [6], and some are shown to be incompatible with their explanation in terms of radicals produced during cavitation.

Experimental procedure

The experimental arrangement has been described in previous publications [2, 7]. We use a 1.5 MHz plane, continuous mode, travelling wave ultrasonic beam, with the sample liquid placed in the acoustic far-field, located at the position of the last axial maximum of the field. This ensures that the beam profile and the acoustic intensity remain relatively constant throughout the sample volume. Figure 1 is a view of the sonication vessel with the sample holder removed. The sample holder has a volume of 50 ml and is shown in Figure 2. Experiments are conducted in a light-tight box, with the photomultiplier tube placed directly over the sample holder. A schematic of the equipment used is shown in Figure 3. The photomultiplier tube is sensitive mainly in the wavelength range, 350–550 nm.

The experimental design allows for simultaneous monitoring of sonoluminescence (S/L), subharmonic emission (S/H), and the voltage applied to the drive transducer. The acoustic output-voltage characteristic of the transducer is determined with a radiation pressure balance. Spatial averages are quoted throughout this work; the peak: average ratio for the beam intensity being ~ 5. Spatial average intensities of 0–10 W/cm² may be easily generated. Triple degassed (by vacuum) distilled water with surfactant added was used as coupling medium, the transducer output checked periodically, and the coupling medium checked to be 'cavitationally' quiet before each experiment.

For all experiments, triple degassed distilled water (checked to be cavitationally quiet) was saturated with the test gas for use as the sample

Applied Scientific Research 38: 45–52 (1982) 0003–0994/82/0381–0045 $01.20.

Figure 1. Sonication vessel with sample holder removed.

Figure 2. View of sample holder.

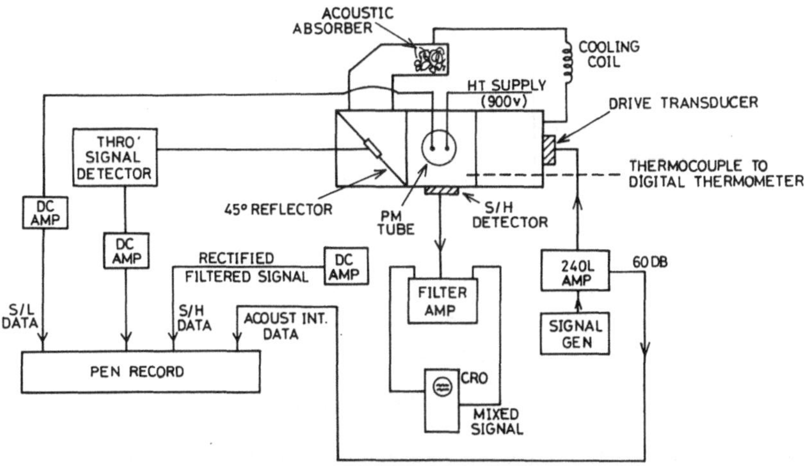

Figure 3. Schematic of equipment.

liquid. All sonications were conducted at 21 ± 1°C but despite the cooling arrangements, sample temperature could rise by up to 5°C during a 5 minute sonication at maximum power (10 W/cm^2). When monitoring the hysteresis effects, the samples are subjected to a series of cycles of increasing acoustic intensity, holding at constant level, then decreasing. Each cycle extends for 4 minutes, with any subharmonication and sonoluminescence produced being chart recorded, together with the voltage to the drive transducer.

When performing the alcohol additions and $Fe^{++} \rightarrow Fe^{+++}$ studies, 5 minute sonications at 10 W/cm^2 were made, and the average subharmonic ($\overline{S/H}$) and sonoluminescence ($\overline{S/L}$) outputs were read from the chart record. For the Fe^{++} oxidation experiments, the test liquid was an appropriately gassed sample from a stock of Fricke solution prepared from degassed distilled water, the composition being 1 m*M* $FeSO_4$, 1 m*M* NaCl, 0.8 N H_2SO_4. The concentration of Fe^{+++} formed was estimated by measuring absorbance at 305 nm with a Beckmann type 25 spectro-photometer. The saturating gases employed are Air, A, O_2, N_2, CO_2 and N_2O. Only a selection of results are reported here.

Results and discussion

The first finding is that of the occurrence of hysteresis. We define Type I hysteresis as the difference between the observed cavitation effect (S/L , S/H) for ascending and descending acoustic intensity, and Type II as referring to any changes in the sample response with repeated insonations.

Figure 4 (H_2-saturated water) shows a typical sonication exhibiting Type I subharmonic hysteresis, which is a marked feature in this case.

A large number of runs have been made with different saturating gases; Figure 6 is typical of such runs.

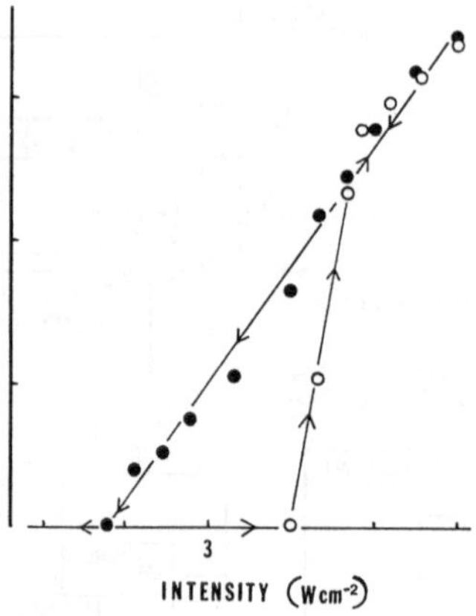

Figure 4. Type I hysteresis for subharmonic emission from H_2-saturated water.

Our experiments show that most of the test gases show Type I hysteresis in distilled water. The effect is more pronounced for subharmonic emission, the sonoluminescence showing a smaller but nonetheless measurable hysteresis. Addition of surfactant increases the extent of hysteresis, an effect which is consistent with the finding of Crum [1] that reducing surface tension raises the cavitation threshold.

We consider that the effect may be explained in terms of frequency-dependent damping coefficients for the bubbles under insonation. Prosperetti's calculations [4] of overall damping coefficient β (for different sized bubbles) explicitly shows such a frequency dependence. Suppose that a bubble has damping coefficient β_1 when forced into pulsation at the drive frequency f_1 and β_2 at frequency f_2. Inspection of Prosperetti's curves, for certain bubble radii, shows that $\beta_1 > \beta_2$ when the frequency f_1 is 1.5 MHz and f_2 is 0.75 MHz. Thus once a bubble has established a subharmonic mode, it will require less excitation energy to maintain this pulsation.

Type II hysteresis is less easily explicable. Figure 5 (N_2-saturated water) shows a clear decline in magnitude of subharmonic with repeated sonications.

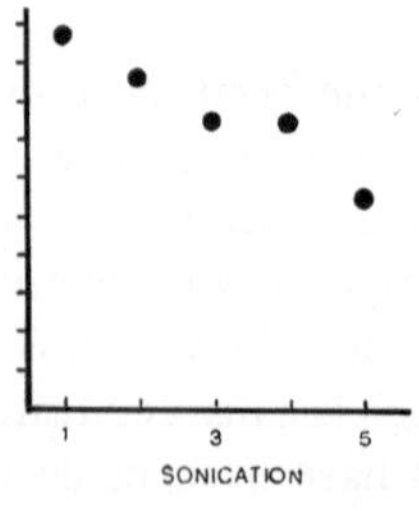

Figure 5. Type II hysteresis for subharmonic emission from N_2-saturated water.

More experiments have to be performed to draw any firm conclusions, but it is difficult to avoid the impression that hysteresis effects are sensitive indicators of bubble dynamic effects.

It is also possible to draw some general conclusions about the cavitational emissions investigated here. Subharmonic activity appears to be a pre-requisite for sonoluminescence. We have not seen a single case of the latter, without the presence of the former, in any sample sonicated to date. The converse is not true: subharmonic thresholds are in general lower than those for sonoluminescence. Moreover, hydrogen saturated water (on its first sonication) shows little or no sonoluminescence but exhibits rather marked subharmonic emission. The usual interpretation of this result [8] is that the high thermal conductivity of this gas prevents high temperatures being reached during the collapse phase of the cavitation bubble. Such explanations obviously give credence to the 'hot-spot' theory of sonoluminescence [3]. However, it is difficult to see how this attractive hypothesis can be upheld by the Type II hysteresis results shown in Figure 6.

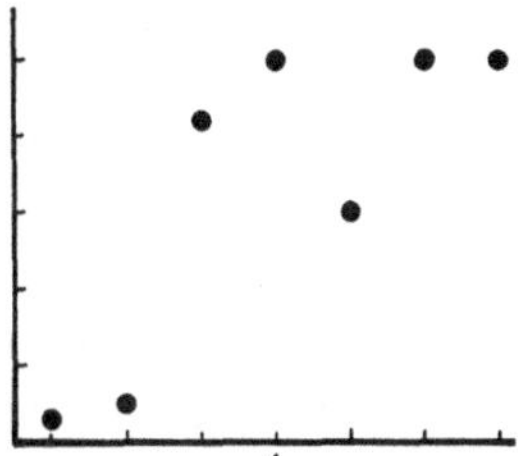

Figure 6. Type II hysteresis for sonoluminescence from H_2-saturated water.

We are following the work of Sehgal et al. [5, 6] when investigating the effects of alcohol addition upon luminescence and upon oxidation of Fe^{++} in gassed Fricke solutions. These authors have argued that, for the alcohol concentrations used, i.e. 0–20 mM, the cavity dynamics would be unlikely to be affected by the presence of alcohol vapour. Thus, observed effects would be of a chemical rather than physical (i.e. bubble dynamical) nature. Sehgal et al. claim that sonoluminescence has its origin in the radicals formed during sonication when cavitation occurs. In the case of Argon-saturated water, the emission peak found in the spectrum of light emitted was attributed to transitions of the OH radical. Aliphatic alcohols are known to scavenge these radicals, and the experimental finding [6] that sonoluminescence in A-saturated water was quenched by these alcohols is adduced as further evidence for the chemical origin of the light emission.

With air-saturated Fricke solution, however, $OH_{\bullet}$ and $HO_{2\bullet}$ radicals are postulated, and a series of reactions proposed to account for the enhancing effect that occurs in the oxidation of Fe^{++} to Fe^{+++} [6].

We have extended the work of these authors, who could not simultaneously monitor the various cavitational effects, as is possible with our experimental arrangement. Our results for air-saturated Fricke solution are shown in Figure 7. The oxidation of Fe^{++} is clearly enhanced with increasing n-Butanol concentration, as Sehgal et al. have found. If their proposed mechanism were valid, the sonoluminescent output would be expected to fall progressively

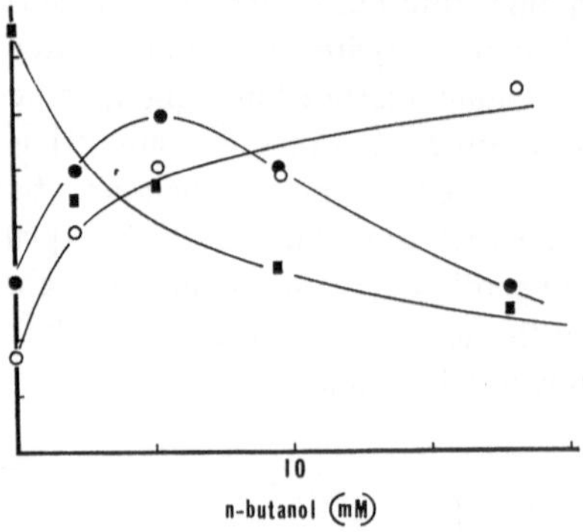

Figure 7. Effect of n-Butanol concentration on Fe^{++} oxidation (○), $\overline{S/L}$ (●), and $\overline{S/H}$ (■) for air-saturated Fricke solution.

with n-Butanol concentration. This is clearly not the case: $\overline{S/L}$ rises to a peak at ~ 5 m*M* alcohol concentration. $\overline{S/H}$, on the other hand, progressively decreases. This suggests that the presence of the alcohol reduces the amplitude of the non-linear bubble pulsations. However, the behaviour of the sonoluminescence indicates that this phenomenon is influenced not only by the bubble dynamics, but also by the chemical environment.

The chain of radical reactions proposed by Sehgal et al., whereby the presence of alcohol enhances Fe^{++} oxidation, depends critically upon the presence of oxygen. By this mechanism, an increase in Fe^{++} oxidation with alcohol addition would not be expected in A-saturated Fricke solution. Our findings (Figure 8) do not substantiate this, not only for A saturated Fricke solution, but also for N_2 (Figure 9) and H_2 (Figure 10) solutions. (H_2-saturated Fricke exhibits strong sonoluminescence at first sonication, while H_2-saturated water shows little or no effect!) Clearly, the radical scheme envisaged by Sehgal et al. needs modification.

Conclusions

We have seen hysteresis effects for both subharmonic and sonoluminescent emissions for a variety of gases dissolved in water. It is clear that an alteration of the chemical environment of the cavitation bubbles can, in certain cases,

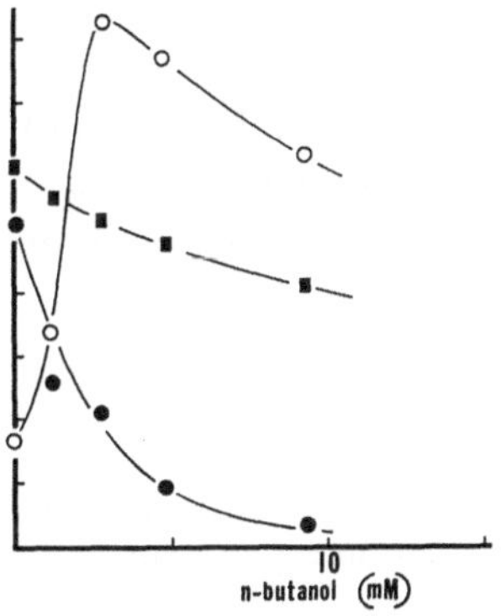

Figure 8. Effect of n-Butanol concentration on Fe^{++} oxidation (○) $\overline{S/L}$ (●), and $\overline{S/H}$ (■) for A-saturated Fricke solution.

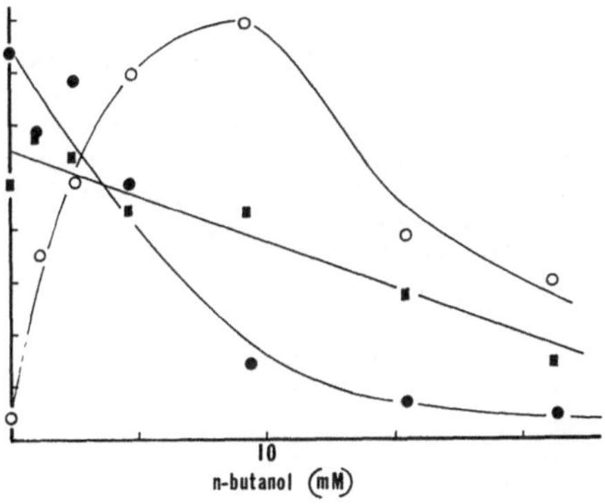

Figure 9. Effect of n-Butanol concentration on Fe^{++} oxidation (○), $\overline{S/L}$ (●), and $\overline{S/H}$ (■) for N_2-saturated Fricke solution.

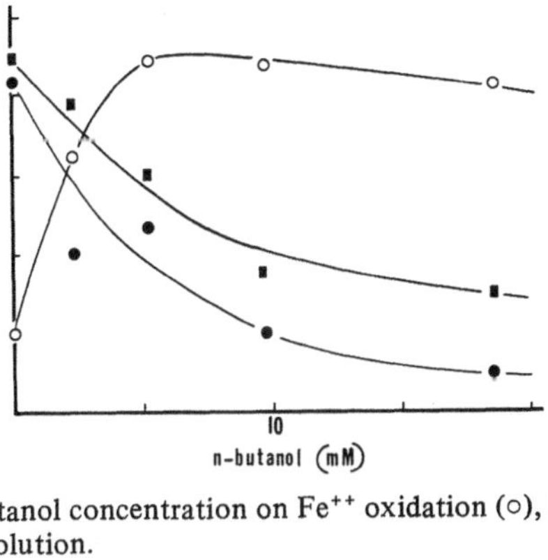

Figure 10. Effect of n-Butanol concentration on Fe^{++} oxidation (○), $\overline{S/L}$ (●), and $\overline{S/H}$ (■) for H_2-saturated Fricke solution.

influence the level of sonoluminescence. The presence of volatile alcohols suppresses cavitational activity, as measured by subharmonic output, but sonoluminescent and chemical (Fe^{++} oxidation) responses depend on the dissolved gas. Our results do not unambiguously support mechanisms proposed for either the chemiluminescent, or the hot-spot, nature of sonoluminescence.

Acknowledgement

We are extremely grateful to Dr Robyn Anderson for her help and encouragement with the Fe^{++} oxidation experiments.

References

1. Crum LA (1979) Tensile Strength of Water. Nature 27B: 148.
2. Graham E, Hedges M, Leeman S and Vaughan P (1980) Cavitational Bioeffects at 1.5 MHz. Ultrasonics 224.
3. Neppiras EA and Noltingk BE (1951) Cavitation Produced by Ultrasonics. Theoretical Conditions for the Onset of Cavitation. Proc Roy Soc B64: 1032.
4. Prosperetti A (1977) Thermal Effects and Damping Mechanisms in the Forced Radial Oscillations of Gas Bubbles in Liquids. J Acoust Soc Am 61: 17.
5. Sehgal C, Steer RP, Sutherland RG and Verrall AE (1977) Sonoluminescence of Aqueous Solutions. J Phys Chem 81: 2618.
6. Sehgal C, Sutherland RG and Verrall AE (1980) Cavitation Induced Oxidation of Aerated Fe^{++} Solutions in the Presence of Aliphatic Alcohols. J Phys Chem 84: 2920.
7. Vaughan P, Leeman S, Hedges M, Graham E and Sutton P (1980) Cavitation Effects at Megahertz Frequencies. In Lauterborn W (ed) Cavitation and Inhomogeneities in Underwater Acoustics
8. Young FR (1976) Sonoluminescence from Water Containing Dissolved Gases. J Acoust Soc Am 60: 100.

PART TWO

Boiling

Acoustic cavitation in cryogenic and boiling liquids

VICTOR A. AKULICHEV

Pacific Oceanological Institute, Far East Science Centre of the USSR Academy of Sciences, Vladivostok, USSR

Abstract. Vapour bubble dynamics in cryogenic and boiling liquids affected by an acoustic field is considered. Linear pulsations and nonstationary growth of vapour bubbles in time due to linear effects of rectified heat and mass transfer are studied. The growth thresholds of vapour bubbles depending on thermodynamic parameters of liquid, static overcompression, and acoustic field frequency are presented. Essential influence of resonance properties of bubbles on the values of growth thresholds is shown. The results for different cryogenic liquids and boiling water are given.

Introduction

Studies of cavitation induced by periodic changes of pressure in cryogenic and boiling liquid are of interest in their association with practical applications in high-energy physics, nuclear energetics and other fields of modern technology. In recent years, special attention has been paid to the study of various phenomena in liquid hydrogen which is assumed to be used in the future as the most perspective fuel.

Cavitation in cryogenic and boiling liquids has the feature that its origin and development is determined by the dynamics of pure vapour cavities and bubbles. Gas inclusions of foreign particles present in usual liquids can not exist, for example, in a liquid as liquid helium. As far as other cryogenic liquids are concerned, they can not contain gas inclusions in considerable amounts due to the technology of their production and storage. The dynamics of cavitation bubbles in such media differ from the dynamics of bubbles in usual underheated water, primarily due to the essential influence of heat and mass transfer processes. These cavitation phenomena occur not only in cryogenic fluids but also in water heated to the boiling state and in molten liquid metals and other boiling or almost boiling media.

Cavitation bubble dynamics

Consider the set of equations describing the behaviour of vapour bubbles in an acoustic field according to Ref. [4]. The set of equations used here is extended and supplemented as compared to those suggested in Refs. [1–3, 5, 6, 8]. We shall consider the inhomogeneity of temperature and pressure inside the bubble and finiteness of evaporation rate and substance condensation, i.e. the nonequilibrium of evaporation and condensation processes. In this case the usual assumptions of spherically symmetrical bubble pulsations will be used. The liquid is assumed to be incompressible and the bubble radius

Applied Scientific Research 38: 55–67 (1982) 0003–6994/82/0381–0055 $01.95.

small in comparison with the acoustic wave length λ in liquid. Hence, dynamic equations will have the form

$$\rho(\dot{U}R + 2U\dot{R} - \tfrac{1}{2}U^2) + p(\infty, t) - p(R, t) = 0, \tag{1}$$

$$\rho c_p \left(\frac{\partial T}{\partial t} + U\frac{R^2}{z^2}\frac{\partial T}{\partial t}\right) = \frac{1}{z^2}\frac{\partial}{\partial z}\left(z^2 \kappa \frac{\partial T}{\partial z}\right) + 12\eta U^2 \frac{R^4}{z^6}, \qquad z > R, \tag{2}$$

where $p(\infty, t)$ is the liquid pressure at infinity, $p(R, t)$ and $U = u(R)$ are the liquid pressure and velocity on the bubble surface, ρ is the liquid density, and c_p, κ, and η are coefficients of heat capacity, heat conductivity and liquid viscosity. Equations (1) and (2) must be supplemented by a set of equations for the vapour phase inside the bubble, which is generally considered compressible,

$$\frac{\partial \rho'}{\partial t} + u'\frac{\partial \rho'}{\partial z} + \frac{1}{z^2}\frac{\partial}{\partial z}(z^2 \rho' u') = 0, \tag{3}$$

$$\frac{\partial u'}{\partial t} + \frac{\partial}{\partial z}\left(\frac{u'^2}{2}\right) + \frac{1}{\rho'}\frac{\partial p'}{\partial z} = 0, \tag{4}$$

$$\rho' c_p' \left(\frac{\partial T'}{\partial t} + u'\frac{\partial T'}{\partial z}\right) = \frac{1}{z^2}\frac{\partial}{\partial z}\left(z^2 \kappa' \frac{\partial T'}{\partial z}\right) + \alpha' T' \frac{\mathrm{d}p'}{\mathrm{d}t}. \tag{5}$$

Designations here are the same as above, only a prime is added. This set of equations must be supplemented by the equation of state for the vapour phase inside the bubble, given by

$$\mathrm{d}\rho' = \rho'(\beta' \mathrm{d}p' - \alpha' \mathrm{d}T'), \tag{6}$$

where

$$\beta' = \frac{1}{\rho'}\left(\frac{\partial \rho'}{\partial p'}\right)_T, \qquad \alpha' = -\frac{1}{\rho'}\left(\frac{\partial \rho'}{\partial p'}\right)_T, \tag{7}$$

Apart from the above equations, the following boundary conditions ensuing from the conservation laws for mass, momentum, and energy must be satisfied at the phase interface

$$\rho(\dot{R} - U) = \rho'(\dot{R} - U') = \mathscr{I}, \tag{8}$$

$$p(R, t) + \frac{2\sigma}{R} - \sigma_{zz} + \frac{\mathscr{I}^2}{\rho} = p'(R, t) - \sigma'_{zz} + \frac{\mathscr{I}^2}{\rho'} \tag{9}$$

$$\kappa\left(\frac{\partial T}{\partial z}\right)_R - \kappa'\left(\frac{\partial T'}{\partial z}\right)_R = \mathscr{I}\left(L + \frac{2\sigma}{R} - \frac{\sigma'_{zz}}{\rho'}\right) + \frac{\mathscr{I}^2}{\rho\rho'\beta' U_{ev}} + \frac{\mathscr{I}^3}{2\rho'^2}$$

$$- \frac{T}{R^2}\frac{d}{dt}\, R^2\left(\frac{d\sigma}{dT}\right). \tag{10}$$

Here $\mathscr{I} = (dM'/dt)(4\pi R^2)^{-1}$ is the vapour mass flux across the bubble surface, L is the latent heat, σ is the surface tension coefficient, σ_{zz} is the radial component of the tensor of viscous stresses in the liquid which is connected with the coefficients of shear and bulk viscosities of η and ζ in the following way

$$\sigma_{zz} = \left(\frac{4}{3}\eta + \zeta\right)\frac{\partial u}{\partial z} + \frac{2u}{z}\left(\zeta - \frac{2}{3}\eta\right).$$

A similar expression can be written for σ'_{zz} in the vapour phase. The rate of substance evaporation into the vacuum U_{ev} is assumed to be finite and for the ideal gas it is expressed by the accommodation coefficient $\bar{\alpha}$, vapour pressure p' and vapour density ρ

$$U_{ev} = \bar{\alpha}\left(\frac{p'}{2\pi\rho'}\right)^{1/2}. \tag{11}$$

In the weak nonequilibrium approximation of the liquid evaporation process on the bubble surface the equality $T(R,t) = T'(R,t)$ is assumed valid. The flux $\mathscr{I}$ in this case can be expressed in the form of the Hertz-Knudsen equation

$$\mathscr{I} = \beta'\rho' U_{ev}[p_\sigma(T) - p'(\rho', T')], \tag{12}$$

where $p_\sigma(T)$ is the saturated vapour pressure associated with temperature T by the Clausius-Clapeyron equation. With small radii of nucleus bubbles, the pressure p'_σ must take into account the correction for the curvature of the phase interface.

The effect of periodic change of pressure on the bubble can be described by the equation

$$p(\infty, t) = p_\sigma(T_0) + \Delta p + p_m \cos \omega t, \tag{13}$$

where p_m is the changing pressure amplitude with frequency $\omega = 2\pi f$, T_0 is the liquid temperature at infinity, Δp is the static overcompression which determines the increase of static pressure p_0 over the saturated vapour pressure p'_σ and is introduced to prevent parasitic boiling of liquid. It should be noted that periodic changes of pressure may be induced by an acoustic or sound field which is fully characterized by the parameters p_m and ω.

In case of periodic effects of hydrodynamic origin one may also emphasize characteristic values of amplitude and frequency.

This set of equations can not be solved analytically. One can obtain numerical solutions by means of modern computers for various particular cases. Figure 1 presents such solutions for vapour bubble pulsations in liquid nitrogen under the influence of a 50 kHz acoustic field for different pressure amplitude of p_m. It is clearly seen that due to pulsations the vapour bubble in liquid nitrogen grows to a typical average size. It reaches this size sooner, the larger the pressure amplitude. Such a regularity becomes apparent at vapour bubble pulsations in different cryogenic and boiling liquids with different parameters of liquid and effecting periodic fields. Figure 2 shows vapour bubble pulsations in liquid nitrogen under the influence of a 10 kHz acoustic field. It is clearly seen that with the decrease of frequency ω the extreme mean radius increases in proportion to $1/\omega$.

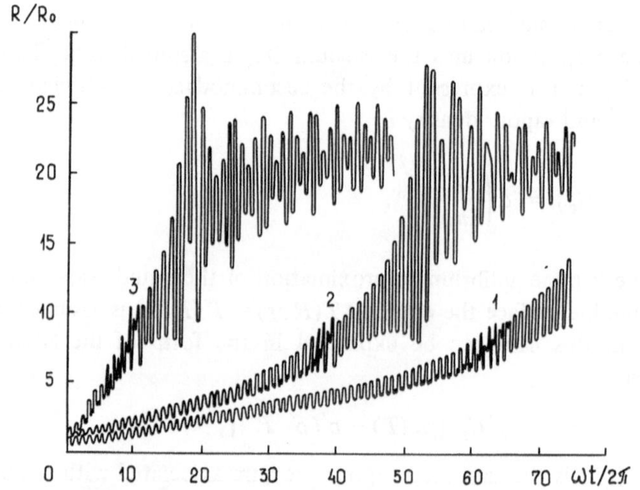

Figure 1. Vapour bubble pulsations in liquid nitrogen affected by a 50 kHz acoustic field for different amplitudes of pressure p_m.

$T_0 = 77.4K; \Delta p = 0.2\,\text{atm}; R_0 = 5.10^{-4}$ cm.

Curve 1 — $p_m = 0.35$ atm, 2 — $p_m = 0.4$ atm, 3 — 0.5 atm.

One can obtain an analytical solution of the complete set of equations only in terms of perturbation theory. If one takes into account that the static pressure in liquid is equal to $p_0 = p_\sigma(T_0) + \Delta p$, then the pressure in the liquid at infinity may be represented as

$$p(\infty, t) = p_0 + p_m\, e^{i\omega t}. \tag{14}$$

A solution for the bubble radius will then be found in the form

$$R(t) = \overline{R(t)} + R_m(t) e^{i\omega t}, \tag{15}$$

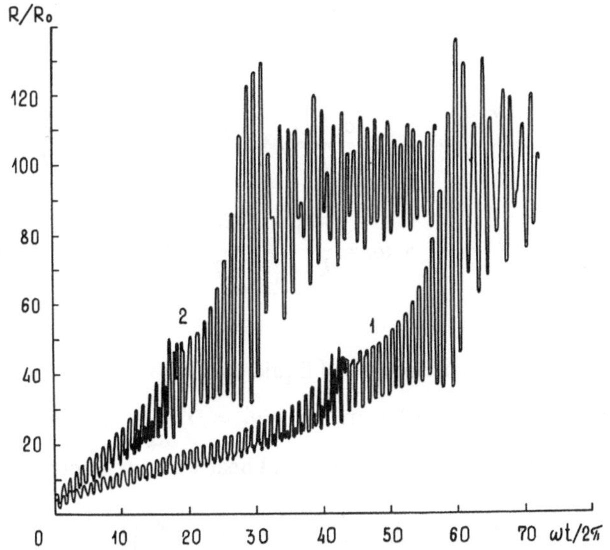

Figure 2. Vapour bubble pulsations in liquid nitrogen affected by a 10 kHz acoustic field for different amplitudes of pressure p_m.

$$T_0 = 77.4\,K; \Delta p = 0.05\,\text{atm}; R_0 = 5.10^{-4}\,\text{cm}.$$

Curve 1 — $p_m = 0.4$ atm, 2 — $p_m = 0.5$ atm.

where the pulsation amplitude of the radius R_m is considered to be small as compared to the average bubble radius $\bar{R}$ while the period of average size changes is considered to be great in comparison with the sound wave period of $2\pi/\omega$. A similar expression may be written for the changes in time of the vapour mass M', the temperature and other liquid and vapour parameters

$$M'(t) = \overline{M'(t)} + M'_m e^{i\omega t}, \tag{16}$$

$$T'(t) = \overline{T'(t)} + T'_m e^{i\omega t}, \ldots \tag{17}$$

In the linear approximation the following expression for R_m is found

$$R_m(t) = -p_m \overline{R(t)} \frac{K}{3Q}, \tag{18}$$

where K and Q are certain functions. Thus the amplitude of R_m bubble pulsations is linearly dependent on the mean radius $\bar{R}$, the pressure amplitude p_m and the function K but is inversely proportional to function Q. K determines the compressibility of the bubble and in the most general form it is as follows

$$K = \frac{3\rho c_p}{2\rho' L}\left(\frac{\partial T}{\partial \rho}\right)_\sigma \mathscr{R}\,\frac{\sqrt{2D/\omega}}{\bar{R}}\left(1 - i - i\,\frac{\sqrt{2D/\omega}}{\bar{R}}\right)$$

$$- \left(1 + \nu_1' + i\omega\frac{\alpha' T' D'}{L}\mathscr{R}\right)\left(\frac{1}{2} - B_\sigma\right)\frac{3c_p' f(k_1'\bar{R})}{\alpha' T'\omega^2 \bar{R}^2}\left(\frac{\partial T}{\partial p}\right)_\sigma$$

$$- \left(1 + \nu_2' + i\omega\frac{\alpha' T' D'}{L}\mathscr{R}\right)\left(\frac{1}{2} - B_\sigma\right)\frac{3c_p' f(k_2'\bar{R})}{\alpha' T'\omega^2 \bar{R}^2}\left(\frac{\partial T}{\partial p}\right)_\sigma, \tag{19}$$

where D and D' are the coefficients of liquid and vapour temperature conductivity and the nondimensional numbers of $\nu_1' = D'\dfrac{k_1'^2}{i\omega}2$ and $\nu_2' = D'\dfrac{k_2'^2}{i\omega}$ are connected with wave numbers of sound and heat waves respectively:

$$k_1' \simeq \frac{\omega}{c'}\left[1 + i\frac{\omega}{2\rho' c'^2}\left(\frac{4}{3}\eta' + \zeta' + \frac{\kappa'}{c_v'} - \frac{\kappa'}{c_p'}\right)\right], \tag{20}$$

$$k_2' \simeq \frac{1 + i}{\sqrt{2D'/\omega}}\left[1 + i\frac{\omega}{2\rho' c'^2}(\gamma' - 1)\left(\frac{4}{3}\eta' - \zeta' - \frac{\kappa'}{c_p'}\right)\right], \tag{21}$$

and $\gamma' = c_p'/c_v'$. Using these nondimensional numbers one determines the coefficient

$$B_\sigma = \left(\frac{\nu_1' + \nu_2'}{2} - \frac{c_\sigma'}{c_p'}\right)(\nu_2' - \nu_1')^{-1}, \tag{22}$$

where $c_\sigma' = T'(\mathrm{d}s'/\mathrm{d}T')_\sigma$ is the vapour capacity along the phase equilibrium curve. The function $f(k\bar{R})$ has the form of $f(k\bar{R}) = k\bar{R}\,\mathrm{cth}\,k\bar{R} - 1$. The calculation of the phase transition nonequilibrium at evaporation and condensation is effected by means of the relaxation factor

$$\mathscr{R} = \frac{1 + i\omega\tau_1}{1 + i\omega\tau_2} \tag{23}$$

in which the relaxation times of τ_1 and τ_2 respectively are

$$\tau_1 \simeq -\frac{\bar{R}}{U_{\text{ev}}}\,\frac{c_p' - c_v'}{c_p' - c_\sigma'}\,\frac{f(k_2'\bar{R})}{k_2'\bar{R}}, \tag{24}$$

$$\tau_2 \simeq \frac{\bar{R}}{U_{\text{ev}}}\left(\frac{\partial T}{\partial p}\right)_\sigma\left[\frac{\rho c_p}{2\rho'\beta' L}\,\frac{\sqrt{2D/\omega}}{R}\,\times\right.$$

$$\left.\left(1 - i - i\,\frac{\sqrt{2D/\omega}}{\bar{R}}\right) - \frac{c_p'}{\beta' L}\,\frac{f(k_2'\bar{R})}{k_2'\bar{R}}\right]. \tag{25}$$

Provided that $\omega \ll \min\left\{\frac{1}{\tau_1}, \frac{1}{\tau_2}\right\}$, the mass exchange process can be considered to be in quasi-equilibrium, whence $\mathscr{R} = 1$. This corresponds to the phase interface equals the saturated vapour pressure. At $|\omega\tau| \gg 1$ mass exchange is absent and the expression for the vapour bubble compressibility (19) takes the form of the expression for gas bubble compressibility.

The function Q in (18) determines the resonance quality of the vapour bubble and is expressed as

$$Q = 1 - \left(\frac{\rho\omega^2\bar{R}^2}{1 + ik_1\bar{R}} + \frac{2\sigma}{\bar{R}} - i4\eta\omega\right)\frac{K}{3} + \frac{\rho' c_p'}{\alpha' T'}\left(\frac{\partial T}{\partial p}\right)_\sigma \tilde{U}\,, \tag{26}$$

$$\tilde{U} = (1 - \nu_1')(\tfrac{1}{2} + B_\sigma) f(k_1'\bar{R}) + (1 - \nu_2')(\tfrac{1}{2} - B_\sigma) f(k_2'\bar{R}).$$

Q shows how much the changing pressure amplitude inside the bubble p_m' differs from the internal pressure amplitude according to the equation

$$Q = \frac{p_m}{p_m'}\,. \tag{27}$$

In order to determine the dependence of the mean radius $\overline{R(t)}$ on time, it is necessary to solve for the nonstationary vapour bubble dynamics in the acoustic field. In the quadratic approximation, it terms of perturbation theory, such a solution has the form of the differential equation

$$\frac{d\bar{R}}{dt} = \frac{\kappa}{\rho' L\bar{R}}\left[\frac{p_m^2}{|Q|^2}\sum_{i=1}^{8} A_i(\bar{R}, \omega) - \left(\Delta p + \frac{2\sigma}{\bar{R}}\right)\left(\frac{\partial T}{\partial p}\right)_\sigma\right] \tag{28}$$

where

$$A_1 = \frac{2}{3}\left(\frac{\partial T}{\partial p}\right)_\sigma [(1 - 3\,\mathrm{Re}\,F_5)\,\mathrm{Re}\,K + 3\,\mathrm{Im}\,F_5\,\mathrm{Im}\,K]\,,$$

$$A_2 = -\frac{1}{4\kappa}\frac{d\kappa}{dT}\left(\frac{\partial T}{\partial p}\right)_\sigma, \qquad A_3 = -\frac{1}{4}\left(\frac{\partial^2 T}{\partial p^2}\right)_\sigma,$$

$$A_4 = \frac{\rho' c_\sigma'}{3\rho c_p}\left(\frac{\partial T}{\partial p}\right)_\sigma \cdot \frac{\bar{R}^2}{2D/\omega}\,\mathrm{Im}\,K,$$

$$A_5 = -\frac{\rho'}{3\rho c_p}\frac{dL}{dT}\left(\frac{\partial T}{\partial p}\right)_\sigma \frac{\bar{R}^2}{2D/\omega}\,\mathrm{Im}\,K,$$

$$A_6 = \frac{\rho'}{3\rho}\left(\frac{\partial T}{\partial p}\right)_\sigma \frac{\bar{R}^2}{2D/\omega}\left[-(1-\operatorname{Re}F_3)\operatorname{Im}K\right.$$

$$\left.+\operatorname{Im}F_3\left(\operatorname{Re}K-\frac{1}{\rho'}\left(\frac{\partial\rho'}{\partial p}\right)_\sigma\right)\right],$$

$$A_7 = -\frac{\rho\omega^2\bar{R}^2}{36}\left(\frac{\partial T}{\partial p}\right)_\sigma |K|^2, \qquad A_8 = \frac{\eta\omega}{3\rho c_p}\frac{\bar{R}^2}{2D/\omega}|K|^2.$$

The functions of F_3 and F_5 are given by the integral

$$F_n(x) = \int_0^\infty \frac{\mathrm{d}t}{(1+t)^n}\,\mathrm{e}^{-(1+i)xt}$$

at $n = 3$ and $n = 5$ respectively, where $x = \bar{R}\sqrt{\omega/2D}$. If in equation 28 the expression in square brackets is larger than zero, the mean radius $\overline{R(t)}$ grows in time due to nonlinear mechanisms determining the phenomenon of rectified heat and mass transfer at vapour bubble pulsations in the field of periodically changing pressure. This phenomenon is somewhat analogous to the phenomenon of the rectified gas diffusion determining the growth of gas bubbles in water with a sound field effect [11]. The physical meaning of the components $A_i(\bar{R}, \omega)$ is treated in Refs. [1–4, 6]. The growth of vapour bubble due to rectified heat and mass transfer is determined by the appearance of a heat flux average in time from liquid to bubbles. This flux appears in the field of periodically changing pressure even in the case when the liquid is statically overpressed, i.e. when the static pressure p_0 is higher than the saturated vapour pressure $p_\sigma(T_0)$ with $\Delta p > 0$.

Resonance frequencies of vapour bubbles

Consider the expression (18) for the amplitude of the bubble radius R_m and introduce the response function

$$\frac{R_m}{R} = -\frac{K}{3Q}\,p_m \tag{29}$$

which appears to be a nondimensional value linearly dependent on the pressure amplitude of the exciting field p_m. Figure 3 shows the dependence of the function $R_m/\bar{R}$ on the mean radius for vapour bubbles in water with temperature of 150°C exposed to an acoustic field with the amplitude of $p_m = 1$ atm. The curves presented correspond to different frequencies of the field in the range of 400 Hz to 1.25 MHz. The given results indicate a resonance nature of the $R_m/\bar{R}$ function dependence on the radius of each excitation frequency. The graphs show that the vapour bubble quality factor is strongly dependent on its size: as the vapour bubble radius decreases, its quality factor decreases too. When the bubble size is less than a definite one, its quality factor reduces to less than a unit and resonance vanishes. In Figure 3 the gently sloping extremum of the response function of small sizes attracts attention. It depends on the effect of surface tension.

On the basis of the $R_m/\bar{R}$ response function analysis, one can build the dependence of the resonance f_0 frequency of vapour bubbles upon the radius $\bar{R}$. Figure 4 presents such $f_0(\bar{R})$ dependencies for vapour bubbles in

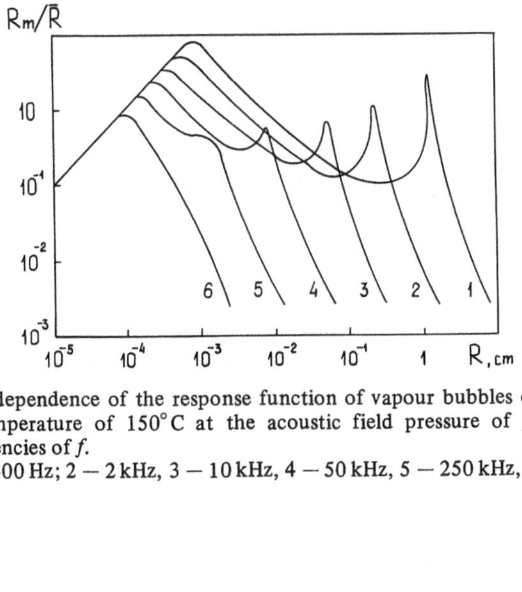

Figure 3. The dependence of the response function of vapour bubbles on their sizes in water with temperature of 150°C at the acoustic field pressure of $p_m = 1$ atm for different frequencies of f.
Curve 1 — $f = 400$ Hz; 2 — 2 kHz, 3 — 10 kHz, 4 — 50 kHz, 5 — 250 kHz, 6 — 1.25 MHz.

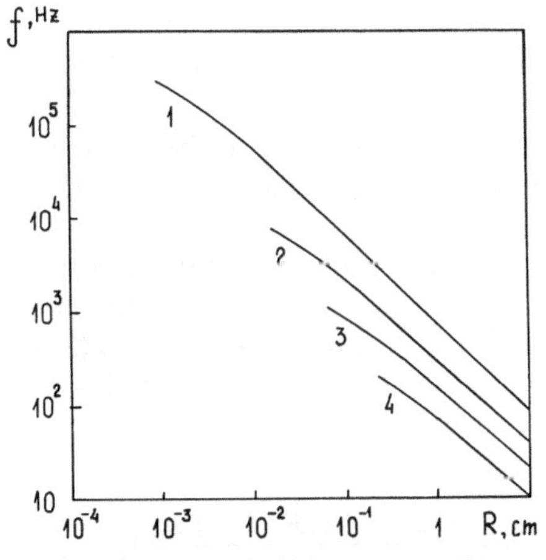

Figure 4. The dependence of the vapour bubble resonance frequencies on their sizes in water at different temperature of T_0.
Curve 1 — $T_0 = 150$°C, 2 — 100°C, 3 — 80°C, 4 — 60°C.

water with different temperature T_0. Left-hand boundaries of the graphs correspond to the vanishing of resonance in the response function due to the reduction of the bubble quality factor.

The vapour bubble resonance frequency may be determined analytically using the condition that the frequency derivative of the module of the complex value in the denominator of the response function is equal to zero (29). When Q is expressed in the most general form (26), it may lead to very cumbersome forms. There is an approximate formula valid at fairly large bubble sizes when the quality factor is sufficiently high [3]

$$\omega_0^2 = \frac{1}{\rho \bar{R}^2}\left(\frac{3\,\mathrm{Re}\,K}{|K|^2} - \frac{2\sigma}{\bar{R}} - 4\eta\omega_0\frac{\mathrm{Im}\,K}{\mathrm{Re}\,K}\right). \tag{30}$$

If the viscosity and surface tension of liquid are disregarded and if it is taken into account that with great bubble radii the compressibility modules $|K|$ is mainly determined by $\mathrm{Re}\,K$, we obtain

$$\omega_0^2 \simeq \frac{1}{\rho \bar{R}}\left(\frac{3}{\mathrm{Re}\,K}\right).$$

It can easily be seen that the formula is similar to some extent to the formula for resonance frequencies of gas bubbles originally obtained by Minnaert [10].

Thresholds of the vapour bubble growth

Consider eqution (28) characterizing the change of the mean vapour bubble radius $\overline{R(t)}$ in time in the field of periodically changing pressure. The first term in square brackets leads to the growth of the vapour bubble due to rectified heat and mass transfer in an acoustic field while the other one is connected with the appearance of static heat transfer and, in case when overcompression $\Delta p > 0$, it may lead to the collapse of the vapour bubble. Depending on which of the terms is predominant, either growth ($\mathrm{d}\bar{R}/\mathrm{d}t > 0$) or collapse ($\mathrm{d}\bar{R}/\mathrm{d}t < 0$) of vapour bubbles occurs.

With a certain amplitude of the acoustic field $p_m = p_m^*$, the condition $\mathrm{d}\bar{R}/\mathrm{d}t = 0$ is fulfilled which determines the vapour bubble growth owing to rectified heat and mass transfer. From equation (28) one can easily obtain a threshold value of the acoustic pressure amplitude p_m^* corresponding to the condition $\mathrm{d}\bar{R}/\mathrm{d}t = 0$

$$p_m^* = |Q|\left\{\frac{(\Delta p + 2\sigma/\bar{R})\,(\partial T/\partial p)_\sigma}{\sum_{i=1}^{8} A_i(\bar{R}, \omega)}\right\}^{1/2}. \tag{31}$$

It is seen that the threshold pressure p_m^* decreases with increase of ΣA_i and with decreasing static overcompression of the bubble. Besides, the value of the threshold pressure p_m^* is linearly dependent on the modulus $|Q|$ of the resonance function which has a minimum value at resonance and increases when moving away from resonance. Therefore, when the sizes of cavitation vapour nuclei coincide with resonance values, the liquid has minimum threshold pressures of the acoustic field.

Figure 5 presents the dependencies of p_m^* growth thresholds on the vapour bubble radius $\bar{R}$ calculated from formula (31) in water with temperature of 150°C in the absence of static overcompression for different acoustic field frequencies. Here the dashed line shows the static thresholds of the liquid with vapour nuclei evaluated by the well-known formula [9] $p^* = \Delta p + 2\sigma/\bar{R}$ valid only for statically stretching liquids. Figure 6 presents the dependencies of p_m^* growth thresholds upon radius $\bar{R}$ with different static overcompressions in water with temperature of 150°C affected by an acoustic 10 kHz field. It is clearly seen that with the absence of static overcompression, the p_m^* threshold pressure of an acoustic field exceeds static thresholds. However, as the static overcompression increases with the associated increase of the quality factor, the possibility to achieve static thresholds arises. At great overcompression values, the cases are possible when p_m^* threshold pressures of the acoustic field are less than static pressures. Physically it is easily explained by the fact that according to equation (27) at resonance with pulsations of high-quality bubbles when $|Q| < 1$, the pressure amplitude p_m' inside the bubble exceeds essentially the pressure amplitude p_m in

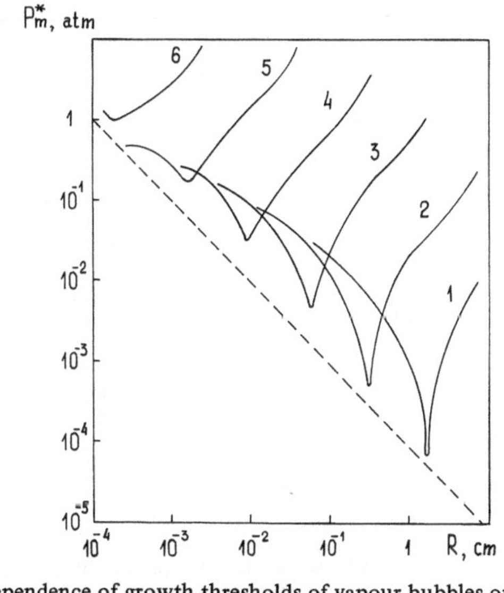

Figure 5. The dependence of growth thresholds of vapour bubbles on their sizes in water with temperature of 150°C at different frequencies of f.
Curve 1 – f = 400 Hz, 2 – 2 kHz, 3 – 10 kHz, 4 – 50 kHz, 5 – 250 kHz, 6 – 1.25 MHz.

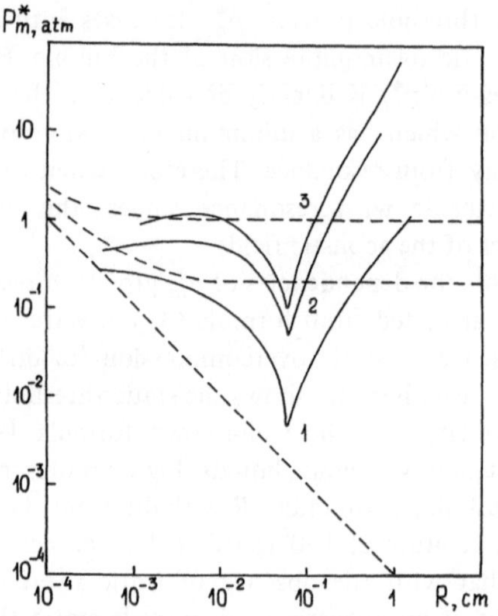

Figure 6. The influence of static overcompression upon growth thresholds of vapour bubbles in water with temperature of 150°C at frequency of 10 kHz.
Curve 1 — $\Delta p = 0$, 2 — 0.2 atm, 3 — 1 atm.

liquid. The ratio p'_m/p_m equals the quality factor of vapour bubbles and in some cases it may prove to be sufficiently large [3].

Figure 7 presents p_m^* threshold pressures of the acoustic field whose excess induces the growth of vapour bubbles in liquid hydrogen with

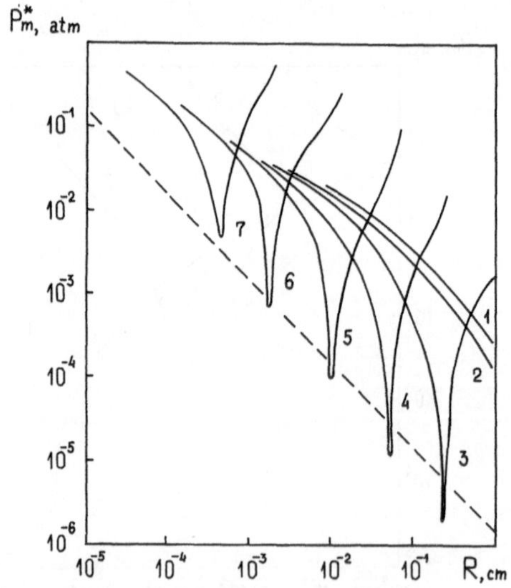

Figure 7. The dependence of growth thresholds of vapour bubbles on their sizes in liquid hydrogen with temperature of 27 K at different frequencies of f.
Curve 1 — f = 400 Hz, 2 — 2 kHz, 3 — 10 kHz, 4 — 50 kHz, 5 — 250 kHz, 6 — 1.25 MHz, 7 — 6.25 MHz.

temperature of 27°K when static overcompression is equal to zero. It is clearly seen that in this case threshold pressure amplitudes may be less than static thresholds even at the absence of static overcompression which depends on the quality factor of vapour bubbles in liquid hydrogen [3].

The results presented indicate the influence of resonance vapour bubble properties upon the thresholds of their growth. This phenomenon has the corresponding analogy in the gas cavitation theory [7]. Characteristically, near the resonance of nucleus vapour bubbles the growth thresholds under the influence of a periodically changing pressure field may prove to be essentially less than the static threshold of the liquid tensile strength.

Acknowledgements

In conclusion the author would like to express his gratitude to his colleagues V.N. Alekseev and V.P. Yushin for their fruitful cooperation in conducting the investigations the results of which are briefly described in this paper.

References

1. Akulichev VA, Alekseev VN and Naugolnykh KA (1971) On dynamics of vapour bubbles in liquid hydrogen ultrasonic bubble chambers. Akusticheskii Zhurnal (Russian) 17 (3): 356–364.
2. Akulichev VA, Zhukov VA and Tkachev LG (1977) Ultrasonic bubble chambers. Physics of Elementary Particles and Atomic Nuclei (Russian) 8 (3): 580–630.
3. Akulichev VA (1978) Cavitation in cryogenic and boiling liquids (Russian). Moscow: 'Nauka'.
4. Akulichev VA, Alekseev VN and Yushin VP (1979) Growth of vapour bubbles in ultrasonic field. Akusticheskii Zhurnal (Russian) 25 (6): 801–809.
5. Alekseev VN (1975) Stationary behaviour of vapour bubble in ultrasonic field. Akusticheskii Zhurnal (Russian) 21 (4): 497–501.
6. Alekseev VN (1976) Nonstationary behaviour of vapour bubble in ultrasonic field. Akusticheskii Zhurnal (Russian) 22 (2): 185–191.
7. Eller AE (1975) Effects of diffusion on gaseous cavitation bubbles. J Acoust Soc America 57 (6): 1374–1378.
8. Finch RD and Nepirras EA (1973) Vapor bubble dynamics. J Acoust Soc America 53 (5): 1402–1410.
9. Flynn HG (1964) Physics of acoustic cavitation in liquids. In Mason WP (ed) Physical acoustics, Vol 1b. New York: Academic Press.
10. Minnaert M (1933) On musical air bubbles and the sounds of running water. Phylos Mag 16 (7): 235–243.
11. Plesset MS and Hsieh DY (1960) Theory of gas bubble dynamics in oscillating pressure fields. Phys Fluids 3 (6): 882–885.

Boiling in a porous bed

V.K. DHIR and IVAN CATTON

University of California, Los Angeles, USA

Abstract. Experimental observations of the heat generation rate at which a porous bed dries out have been made for bed particle sizes ranging from 245 to 4,783 microns in beds up to 40 cm deep with different coolants. The governing mechanisms are identified and used to develop models of the observed phenomena.

Introduction

A great deal of attention has been given to answering the question of whether or not a given porous bed will dry out when the volumetric heating rate is at some prescribed value. The governing mechanism leading to dryout is buoyancy driven, countercurrent flow in a porous medium with phase change. Dryout is simply the condition where the liquid phase cannot reach all portions of the bed.

Disagreement exists concerning some of the mechanisms governing various aspects of the process. Some believe surface tension is important [14, 15] whereas others do not. In their modeling some have tried to account in an approximate way for vapor flux as well as liquid flux flow resistance [8, 11, 12] while others have neglected it [1–5, 9]. A considerable body of experimental data has been collected for a large number of material combinations as well as for several different methods of heating.

Analysis

Visual observations of boiling in porous layers showed that for particles less than 1 mm in diameter vapor moved in the porous layer in the form of channels. These vapor channels were free of any particles. In deep beds the vapor channels extended only part way into the bed while in shallow beds the channels were well formed throughout the bed. In beds of large diameter particles, vapor could not push the particles aside and thus appeared to move through the interstitial spaces between the particles. Keeping these physical differences in mind, separate models for dryout heat fluxes in deep and shallow beds of small particles and deep beds of large particles are developed.

Small particle diameter deep bed

Figure 1 shows the flow model proposed in [3] for a volume heated bed. The dryout heat flux in a deep bed is controlled by the maximum coolant flow rate through the lower packed region of the bed. The liquid encounters minimal

Applied Scientific Research 38: 69–76 (1982) 0003–6994/82/0381–0069 $01.20.

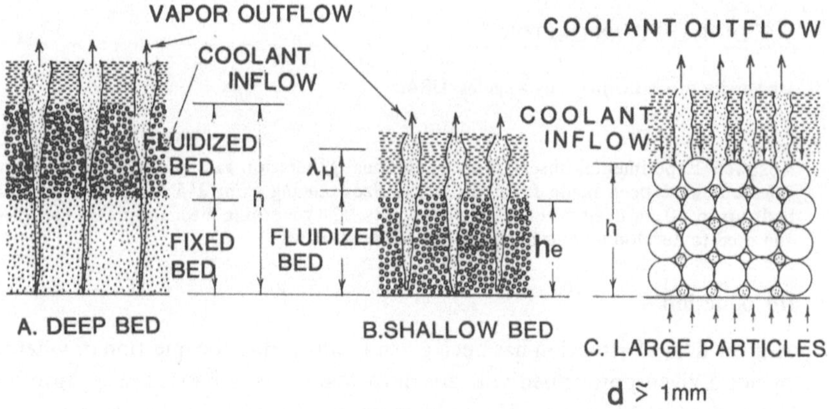

Figure 1. Models for dryout heat flux.

resistance in the upper levitated region of the bed. The bed dries out at a particular location when gravity can no longer maintain the necessary flow rate to compensate for evaporation rate.

Under the assumptions of (1) no liquid inertia, (2) inviscid vapor, (3) small cross-sectional area occupied by the vapor, and (4) spherical particles of near uniform size, the dryout heat flux, normalized with Zuber's value of the critical heat flux [16], is given by

$$\frac{q_{vd}}{q_z} = \frac{Q_v h(1-\epsilon)}{q_z} = C_1 K d^{*2} \Gamma^{1/2} M, \tag{1}$$

where $Q_v h$, ϵ and q_z are the particle heat generation rate per unit volume, bed height, bed void fraction and Zuber's value of the critical heat flux [16] with

$$K = \frac{\epsilon^3}{(1-\epsilon)^2}, \qquad \Gamma = \rho_f/\rho_g,$$

$$M = \left[\frac{\rho_f^4 \sigma^3}{\mu_f^4 g(\rho_f-\rho_g)^3}\right]^{1/4}, \qquad d^* = \frac{d}{[\sigma/g(\rho_f-\rho_g)]^{1/2}},$$

and C_1 a constant to be determined from experiment. The variables ρ_f, ρ_g, σ, μ_g, g and d are coolant density, vapor density, surface tension, viscosity, gravitational acceleration and particle diameter respectively.

Small particle diameter shallow bed

In shallow beds, the particles are in a fluidized state and the coolant has easy access to the particles throughout the bed. The flow of vapor in the channels is similar to that in a tube. A hydrostatic head equal to the height of the bed drives the vapor in the channels. The flow configuration in the channels within the bed is very stable. If, however, the vapor velocity exceeds a certain critical value, the vapor jets in the overlying liquid become unstable and block the entering coolant. A deficiency in liquid inflow leads to dryout. The dryout heat flux in shallow beds is, thus, limited by the maximum or critical velocity of the vapor jets.

To determine the dryout heat flux, the approach followed is that of Zuber [16]. The vapor velocity out of the bed and relationships between critical wavelength and critical velocity are used to obtain

$$\frac{q_{vs}}{q_z} = 1.84 \frac{24}{\pi} \frac{nA_c}{A_v} \frac{1}{f^{1/4}},$$

where n, A_c, A_v and f are the number of vapor channels per unit area, the average vapor channel cross-sectional area at the top of the bed, the bed cross-sectional area and a friction factor for the vapor channel respectively.

The area ratio, noting that the cross-sectional area of the bed occupied by vapor is $\pi/16$ when $h \to 0$, is

$$\frac{nA_c}{A_v} = \frac{\pi}{16} \epsilon_e \left[1 - \frac{h(1-\epsilon)}{h_e(1-\epsilon_e)}\right],$$

where the subscript e refers to the expanded bed. With this expression and the assumption that $\epsilon_e/f^{1/4}$ depends only on the dimensionless channel diameter allows one to write

$$\frac{q_{vs}}{q_z} = 1.84 C_2 \left[1 - \frac{h^*(1-\epsilon)}{C_3}\right], \tag{2}$$

where h^* is bed depth scaled by $[\sigma/g(\rho_f - \rho_g)]^{1/2}$. The constants C_2 and C_3 are to be determined from experiment.

Large particle diameter deep bed

In porous layers composed of particles typically of diameter on the order of 1 mm or more, vapor moves through interstitial spaces between the particles. Under such a situation both liquid and vapor resistances need to be considered. Also, as the particle size becomes large the flow changes from laminar to turbulent. Now neglect of inertia term in the Kozeny-Carman relation is not appropriate as was done earlier for small particles. In fact, as the particles become large, the flow resistance mainly comes from the term involving square of the superficial velocity of both liquid and vapor. Going through mechanistic details of countercurrent flow of two immiscible fluids in a porous layer, an expression for limiting fluxes of two fluids was obtained in [7] as

$$j_f^{*2} + j_g^{*2} = 0.33,$$

where

$$j_f^* = j_f \sqrt{\rho_f} \,/\, \sqrt{gd\epsilon^3(\rho_g - \rho_g)/[6(1-\epsilon)]} \;,$$
$$j_g^* = j_g \sqrt{\rho_g} \,/\, \sqrt{gd\epsilon^3(\rho_f - \rho_g)/[6(1-\epsilon)]} \,.$$

Using the continuity equation and the energy equation

$$j_f\rho_f = j_g\rho_g; \qquad q = h_{fg}\rho_f j_f.$$

An expression for the dryout heat flux is obtained as

$$q = \frac{0.234 h_{fg} \sqrt{\rho_g(\rho_f - \rho_g) gd\epsilon^3/(1-\epsilon)}}{[1 + 0.69\rho_g/\rho_f]^{1/2}} \,. \tag{3}$$

Equation (3) should be applicable to a bottom heated porous layer unless limiting conditions occur at the plate. For a volumetrically heated porous layer, equation (3) should hold good but with a different numerical constant.

Experiment

Experiments were performed by inductively heating metallic particles using the apparatus shown schematically in Figure 2. Calibration of the heat generation rate in the particles as a function of the frequency generator power was made before and after each dryout heat flux measurement. The calibration was accomplished by measuring the initial rate of temperature rise and multiplying by the thermal capacity of the bed (particles and coolant). A detailed description of the experimental apparatus and procedure can be found in [2].

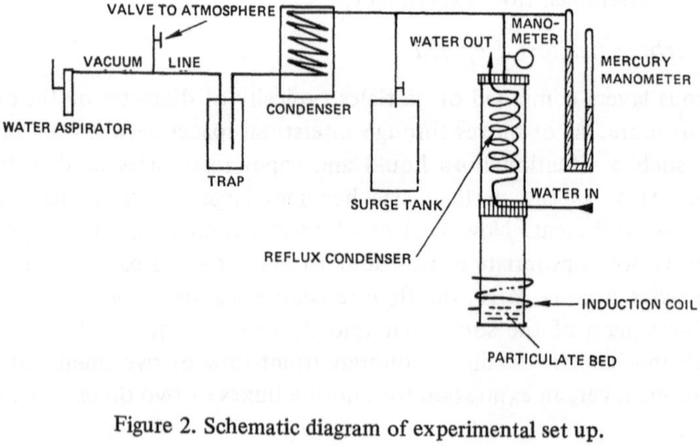

Figure 2. Schematic diagram of experimental set up.

Results

Measurements have been made for particle sizes in the range from 295 to 4,173 microns and for three different coolants. Bed heights were varied from 1.9 to 40 cm and the bed porosity from 0.38 to 0.45. A number of dryout heat flux observations for shallow beds were made.

The data for volume heated deep beds of small particles are plotted in Figure 3 with the correlation equation (1). The data of Gabor et al. [9] are for UO_2 particles with water and sodium as coolants. The data representing a three-fold variation in d^*, a seven-fold variation in M and K and a three-fold variation in Γ are correlated reasonably well when C_1 is taken to be 7.5×10^{-4}. The data are tabulated in [2]. To insure that the dryout heat flux was independent of bed depth, measurements were made for beds up to 40 cm deep [6]. No bed depth dependence was evident.

Dryout heat flux data for shallow beds are shown in Figure 4 as a function of dimensionless bed height for different particle sizes and coolants. The data are correlated to within 25% when the empirical constants C_2 and C_3 in equation (2) are taken to be 1 and 10.87 respectively. The dryout heat flux is seen to increase linearly with decreasing bed depth as anticipated.

Dryout heat flux measurements for beds of large diameter particles are given in Figure 5. It appears that within experimental uncertainty the heat flux ratio is independent of bed depth. Also, as the particle diameter increases, the dryout heat flux increases. For all of the data shown, the overlying fluid layer depth was one tenth of the bed depth.

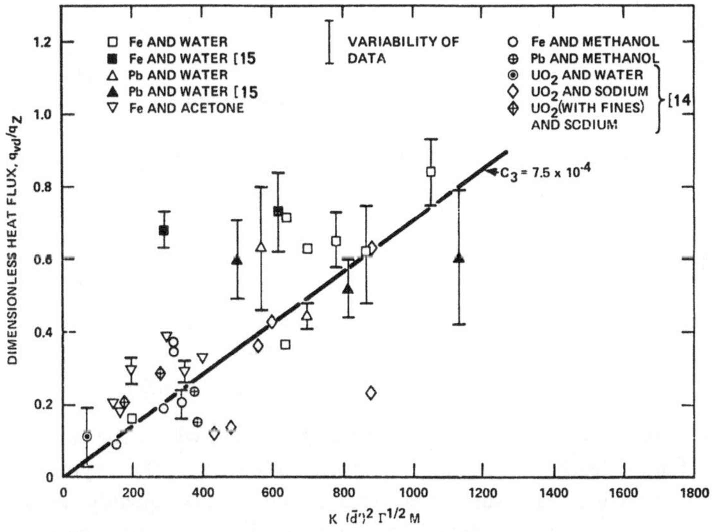

Figure 3. Dimensionless dryout heat flux in volume heated deep beds.

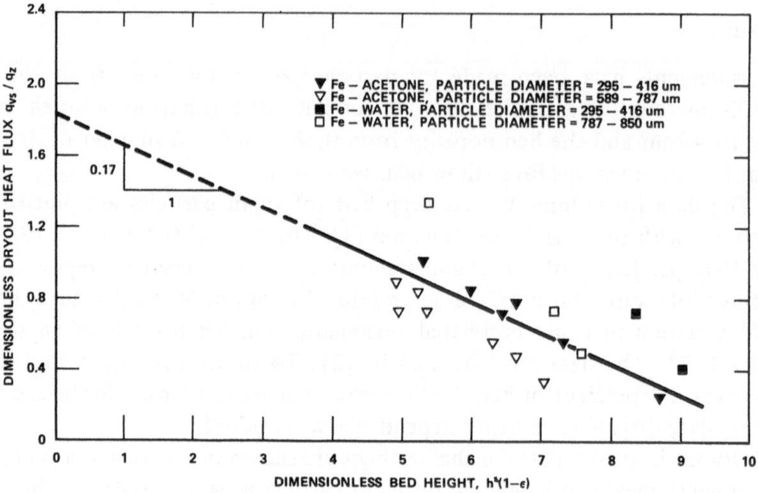

Figure 4. Dimensionless dryout heat flux in volumetrically heated shallow beds.

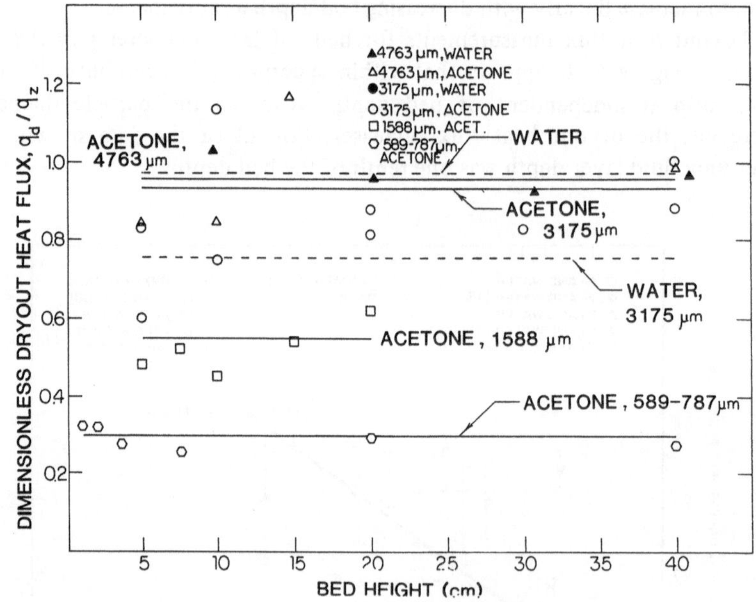

Figure 5. Experimental results for large particles.

In contrast with the small particle experiments where the flow processes were fairly steady, a surging phenomena initiated as the dryout heat flux was approached. The surging is periodic with the frequency decreasing to zero as the dryout heat flux is approached. The bed temperature was also periodic with the maximum value occurring when the liquid was expelled.

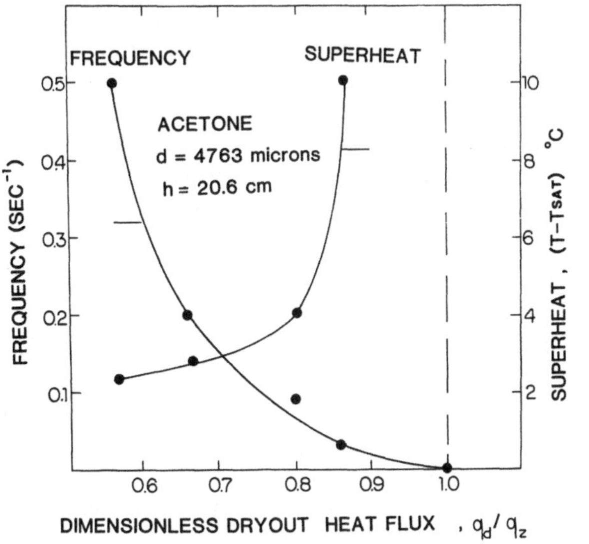

Figure 6. Surging frequency and bed superheat.

The magnitude of the temperature oscillation increased as the dryout heat flux was approached. Results for acetone are shown in Figure 6.

Figure 7 compares the observed dryout heat fluxes in volumetrically heated porous layers with the models developed for porous layers of small and large diameter particles. The data plotted in Figure 7 are for deep beds saturated with acetone. It is seen that for particles less than 1000 μm, our [3] equation (1) is in good agreement with the data. Prediction of Hardee and Nilson [10] gives much higher heat fluxes while that of Gabor et al. [8] tends to give much lower heat fluxes. Equation (3), although developed for bottom

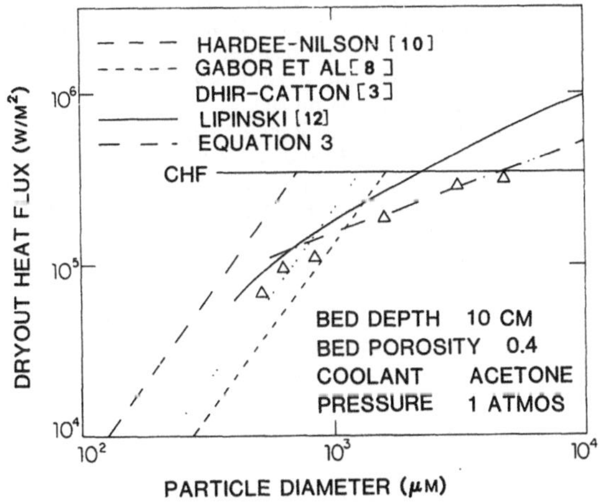

Figure 7. A comparison of dryout heat flux predictions with experimental data.

heated or porous layers, is seen to compare well with volumetrically heated large particle data. Lipinski's model predicts higher dryout heat fluxes. Both models predict that dryout heat fluxes for porous layers of very large particles can be greater than critical heat fluxes on flat plates. However, present data are insufficient to provide a confirmation of the limiting values of the dryout heat fluxes on porous layers of very large particles.

References

1. Dhir VK and Catton I (1976) Prediction of Dryout Heat Fluxes in Beds of Volumetrically Heated Particles. Proceedings of Fast Reactor Safety Conference, Chicago, CONF 761001, p. 2026.
2. Dhir VK and Catton I (1977) Dryout Heat Fluxes in Beds of Inductively Heated Particles NUREG-0262.
3. Dhir VK and Catton I (1977) Dryout Heat Fluxes for Inductively Heated Particle Beds. J Heat Transfer 99: 251–256.
4. Dhir VK and Catton I (1978) The effect of Particle Size Distribution and Nonheated Constituents on Debris Bed Dryout ANL-78-10.
5. Dhir VK (1979) Dryout Heat Fluxes in Debris Beds containing Particles of Different Size Distributions. Proceedings of the International Meeting on Fast Reactor Safety Technology, Seattle, p. 770.
6. Dhir VK and Catton I (1980) Dryout Heat Fluxes in Very Deep Debris Beds. Transactions of the ANS/ENS International Conference Washington DC.
7. Dhir VK and Barleon L. On Counter Current Flooding in Porous Layers (submitted to Letters in Heat and Mass Transfer).
8. Gabor JD, Epstein M, Jones SW and Cassulo JC (1980) Status Report on Limiting Heat Fluxes in Debris Beds. Argonne National Laboratories Report ANL/RAS 80–21.
9. Gabor JD, Sowa ES, Baker Jr L and Cassulo JC (1974) Stuides and Experiments on Heat Removal from Debris in Sodium. Proceedings of ANS Fast Reactor Safety Conference, Beverly Hills, CONF 740701, p. 823.
10. Hardee HC and Nilson RN (1977) Natural Convection in Porous Media with Heat Generation. Nuclear Science and Engineering 63:119–132.
11. Lipinski RJ and Rivard JB (1979) Debris Bed Heat Removal Models: Boiling and Dryout with Top and Bottom Cooling. Proceedings of the International Meeting on Fast Reactor Safety Technology, Seattle, p. 757.
12. Lipinski RJ (1980) Particle Bed Dryout Model. Transactions of the ANS/ENS International Conference Washington DC.
13. Ostensen RW (1980) Advanced Reactor Safety Research Program. Quarterly Report, October-December 1979 Sandia, Albuquerque, New Mexico.
14. Shires GL and Stevens GF (1980) Dryout During Boiling in Heated Particulate Beds. United Kingdom Atomic Energy Authority Report AEEW-M1779.
15. Trenbert R and Stevens GF (1980) An Experimental Study of Boiling Heat Transfer and Dryout in Heated Particulate Beds. AEE Winfrith Dorchester Report AEWW-R 1342.
16. Zuber N (1959) Hydrodynamic Aspects of Boiling Heat Transfer. AEC Report No. AECU-4439, Physics and Mathematics.

Growth and departure of individual bubbles at a wall

M.G. COOPER

Department of Engineering Science, Oxford University, UK

1. Introduction

This paper is intended to describe in outline recent work with several research students to improve fundamental understanding of boiling, through a largely experimental study of individual bubbles of vapour growing at a wall. The outline plan is indicated in Table 1, which shows the selection of operating conditions. The aim was to start with the simplest possible situation (liquid initially stagnant and isothermally supersaturated, at zero gravity). Then, having understood that, aiming to proceed by adding complications successively.

Table 1.

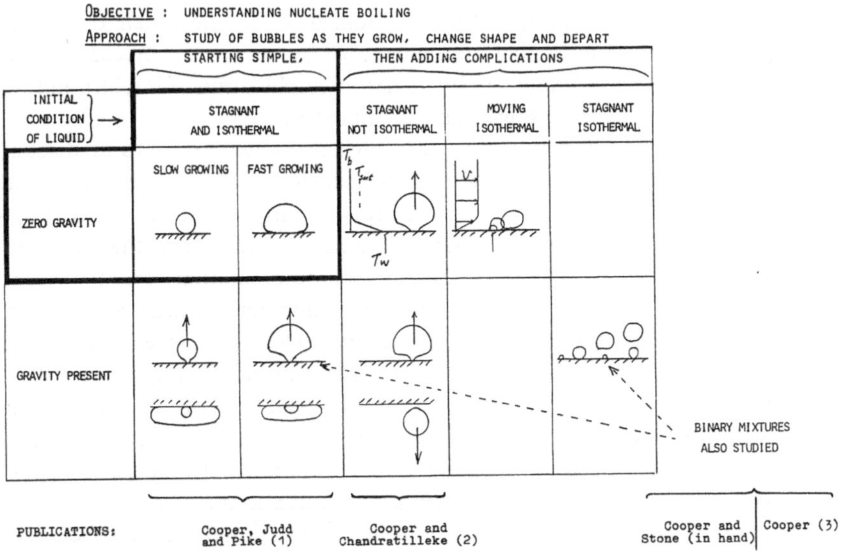
OBJECTIVE : UNDERSTANDING NUCLEATE BOILING
APPROACH : STUDY OF BUBBLES AS THEY GROW, CHANGE SHAPE AND DEPART
STARTING SIMPLE,
THEN ADDING COMPLICATIONS
INITIAL CONDITION OF LIQUID
STAGNANT AND ISOTHERMAL
STAGNANT NOT ISOTHERMAL
MOVING ISOTHERMAL
STAGNANT ISOTHERMAL
SLOW GROWING
FAST GROWING
ZERO GRAVITY
GRAVITY PRESENT
BINARY MIXTURES ALSO STUDIED
PUBLICATIONS:
Cooper, Judd and Pike (1)
Cooper and Chandratilleke (2)
Cooper and Stone (in hand)
Cooper (3)

Those aims have been achieved, as we can claim a close understanding of that simplest case (described briefly in Section 3.1 below, and more fully in [1]) and we have since restored in succession various complications, several of which will also be described briefly below. Regarding that simplest case as category (1), we can categorise others below by listing the complications restored in each case (unless stated otherwise, the liquid is initially stagnant and isothermal, and gravity is absent); (2) positive, negative or fractional gravity restored, normal to the wall (Section 3.2 below and [1]); (3) known initial thermal boundary layer introduced in the liquid (Section 3.3 below

Applied Scientific Research 38: 77–84 (1982) 0003–6994/82/0381–0077 $01.20.

and [2]; (4) liquid initially in known motion parallel to the wall (Section 3.4 below and [3], work continuing).

Due to shortage of time and space, a further two categories of tests will not be discussed below, though they have been or are being carried out. They are: (5) tests with liquids which are binary mixtures ([3, 7]), (6) tests with pairs of bubbles, in which one bubble is created in initially stangant isothermal liquid with buoyant gravity, followed after a selected interval by another bubble from the same site ([7], work continuing).

In each category, these tests therefore explored, or at least entered at one point, an unknown regime of behaviour of bubbles growing at a wall. Tests have also been done with two or more of these complications, such as a combination of categories 2 and 3, involving gravity with initial thermal boundary layer ([2]).

Of course, in an actual boiling situation, many of these complications will be present simultaneously; our work is beginning to indicate which complications are significant in any given situation. However, boiling may not be overwhelmingly influenced by these phenomena of bubble dynamics: broadly speaking, the stage in bubble life represented by our work is preceded by the nucleation stage and succeeded by stages of combination of bubbles and two phase flow. Any of these stages may be important.

Faced with such complexity, an engineer often seeks to understand enough of the underlying phenomena to enable him to look sensibly at the complex problem as a whole. In the present context that means seeking to understand bubble dynamics as far as establishing trends and the interaction of simultaneous influences of growth and other motion, gravity, surface tension, etc. Specifically, we have established non-dimensional groups (t^*, g^*, δ_s^*, t^+, z^+, . . .) whose values indicate, for example, the relative magnitudes of stresses from bubble inertia ($\frac{1}{2}\rho\dot{R}^2$), surface tension ($2\sigma/R$), gravity ($\rho g R$), mainstream inertia ($\frac{1}{2}\rho v^2$) or indicate relative magnitudes of thermal boundary layer thickness (δ_s) compared with a characteristic length in bubble history. By suitably chosen experiments we have shown, not only the general importance of these stress ratios and other groups, but also the specific dependence of required parameters on specific groups, such as unexpected effects upon departure time, arising from surface tension and thickness of the thermal boundary layer.

The non-engineering, theoretical member of I.U.T.A.M. may aim instead to derive the behaviour of these bubbles by solving or computing the known equations of fluid mechanics and energy. Since this is a coupled problem of energy and fluid mechanics with unknown moving free surface showing insignificant effects of inertia, surface tension and viscosity, the computation would be difficult. It would be somewhat less difficult if the energy phenomena can be hived off, leaving a purely fluid mechanics problem. That can be done in the simplest case (category 1 above). The result is a purely fluid mechanics problem which is formulated in the appendix in terms of stream function and vorticity, but not solved. Instead the answer is given from our experiments, and the solution is left for the interested reader.

2. Experimental techniques and limitations

Experimental techniques include: producing a low gravity environment by use of a freely falling table; measuring the 'effective *g*' by a fly-ball accelerometer; producing a stagnant isothermal liquid by surrounding an inner test vessel with a heated jacket, reducing pressure in the test vessel to cause supersaturation just before the test; triggering bubbles, either by creating hot spots in thin film electrical circuits deposited on the surface of the wall, or by an electrolytic triggering bubble; observing rapid variations in the temperature at that surface by other thin film circuits; creating a known thermal boundary layer in the liquid by transient electrical heating in further thin film circuits, just before the test; creating relative motion between liquid and wall, either by operating a piston in the liquid, or by moving the wall, just before the test; automatic sequencing of these operations; high speed recording of temperature and pressure by ultra violet recorder or by microprocessor-controlled digital data acquisition; high speed cine photography. More details are given in the references.

Certain limitations were imposed by the size of the apparatus and the nature of the experiments. Due to the initial size of the triggered bubble, attainable pressures and supersaturation, etc., bubbles were typically observed for sizes from 1 mm to a few cm in diameter and for times up to 250 ms. Thermal boundary layers were normally a few hundred μm in thickness, but could be produced up to 4 mm thick. Mainstream flow was laminar with velocity up to 0.2 m/s and velocity boundary layer similarly up to a few mm thick. If thin film circuits were in use, the liquid must be of very low electrical conductivity. A range of organic liquids were used, and water when thin films were not in use. It would be desirable to use a liquid with radically different Prandtl number, but there do not seem to be any which are transparent and boil at reasonable temperature and pressure.

It was vital to produce the bubbles at the required time and place, so we used the triggering methods mentioned above. The penalty was that these experiments shed no light on nucleation, and also the earliest stages of bubble life (below 1 mm diameter) may not be representative of boiling.

For our bubbles, the rate of subsequent growth was controlled by rate of diffusion of heat, for virtually all of their lives. Any 'inertia controlled' stage, when liquid inertia significantly affected growth through affecting saturation temperature in the bubble, was confined to the first millisecond.

In general, our bubbles correspond directly to boiling at pressures below one atmosphere. It is therefore important to develop from them sufficient understanding to be able to apply the results to boiling at the higher pressures normally met in industry, where bubbles grow more slowly and are therefore generally nearly spherical, as discussed below.

3. Experimental results

3.1. Liquid initially stagnant and isothermal, zero gravity

It was found, not for the first time, that slow-growing bubbles tend to be spherical and faster-growing bubbles tend to be hemispherical. The growth rates were observed to be, for spherical bubbles about $2\,\mathrm{Ja}\sqrt{\alpha t}$, similar to diffusion based predictions for reasonably large Ja such as Plesset and Zwick [6] $2\sqrt{3/\pi}\,\mathrm{Ja}\sqrt{\alpha t}$. For hemispherical bubbles, it was close to $R = 3\,\mathrm{Ja}\sqrt{\alpha t}$, as predicted by analyses allowing for evaporation from a thin layer of liquid (microlayer) beneath the bubble [4].

As described in [1], it was noted that these two growth rates were both consistent with one simple expression for growth, namely

$$\frac{\mathrm{d}}{\mathrm{d}t}\,(\text{bubble volume}) = \mathrm{Ja}\sqrt{\alpha t}\;(\text{total bubble area}),$$

where the area includes the flat base (microlayer) as well as the curved top. Experimentally, it was found to apply quite accurately throughout the change in shape from hemispherical to spherical.

That enables us to de-couple the energy and flow equations, in the sense that $\mathrm{Ja}\sqrt{\alpha}$ is now seen to be the only input required from the energy equation. Given that value, the remaining problem is purely fluid mechanics. Hence, as discussed in [1], we can say that any quality of interest, X, must be a function only of $\mathrm{Ja}\sqrt{\alpha}$ (renamed b for brevity) and relevant fluid mechanics quantities, ρ, μ, σ and time t. Dimensional analysis reduces those five quantities to two, conveniently taken as a dimensionless time containing no viscosity and a group involving viscosity but not time, giving

$$X^* = f\left(\frac{b}{\sqrt{\nu}}, \frac{t}{b^6\rho^2/\sigma^2}\right) = f\left(\frac{\mathrm{Ja}}{\sqrt{\mathrm{Pr}}}, t^*\right).$$

The group $\mathrm{Ja}/\sqrt{\mathrm{Pr}}$ is involved in the thickness of the microlayer, but it apparently does not strongly influence the overall behaviour of the bubble, such as change of shape. We found that the shape parameter D/H appeared to depend only on the other group, t^*, though its independence from $\mathrm{Ja}/\sqrt{\mathrm{Pr}}$ is not fully established and a check by analysis or computation would be of interest, since quantities like Pr could then be varied more readily in the computation than in experiments.

The well-known change of shape from hemispherical ($D/H = 2$) to spherical ($D/H = 1$) could now be quantified and seen to occur as t^* changed from about 0.1 to 100. Having derived t^*, we could see its physical significance because $\sqrt{t^*}$ arises in the ratio of stresses from inertia of growth ($\frac{1}{2}\rho\dot{R}^2$) and from surface tension ($2\sigma/R$). Predominance of the former implies small t^*. Since t^* involves b^6, hence Ja^6, it is very dependent on Ja, which can change widely. In fact the modest change from Ja = 30 to Ja = 10 causes nearly a thousandfold change in t^*, making the difference between a bubble which is hemispherical throughout its observed life to one which is spherical throughout its observed life.

Bubbles did not appear to depart from the wall, except perhaps intermittently due to oscillations at very large t^*.

3.2. Liquid initially stagnant and isothermal, gravity present

With buoyant gravity and in many cases with non-buoyant gravity, experiment showed that the simple growth law above was still highly accurate. Hence the dimensional analysis still applied, with the addition of g, leading to g^* ($= gb^8\rho^3/\sigma^3$). We could, therefore, expect a variant of the category 1 behaviour described in 3.1 above, variations being controlled by the single parameter g^*. Experiment confirmed that, showing that bubbles initially behaved as in category 1, diverging from that pattern at a time t^* dependent on g^*. For positive (buoyant) g^*, the bubble would depart at a time t_D^* related uniquely to g^*. The relation between t_D^* and g^* for large g^* showed the interaction of inertia ($\frac{1}{2}\rho\dot{R}^2$) and gravity (ρgR). For smaller g^* the effect of $2\sigma/R$ could be seen, and it appeared that surface tension reduced t_D^*, assisting departure, as discussed in [1].

3.3. Liquid initially stagnant with thermal boundary layer [2]

With a thermal boundary layer, even with zero gravity, growth did not follow the simple law described above, nor has any simple and accurate extension of that law emerged. This is partly because observation, particularly of the bubble base, is hindered by a 'mirage' effect. Hence there is no formal reason for applying the same dimensional argument, but it was tried, leading to

$$X^* = f\left(\frac{\mathrm{Ja}}{\sqrt{\mathrm{Pr}}},\, t^*,\, \delta_s^*\right), \qquad \text{where} \qquad \delta_s^* = \frac{\delta_s}{b^4\rho/\sigma}.$$

That proved to be helpful, as the experimental results could again be seen to fall into a pattern which was a one-parameter variant of the behaviour with isothermal liquid, category 1, Section 3.1 above, here regarded as $\delta_s = \infty$.

The chief result was that the bubbles depart in zero gravity (which they did not do in isothermal liquid) and even in negative (non-buoyant) gravity. That phenomenon had long been known, but we were able to produce it under controlled conditions and show that it is associated with thermal boundary layer, and encouraged by subcooling. Curious empirical relationships emerge between parameters, some of which are simple enough to offer hope that they can usefully be incorporated into the complex phenomenon of boiling as a whole. One such relationship is between diameter and time at departure in saturation boiling in zero gravity, where

$$D_D = 1.5\, t_D^{2/3} \left(\frac{\sigma}{\rho}\right)^{1/3}$$

in which it will be noted that $(\sigma/\rho)^{1/3}$ varies little for a wide range of fluids, and there is no dependence on growth rate and thermal boundary layer thickness. When gravity was restored it could influence the departure time, but it is interesting to note that in our experiments, if there was a thermal boundary layer, then gravity was never the predominant influence on departure.

3.4. Liquid initially isothermal but in motion, zero gravity

With the liquid in motion there were increased problems of avoiding unwanted bubbles (familiar in cavitation), but enough results were obtained [5] to enable design of improved apparatus. Growth rate was augmented because the convective effects increase heat flows. The results indicated strongly an interaction between inertia from growth ($\frac{1}{2}\rho\dot{R}^2$) and from mainstream flow ($\frac{1}{2}\rho v^2$), with very little effect from surface tension. The bubbles slid or rolled along the wall in a manner governed by b and v, so non-dimensionalising time and distance using b and v ($t^+ = t/(b/v)^2$, $z^+ = z/(b^2/v)$) produced a common curve. The work is continuing and, as always, strenuous efforts will be made to extend the range of conditions to produce results relevant to industrial boiling. Our experiment was specifically designed to explore the range around $\frac{1}{2}\rho\dot{R}^2 \simeq \frac{1}{2}\rho v^2$ ($\dot{R} \simeq v$) because that was thought likely to prove interesting. It is interesting. but we must recognise that surface tension may well prove important under other conditions.

4. Conclusions (so far) and continuation

These experimental techniques are enabling us to develop an increasing understanding of the behaviour of bubbles relevant to boiling. In principle the results could be obtained by computation, but not easily, since it involves a coupled problem of energy and fluid mechanics with phase change at a moving unknown interface which shows significant effects of viscosity and surface tension. The simplest experiment corresponds to a problem in fluid mechanics alone, which is formulated here and may be at about the current borderline of feasibility in computational fluid mechanics.

Experimentally, that simplest case is readily understood, and we have made and understood various one-parameter variants of that. Two-parameter variants are less comprehensible, but are being or will be tackled, selected chiefly for their importance for boiling, such as combined effects of relative motion and thermal boundary layer.

Acknowledgements

This work was supported initially by the Science Research Council of the UK, and latterly by the Engineering Sciences Division of the Atomic Energy Research Establishment, Harwell, UK.

References

1. Cooper MG, Judd AM and Pike RA (1978) Shape and departure of single bubbles growing at a wall. Paper PB-1, 6 Internat Conf on Heat and Mass Transfer, Toronto, August 1978
2. Cooper MG and Chandratilleke TT (1981) Growth of diffusion-controlled vapour bubbles at a wall in a known temperature gradient. Int J Heat Mass Transfer 24: 1475–1492
3. Cooper MG (1982) The binary microlayer – a double diffusion problem. Chem Eng Sci 37: 27–35
4. Cooper MG and Merry JMD (1973) A general expression for the rate of evaporation of a layer of liquid on a solid body. Int J Heat Mass Transfer 16: 1611–1815.
5. Mori K (1980) Behaviour of vapour bubbles growing at a wall with forced flow. MSc Thesis, Oxford University, England
6. Plesset MA and Zwick SA (1952) A non-steady heat diffusion problem with spherical symmetry. J Appl Phys 23: 95–
7. Stone CR (1980) Boiling: bubble growth in pure and binary liquids. DPhil Thesis, Oxford University, England

Appendix

Mathematical formulation in initially stagnant isothermal liquid

The problem can usefully be put in non-dimensional form with $r^x = r/b\sqrt{t}$. In zero gravity, (category 1, Section 3.1 above), the behaviour of the bubble is then represented by a change from approximately a hemisphere with radius $R^x = 3$, centred on the wall to approximately a sphere with radius $R^x = 2$, touching the wall.

It can also be formulated in terms of stream function and vorticity, non-dimensionalised as

$$\psi^* = \frac{\psi}{b^3\sqrt{t}} \qquad \omega^* = \frac{\omega}{1/t}$$

and using t^* defined earlier.

Boundary conditions for ψ^* at $\theta = 0$ (the axis) and $\theta = \pi/2$ (the wall) can be derived from the growth law

$$\frac{\mathrm{d}V}{\mathrm{d}t} = A\frac{b}{\sqrt{t}}$$

which yields $\psi^* = 0$ at $\theta = \pi/2$, $\psi^* = A^*/2\pi$ at $\theta = 0$.

Boundary conditions for ψ^* at large r^x can be derived from assuming generally radial flow there, with a velocity boundary layer at the wall. If the boundary is sufficiently remote from the bubble, any fault in that assumption should not affect the bubble.

Boundary condition for ψ^* on the bubble can be derived from the Navier-Stokes equation parallel to the interface. Pressure gradient there is determined by the gradient of the sum of principal curvatures multiplied by surface tension. If gravity is present, (category 2, Section 3.2 above) it can be expressed by an additional term in this equation.

Boundary conditions for ω^* on the axis and wall are derived respectively from symmetry, $\omega^* = 0$ on the axis, and from the condition for zero slip at the wall.

Boundary conditions for ω^* at large r^* can also be derived from the assumed flow pattern there.

Boundary conditions for ω^* on the bubble must reflect the fact that the bubble surface is a principal plane of stress, which rotates at angular velocity $\omega/2$, hence $\omega = 2D\beta/Dt$ where β is the angle of inclination of the surface and D/Dt is the usual Lagrangian (convected) derivative.

The initial condition (virtually impossible to observe experimentally) at short t^*, is nearly hemispherical with curvature near the rim in a region of extent similar to the velocity boundary layer. The condition for the surface to be a principal plane of stress must be met or reached before proceeding to large t^*.

Evaporative instability at the superheat limit

B. STURTEVANT and J.E. SHEPHERD

Graduate Aeronautical Laboratories, California Institute of Technology,
Pasadena, CA 91125, USA

Abstract. The explosive vaporization of a single bubble inside a droplet of butane heated to the limit of superheat has been investigated experimentally using short-exposure photographs and fast-response pressure measurements. An interfacial instability driven by rapid evaporation has been observed on the surface of the bubbles. It is proposed that the Landau mechanism of instability, originally described in connection with the instability of laminar flames, also applies to rapid evaporation at the superheat limit. Calculations suggest that other technically important fluids may be even more unstable when boiling at the superheat limit. The rate of evaporation after the onset of instability is estimated from the experimental measurements to be two orders of magnitude greater than would be predicted by conventional bubble-growth theories that do not account for the effects of instability. An estimate of the mean density within the bubbles during the evaporative stage indicates that it is nearly equal to the critical density of butane.

1. Introduction

Vapor bubble growth in slightly superheated liquids has been investigated extensively both theoretically and experimentally, beginning with the work of Plesset and Zwick [11] and Dergarabedian [1]. In those and subsequent investigtions the evaporating bubble surface has been assumed to be smooth and free from instability in the theoretical analysis and, indeed, no previous observations of growing vapor bubbles have shown unstable evaporating surfaces. All vaporizing bubbles observed so far have had smooth and regular surfaces. In the experiments reported here, we have observed a large-amplitude, small-scale roughening of the liquid-vapor interface on bubbles growing in droplets of liquid butane heated to the limit of superheat. This roughening is believed to be the manifestation of a previously undiscovered interfacial instability which is driven by rapid evaporation. The previous experiments were done at relatively low superheats, while in the present experiments the superheat is 105°C, so the evaporative mass flux is much larger. Therefore, it is not surprising that new phenomena should appear, and that the results of the present experiments might not be predicted by extrapolation from the previous experiments or from theories developed for near-equilibrium evaporation, such as the classical theory of bubble growth.

The theory referred to here as the *classical theory of bubble growth*, the most complete version of which is due to Prosperetti and Plesset [12], treats the growth of a smooth spherical vapor bubble from a critical nucleus in a uniformly superheated liquid, accounting for heat-transfer effects in the liquid but assuming thermodynamic equilibrium in the vapor. The theory shows that bubble growth develops in three stages: first, a surface-tension-

Applied Scientific Research 38: 85–97 (1982) 0003–6994/82/0381–0085 $01.95.

controlled stage in which the bubble grows from a critical radius; second, an inertia-controlled stage in which the bubble surface grows with a constant velocity $\dot{R}$ determined by the vapor pressure and density of the superheated fluid; and third, an asymptotic stage in which bubble growth is dominated by heat transfer and follows a $t^{1/2}$ dependence. The stability of the evaporating surface of a bubble described by this theory has been investigated, and some results for various liquids boiling at the superheat limit, together with a comparison to experiment in the case of butane, are presented in this paper. The experimental observations reported here were obtained during the course of an investigation of the transient processes that take place just after a single droplet of metastable liquid at the superheat limit begins to boil.

In the absence of bubble-forming nuclei, liquids may be heated to temperatures far above their boiling points. There is, however, an absolute limit of superheat, T_s, defined by the limit of mechanical stability of fluids. T_s is only about 10% below the critical temperature T_c of many substances, so superheats of more than 100°C can in principle be attained. Heating to the superheat limit is made possible by suppressing heterogeneous nucleation and ordinary boiling by, for example, heating or depressurizing the liquid very rapidly (on a microsecond time scale), or by immersing the volatile liquid in another liquid, thus isolating it from rough solid surfaces containing gas nuclei. The latter method was first used by Dufour [2], who 120 years ago was able to superheat water by as much as 75°C. When such extreme superheats do occur, and boiling begins spontaneously by homogeneous nucleation, the ensuing evaporative fluxes, fluid accelerations and departures from thermodynamic equilibrium are orders of magnitude greater than in ordinary boiling. The resulting explosive process is known as a vapor explosion, and, when it occurs accidentally in industry or in nature, it can be very destructive.

One simple configuration in which rapid evaporation can effectively be studied in relative isolation from other complicating factors is the vapor explosion of droplets in the so-called bubble-column apparatus [15, 9]. In this device the volatile liquid is isolated from possible nucleation centres by immersion in another liquid. Despite the widespread use of this technique for many decades, and despite considerable speculation in the literature about the miniature explosions that are always observed when the droplets vaporize, the fundamental behavior of the explosion has not previously been studied. In the present work the vaporization process and resulting blast-generated pressure field is studied during the explosion of butane droplets immersed in ethylene glycol. Short-exposure photographs and fast-response pressure measurements have been used to obtain a complete description of the explosion process. In this paper results pertaining to the evaporative phase of the explosion are presented. A more complete description and further results will be presented in a forthcoming publication [14].

2. Evaporative instability

It is proposed that the ditortion and roughening of the liquid-vapour interface observed in the present experiments is attributable to an inertial instability first introduced to explain the instability of laminar flames by Landau [5, 6] * and subsequently discussed, together with several other mechanisms, in connection with evaporation by, e.g., Miller [8] and Palmer [10]. The important physical processes influencing stability that are included in the theory are mass flux across the liquid-vapor interface, acceleration of the fluid normal to the interface and surface tension. The Landau instability has been treated for spherical flames growing at constant speed by Istratov and Librovich [4], but the effects of acceleration and surface tension have not yet been included in an analysis applicable to rapid evaportion in spherical bubbles. Acceleration may be destabilizing to growing bubbles only during the initial surface-tension-controlled stage of growth when the acceleration is outward. Though for evaporation at the superheat limit the outward accelerations during this phase are enormous ($\sim 10^9$ g), the radii are very small, so surface tension actually stabilizes the bubbles against the Rayleigh-Taylor instability. On *observable* time scales (say, a few microseconds) the growth before the onset of instability is diffusion limited and the consequent deceleration of the fluid motion, which for the conditions of the present experiments is of order 10^5 g, is stabilizing.

Thus, it must be the mass flux across the liquid-vapor interface that drives the evaporative instability observed in the present experiments. This instability has not been observed before in other evaporative systems because the rates of evaporation in those cases were orders of magnitude smaller than those which occur at the superheat limit. The presence of substantial mass flux across a distorted liquid-vapor interface has the important consequence that, though the incoming flow upstream of the discontinuity (i.e., in the liquid phase) may be irrotational, vorticity is generated by the flow transition at the interface and appears in the downstream flow (the vapor phase). It is this interaction that drives the Landau instability and that distinguishes it from other instabilities.

Calculations have been made of the stability limits and growth rates given by the dispersion relation of the Landau theory, using the mass flux and acceleration predicted as a function of time by the classical theory of bubble growth for the conditions of the present experiments. The results of these calculations for butane are shown in Figure 1. The range of unstable wave numbers is plotted as a function of time (increasing from right to left). Also shown on the abcissa are (i) the ***inertia number,*** the ratio of the destabilizing force of the mass flux to the stabilizing force of surface tension,

*The authors are indebted to Professor E. Marble of the Jet Propulsion Center for suggesting that the Landau instability might be relevant to our experiment.

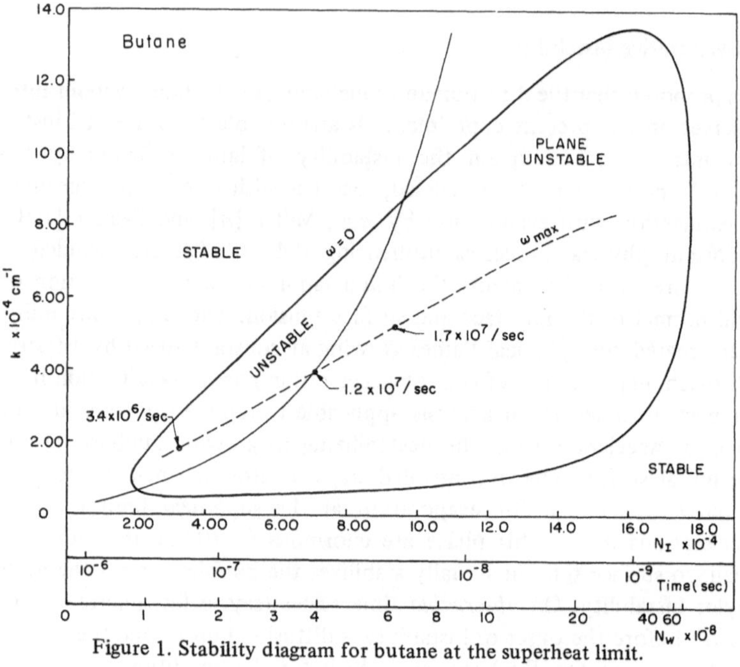

Figure 1. Stability diagram for butane at the superheat limit.

$$N_I = \delta \frac{\dot{m}^2}{\sigma}\left(\frac{1}{\rho_v} - \frac{1}{\rho_l}\right), \tag{1}$$

where $\dot{m}$ is the evaporative flux, σ is the surface tension, ρ_v and ρ_l are the vapor and liquid densities, and δ is some length scale, say 1 cm, and (ii) the *Weber number,* the ratio of the stabilizing acceleration forces to the surface tension forces,

$$N_W = g \frac{\Delta\rho\delta^2}{\sigma}, \tag{2}$$

where $\Delta\rho = \rho_l - \rho_v$ and in this application $g = -\ddot{R}$, where R is the bubble radius. At small times ($t < 10^{-10}$ sec) accelerations stabilize the interface absolutely and no wave numbers are unstable. For large times ($t > 10^{-6}$ sec for conditions of the present experiments) a classically growing bubble is again stable, because the evaporative mass flux decreases below a critical value. At *intermediate* times the liquid-vapor interface of butane exploding at the superheat limit is indeed predicted to be unstable over a large range of wave numbers.

In a spherical system, however, another limitation on instability arises from the fact that disturbances of wave length greater than the diameter of the bubble can not exist. Therefore, for the purpose of these calculations we have imposed an ad hoc low wave number limit of instability for spherical bubbles, $k = 2\pi/R$. This limit is shown in Figure 1 as a curve which traverses the plot from the top center to the lower left. Only those wave numbers above this curve can actually be unstable according to the ad hoc criterion given above. Actually, as shown by Istratov and Librovich [4], since disturbances on growing bubbles must grow more rapidly than the bubble itself before the behavior can be termed unstable, a finite number of the lowest modes of oscillation of the bubble may be stable. Furthermore, spherical growth of the perturbed system leads in some cases to slower, algebraic, growth of instabilities rather than the exponential rates predicted for planar systems. Therefore, growing bubbles may be somewhat more stable than implied by the estimate obtained here, but, in view of the uncertainties of the values of the parameters used in these calculations, the added complexity of a more sophisticated criterion for the low wave number limit of instability does not seem justified. The locus of wave numbers at which the maximum growth rate of disturbances on a plane interface occurs (shown as a dashed line in Figure 1) happens to fall within the band of unstable wave numbers predicted for the *spherical* surface. We adopt these maximum growth rates as an estimate of the upper bound of the growth rate of disturbances on spherical bubbles.

A simple quantitative measure of the relative susceptibility of a substance at its superheat limit to evaporative instability can be defined as the product of a typical value ω of the maximum growth rate calculated to occur within the region of instability and the time interval τ during which a classically growing bubble is predicted to be linearly unstable. The resulting figure of merit $F = \omega\tau$ is the amount by which disturbance amplitudes are exponentiated during instability, according to linear stability theory for plane interfaces. We take for ω the *largest* value of maximum growth rate calculated within the region of instability, which, incidentally, always occurs near the beginning of the period of instability. The result $F = 2.9$ for butane indicates that that substance is marginally unstable, and that the discovery of the instability during a study of the vapor explosion of butane droplets was a matter of some luck!

The susceptibility of other substances to instability can be estimated by inserting the appropriate physical properties into the above-described calculations. This has been done for two technically important fluids, water and liquid sodium, though the properties and behavior of sodium at the limit of superheat are not well known, so that estimate is very crude. The resulting stability diagrams for these two fluids are shown in Figures 2 and 3. The figure of merit F for water is found to be 29, and for sodium is 2900. Clearly, these two fluids are very much more unstable than butane.

3. Experimental results

3.1. Photographs

Figure 4 shows representative examples of photographs of exploding droplets at the earliest time ($\sim 10\,\mu$sec) that pictures could be taken after initiation of

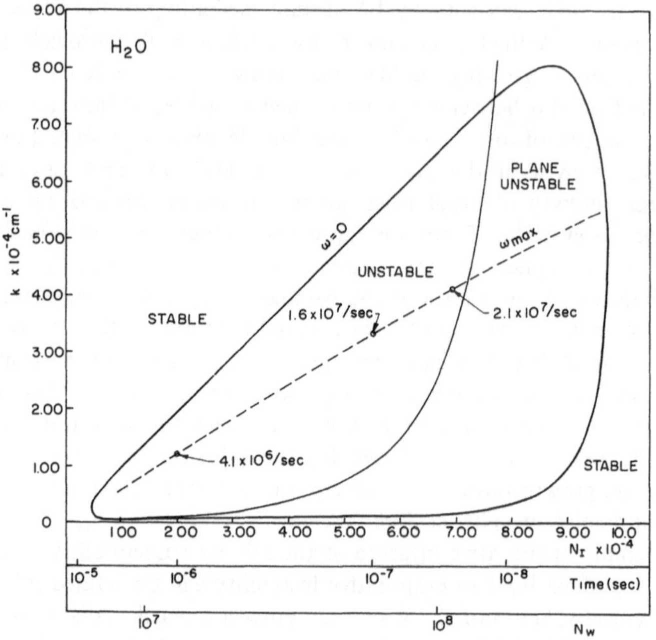

Figure 2. Stability diagram for butane at the superheat limit.

the vapor explosion by homogeneous nucleation. The drops are shown immersed in host liquid, which appears pure white in the photographs. In none of the nearly 500 photographs taken in this set of experiments has ever more than one bubble seen within a vapor-exploding drop. As is apparent from Figure 4, the bubbles form at random asymmetric locations within the drops.

Of particular note is the somewhat regular pattern seen on the image of the bubbles in Figure 4, especially Figure 4a. It is hypothesized that this pattern is caused by variations of transmitted light through a regularly wrinkled liquid-vapor interface. At slightly later times (Fgiure 5) the bubble surface is extremely rough on a very small scale. When viewed in profile, the distortion is seen to be random and of large amplitude.

The regular pattern on the bubble surface observed in Figure 4 has been examined in more detail in one case by enhancing the image of the bubble by computer. The original photographic negative was digitized and was processed numerically using linear interpolation to expand the data base, median filtering to sharpen the edges, high-pass filtering to eliminate large-scale variations of photographic density, and contrast enhancement.* The results are shown in Figure 6. The enhanced image shows that the pattern

*The authors are indebted to Dr B. White and Mr D. Madura of the Medical Image Analysis Facility, Jet Propulsion Laboratory, for their aid in this effort.

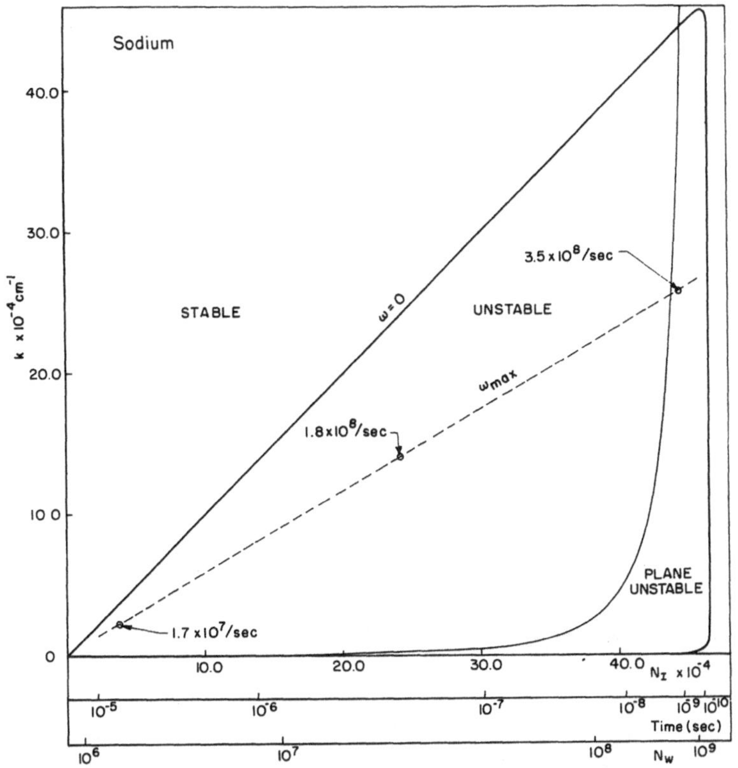

Figure 3. Stability diagram for sodium at the superheat limit.

covers the entire bubble. Though the regularity of the pattern is much more striking in the enhanced image, there is still substantial asymmetry and randomness. The bubble in this figure has a diameter of approximately 150 μm and the length scale of the pattern is 20–40 μm. This scale is 2–4 times larger than the largest scale in the unstable band of wave numbers in Figure 1, but the observations are made at times an order of magnitude larger than those during which the linear theory predicts the bubbles to the unstable. Thus, these experiments are not carried out under conditions that would provide definite verification that the Landau mechanism is indeed the one that drives the instability, but the results are consistent with the Landau mechanism, so long as the scale of the disturbances that develop during the linear phase of the instability at very small times remain frozen in the flow during the initial development of the nonlinear stage. Further experiments should be carried out to investigate this possibility.

Evidently, after the initial development of the instability, the wrinkling of the liquid-vapor interface saturates and persists at a nearly constant amplitude for the remainder of the evaporation process (Figure 5). The significance of

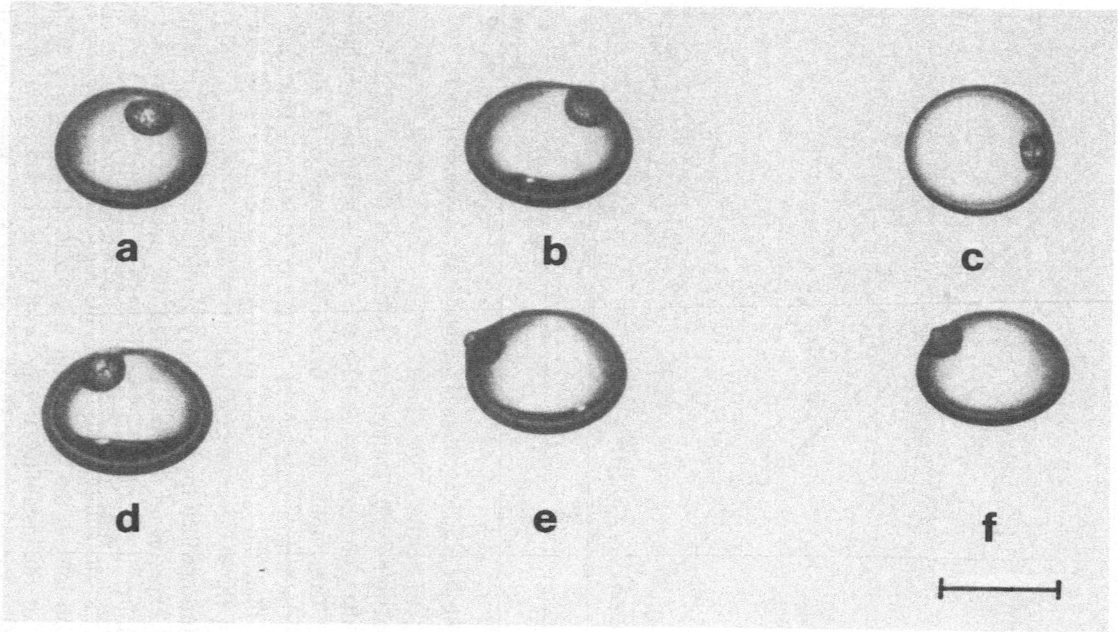

Figure 4. Vapor bubbles in drops at the earliest observed times (9–12 μsec). Scale indicates 1 mm.

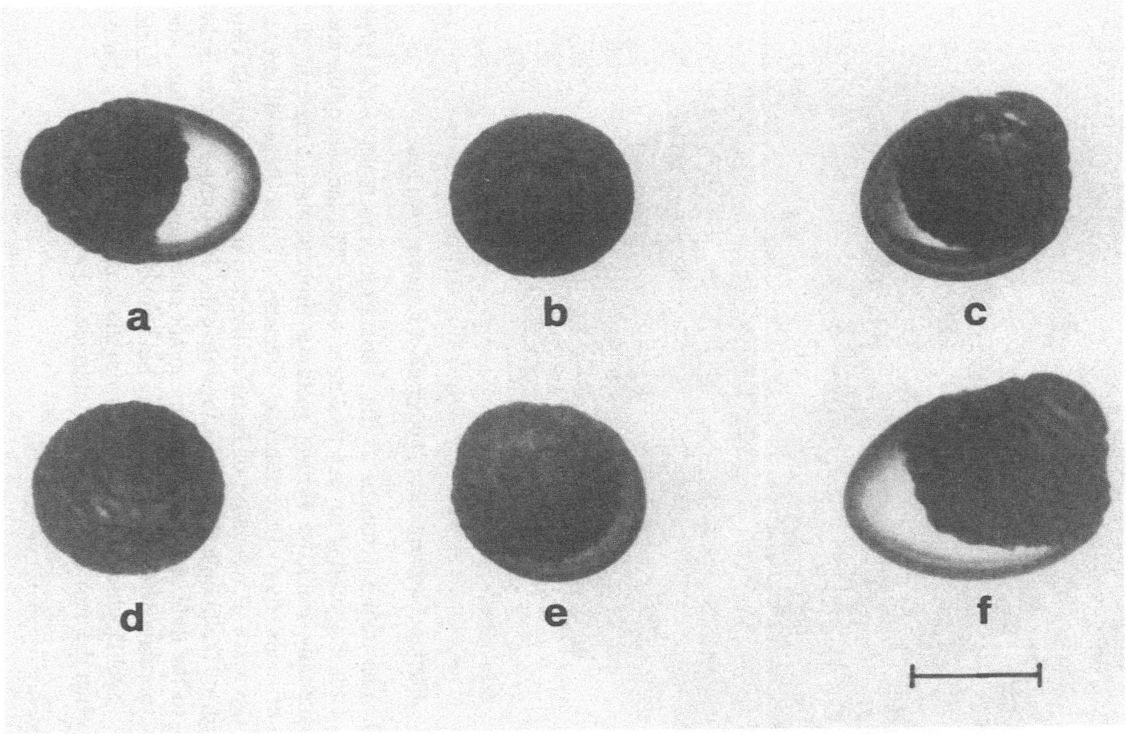

Figure 5. Growing vapor bubbles, showing developing jet structure and roughened evaporating surfaces (55–65 μsec). Scale indicates 1 mm.

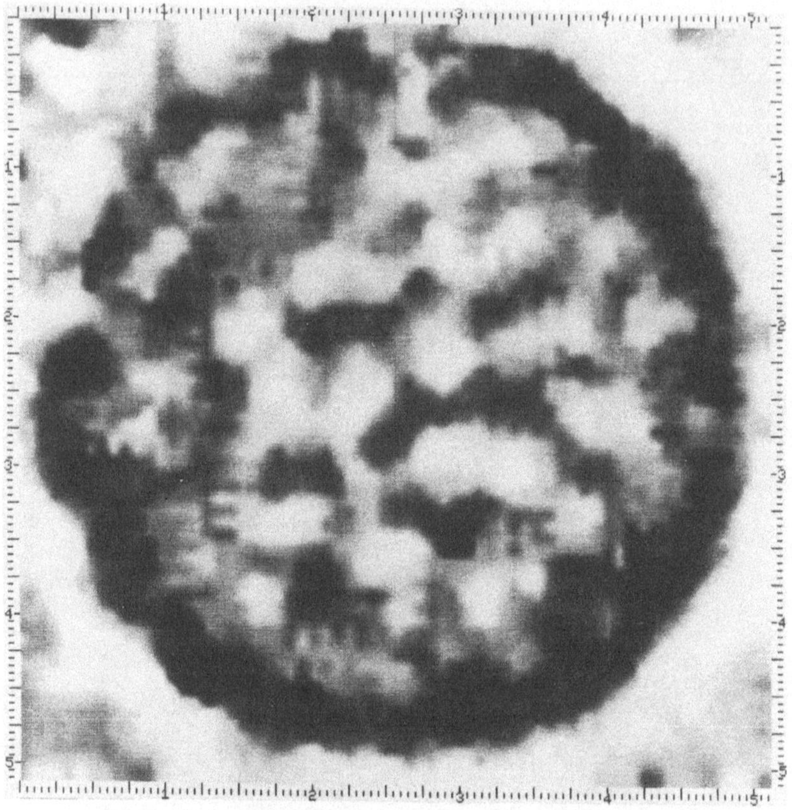

Figure 6. Image-enhanced photograph of bubble at $t = 12\,\mu$sec.

the roughening to the dynamics of the evaporation process lies in the large increase of the evaporating surface that must occur. The increase of the area of the randomly distorted surface yields a proportional increase in the evaporation mass flux. Unfortunately, because of the large difference between the index of refraction of liquid and vapor, it is impossible to view the interior of the bubble during the time that the evaporative surface is the roughest, so the fluid *inside* the bubble can not be examined. Therefore, one can only speculate whether there might be a substantial amount of *liquid* within the bubble which has been torn from the liquid-vapor interface during the time when the mass flux across the interface is the largest.

3.2. Estimate of evaporative mass flux

Using the data obtained in these experiments, an order-of-magnitude estimate of the evaporative flux can be calculated by inserting the photographically-determined bubble growth rate and the acoustic source strength obtained from the pressue measurements into the global conservation equation for mass in the bubble,

$$\frac{\mathrm{d}M}{\mathrm{d}t} = \rho_l \frac{\mathrm{d}V}{\mathrm{d}t} - \rho_l Q, \tag{3}$$

where M is the total mass in the bubble, ρ_l is the liquid density, V is the volume of the bubble and Q is the acoustic source strength, that is, the rate of volume outflow in the fluid outside the bubble. The effective evaporative mass flux $\dot{m}$ is calculated from $\mathrm{d}M/\mathrm{d}t$ dividing by the measured mean evaporating area A.

The shape of twenty-two bubbles in representative photographs at times ranging from 8.3 to 91 μsec has been digitized, and V and A have been calculated treating the bubbles as bodies of revolution. *No correction to A for the roughness of the interface was made.* The *effective* radius of a spherical bubble implied by the volume data is plotted versus time in Figure 7. It is remarkable that the growth rate appears to be linear, with a mean velocity of approximately 14.3 m/sec. Since the present data are obtained after onset of instability, it is not surprising that the growth rate disagrees with the predictions of the classical diffusion-limited theory. Indeed, the observed rate is substantially *less* than the linear growth rate (dashed line; 40 m/sec) predicted by the theory for the inertially-dominated stage at very small times, but is *larger* than the prediction (solid line) for later times during the actual period of observation in these experiments, when the limiting effects of heat transfer are supposed to dominate. As will be seen below, the fact that the observed growth rate is less than the predicted inertially-dominated rate does not mean that evaporation rates are particularly small.

The source strength Q is calculated using the equation from acoustics theory relating far-field pressure to the behavior of a spherical source,

$$Q(t) = \frac{4\pi r}{(1+\alpha)\rho_\infty} \int_0^t p(t' + r/c)\,\mathrm{d}t' \tag{4}$$

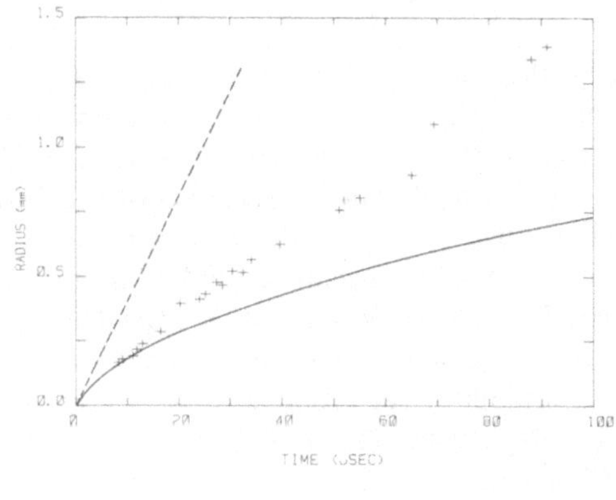

Figure 7. Effective bubble radius as a function of time: +, experiment; ———, classical theory; – – – –, classical inertial growth rate, 40 m/s.

where α is the coefficient of reflection of waves normally incident on the transducer face (taken to be 0.83 for an aluminum-glycol interface), and p, ρ_∞ and c are the pressure measured by the transducer, the host liquid density and sound speed, respectively. Seven pressure signals were averaged for the computation of an average source strength from (4).

The calculations show that the effective evaporative mass flux calculated from equation (3) rises very rapidly and after some fluctuation levels off at about 400 gm/cm^2 sec. This value is to be compared with mass fluxes typical of evaporation at low superheat, which are of the order of 10^{-3} – 10^{-1} gm/cm^2 sec. Only in extreme situations such as vacuum distillation [3] or laser ablation [13] does the mass flux reach the magnitude observed in these experiments.

An estimate of the mean vapor density ρ_{eff} within the bubble has been calculated by dividing $M(t)$ by $V(t)$. A plot of ρ_{eff} versus time shows that, after the transients, it settles down to a nearly constant value, of about 0.2 gm/cm^3. This value is nearly equal to the *critical* density of butane, 0.23 gm/cm^3!

Calculation of bubble growth by the classical theory (as in Figure 7) has been done using approximate analytical expressions for R and T_s suggested by Mikic et al. [7], rather than computing the exact solutions of Prosperetti and Plesset [12]. However, the modified method for evaluating the scaling constants suggested by Prosperetti and Plesset was used. Taking the predicted evaporative mass flux $\dot{m}$ to be $\rho_v(T_l)\mathrm{d}R/\mathrm{d}t$, where ρ_v is the density of saturated vapor at the temperature T_l of the liquid interface, the value predicted at 10 μsec for the conditions of the present experiments is two orders of magnitude less than observed and becomes even smaller with increasing time.

4. Conclusions

An interfacial instability driven by rapid evaporation has been observed on the surface of single, rapidly growing bubbles in droplets of liquid butane boiling at the superheat limit. The instability is observed during relatively early stages of its development, when the pattern of the disturbances on the bubble is relatively regular, and during the later stages, when the wrinkling of the surface saturates at nearly constant amplitude, is random and occurs on many length scales. The rate of evaporation is estimated from the data to be two orders of magnitude greater than would be predicted without accounting for the effects of instability. An estimate of the mean density within the bubbles during the evaporative stage indicates that it is nearly equal to the critical density of butane.

It is proposed that the theory developed by Landau for explaining the instability of laminar flames, may be applicable to the evaporating systems studied in this work. The experimental observations are consistent with this hypothesis, so long as the disturbances which develop during the linear stage of the instability (at times predicted to be much smaller than are observable in the experiments) remain frozen in the flow during the early part of the nonlinear phase of the instability. The disturbance growth rate and the time interval during which the bubbles are predicted to be linearly unstable imply that butane is marginally unstable, but that other technically important liquids (e.g., water and sodium) may be more unstable.

Acknowledgments

This research was supported by the United States Department of Energy, Division of Chemical Sciences, under Project Agreement DE-AT03-80ER10634.

References

1. Degarabedian P (1953) The rate of growth of vapor bubbles in superheated water. J App Mech 20: 537.
2. Dufour L (1861) Sur l'ebullition des liquides. Comptes Rendus 52: 986.
3. Hickman K (1972) Torpid phenomena and pump oils. J Vac Sci Tech 9: 960.
4. Istratov AG and Librovich VB (1969) On the stability of gasdynamic discontinuities associated with chemical reactions. The case of spherical flame. Astron Acta 14: 453.
5. Landau LD (1944) On the theory of slow combustion. Acta Physiochimica URSS 19: 77.
6. Landau LD and Lifshitz EM (1959) Fluid Mechanics, p. 479. New York: Pergamon Press.
7. Mikic BB, Rohsenow WM and Griffith P (1970) On bubble growth rates. Int J Heat Mass Trasnfer 13: 657.
8. Miller CA (1973) Stability of moving surfaces in fluid systems with heat and mass transport – II. Combined effects of transport and density difference between phases. AIChe Journal 19: 909.
9. Moore GR (1959) Vaporization of superheated drops in liquids. AIChE Journal 5: 458.
10. Palmer HJ (1976) The hydrodynamic stability of rapidly evaporating liquids at reduced pressure. J Fluid Mech 75: 487.
11. Plesset MS and Zwick SA (1954) The growth of vapor bubbles in superheated liquids. J App Phys 25(4): 493.
12. Prosperetti A and Plesset MS (1978) Vapor bubble growth in a superheated liquid. J Fluid Mech 85: 349.
13. Ready JF (1965) Effects due to absorption of laser radiation. J App Phys 36: 462.
14. Shepherd JE and Sturtevant B (1982) Rapid Evaporation at the Superheat Limit (to be published in J Fluid Mech).
15. Wakeshima H and Takata K (1958) On the limit of superheat. J Phys Soc Japan 13: 1398.

PART THREE

Nucleation

Nucleation and stabilization of microbubbles in liquids

LAWRENCE A. CRUM

Department of Physics and Astronomy, University of Mississippi, Oxford, MS 38677, USA

Abstract. An important aspect of the processes of cavitation and boiling is the concept of a nucleus that acts as a preferential site for the inception of these events. It is commonly thought that except for rare instances or specially controlled experiments, all cavitation and boiling sites originate at the location of such a nucleus. In order to study these important phenomena, then, it is imperative that as much as possible be known about nucleation in cavitation and boiling. It is generally accepted that free air bubbles normally do not act as nucleation sites because they are inherently unstable to dissolution due to surface tension. Thus, the study of nucleation is necessarily associated with mechanisms for stabilizing microbubbles or pockets of gas within the liquid. In this paper, various stabilization models that have been proposed are reviewed as well as the experimental evidence that supports the specific models. One particular model, the crevice model, is examined in some detail, and its predictions are used to explain several different measurements of boiling and cavitation inception. Finally, some evidence that has recently become available concerning the damaging aspects of high intensity ultrasound is examined. Many aspects of this evidence point to the existence of cavitation as the damage mechanism. Also given in this paper are preliminary explanations of these effects due to the growth of microbubbles or cavitation nuclei by rectified diffusion.

1. Introduction

The measured tensile strength of liquids has long been known to be significantly less than theoretical predictions. This reduced strength has historically been attributed to the presence of inhomogeneities in the liquid that serve as preferential sites for liquid rupture. We shall for the purpose of communication give these inhomogeneities the generic term of 'nuclei'. Further, since the most popular kind of nuclei take the form of collections of gas molecules in a localized region of small extent we shall occasionally also designate these nuclei as 'microbubbles', primarily for lack of a more descriptive term.

In this paper we wish to examine the role that microbubbles play in the nucleation of boiling, in transient acoustic cavitation inception, and finally in the stable cavitation associated with the oscillation and growth of gas bubbles in insonated tissue.

The nuclei that we shall deal with in this paper are of microscopic dimensions (perhaps even smaller) and play such an enigmatic role that an actual physical description of one is currently not possible. There have been, however, a variety of models proposed to explain certain aspects of their behaviour and we now examine some of these models, their characteristics, and their predictions. Since this paper is meant to review the state of the art, the author will lean heavily on the work of others as well as introduce some new results of his own.

Applied Scientific Research 38: 101–115 (1982) 0003–6994/82/0382–0101 $02.25.

2. Historical review

The obvious and simplest model for a nucleus is simply that of a small gas bubble present in the bulk of the liquid. Unfortunately for this model, all liquids possess some level of surface tension and thus a free gas bubble present in a liquid is unstable. That is, it will tend to dissolve due to the equivalent external pressure generated by surface tension that will force the gas out of the bubble into solution in the liquid. Epstein and Plesset [13] have examined the expected lifetime of air bubbles in water and using their equations we can calculate that air bubbles of size on the order of a few microns will dissolve in a few seconds. Of course, larger bubbles will live longer, but these larger bubbles will overcome Brownian motion and rise to the surface, also in a relatively short time [14]. There is, of course, a continual production of extremely small microbubbles created by thermal fluctuations but the measured tensile strengths of liquids can not be explained by homogeneous nucleation except for extremely small samples [4] wherein there are none of the larger (and consequently weaker) inhomogeneous nuclei.

It should be noted that the presence of nuclear radiation can induce microbubbles momentarily, by interacting with the molecules of the liquid, and these 'bubble chamber' type cavities can serve as nucleation sites. There have been several studies of radiation-induced nucleation and the reader is referred to the excellent report of Greenspan and Tschiegg [16].

It is perhaps now appropriate to distinguish between nucleation and stabilization. If a neutron interacts with an oxygen nucleus and creates a cylindrical vapor cavity in a liquid, then this cavity can serve as a nucleation site only if the local environmental conditions are such as to cause the cavity to grow instantaneously. Otherwise, the cavity will soon collapse and no permanent nuclei will have been formed. It is obvious then that free bubbles can not serve as permanent nucleation sites unless there is some stabilization mechanism to prevent their rapid disappearance. We now examine models for the stabilization of microbubbles in liquids.

2.1. Rigid organic skin model

Since organic compounds are abundant in our environment, and since these compounds are often surface active, Fox and Herzfeld [15] proposed that these compounds could form a rigid skin about the surface of a free gas bubble and prevent it from dissolving. This skin was necessarily impermeable to gas diffusion and must have sufficient mechanical strength to overcome the hydrostatic and surface tension forces tending to collapse it. Herzfeld [19] subsequently abandoned the hypothesis because it apparently was inconsistent with the experiments of Strasberg [27] who examined the acoustic cavitation threshold as a function of static pressure and found a linear dependence. If the rigid skin hypothesis were appropriate, Herzfeld reasoned that there would be a threshold crushing pressure that would eventually destroy this linear dependence. This model is discussed here because modifications of this model are still being considered.

2.2. Ionic skin model

It has been known for some time that air bubbles in water possess some excess electrical charge [2] and thus is seems reasonable to assume that the repulsive nature of like electrical changes on the surfaces of an air bubble could stabilize it from dissolution. Akulichev [1] has examined this model in some detail and has hypothesized that certain ions such as Cl^- and F^- are hydrophobic and will migrate to the surface of an air bubble in water. Other ions such as OH^- are hydrophilic and will not migrate to the surface. Accordingly, one would expect that there would be a dependence of the cavitation threshold on the concentration of such salts as NaCl or NaF, but not of NaOH. Figure 1 shows the results of measurements by Akulichev of the acoustic cavitation threshold as a function of concentration for the three compounds mentioned above. The expected effects are observed.

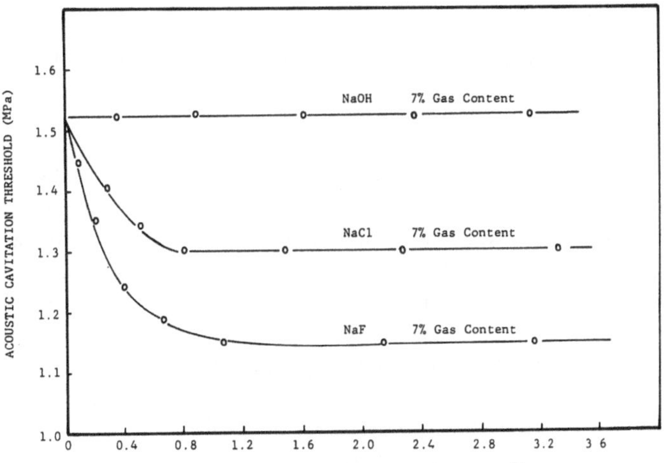

Figure 1. Effect of dissolved ions on the acoustic cavitation threshold of water. After Akulichev [1].

2.3. Film of surface-active substances model

In a modified version of the rigid-skin hypothesis of Fox and Herzfeld, Sirotyuk [24, 25, 26] has examined the possible stabilization of microbubbles by a film of surface-active substances. Detergents and soaps tend to be composed of an oxygen-rich group on one end and a long hydrocarbon tail on the other. For an air bubble in water, the polar end will bond to the water surface and the tail will extend outward into the air. This 'picket-fence'

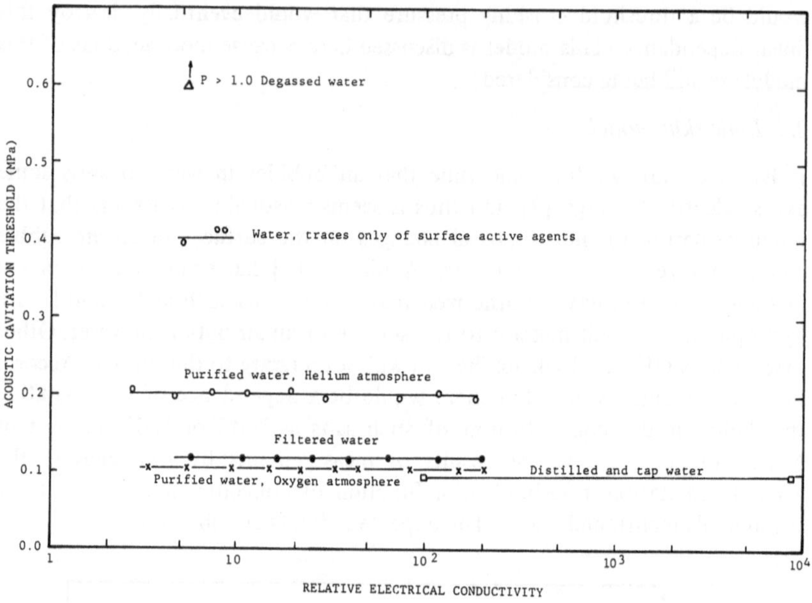

Figure 2. Dependence of the acoustic cavitation threshold of water on different degrees of purification. After Sirotyuk [24].

configuration has an intrinsic elasticity that may stabilize the microbubble. Sirotyuk has measured the acoustic cavitation threshold of water in a device designed to maintain a high degree of purity. He has examined the threshold as a function of the electrical conductivity, to test the model of stabilization by an ionic skin, and as a function of concentration of surface-active substances, to test the model of stabilization by a surface-active film. His results are reproduced here for reference and are shown in Figure 2, which plots acoustic cavitation threshold versus electrical conductivity for a variety of conditions. His results indicate that preparing water in an oxygen atmosphere only slightly increases the threshold, as does filtration, which increases the threshold by a factor of approximately 1.3. If purified water is produced in a helium atmosphere, the increase in threshold is more significant, being near a factor of 2. The most significant result, however, is that when all but traces of surface-active substances are removed, as indicated here by a reduced range of electrical conductivity, the threshold increases by a factor of nearly 4. It is also important to note that when the liquid was degassed, the threshold increased above the highest pressure attainable by the apparatus, approximately 1.0 MPa (10 bars).

Sirotyuk derives several conclusions from these results. First, it is noted that there is essentially no variation in the threshold with electrical condictivity, thus questioning the model of stabilization by an ionic skin. Second, there is little increase in the threshold after filtration, thus questioning models based upon stabilization by crevices in particulate matter (to be discussed below). Finally, there is a significant increase in the threshold when all but traces of surface-active substances are removed, thus supporting the model of stabilization by surface-active films.

In this review of the various stabilization models, it is appropriate to point out that the measurements made by Sirotyuk were obtained for water containing near-saturation amounts of gas. Akulichev has found, see Figure 1 in [1], that a change in the cavitation threshold with concentration of ionic solutes could not be detected until the amount of dissolved gas was reduced below approximately 45% of saturation. Thus, the evidence presented by Sirotyuk does not rule out the ionic skin stabilization model. It is also appropriate to point out that Greenspan and Tschiegg [16] detected a more than 100-fold increase in the cavitation threshold of water with extensive and protracted filtration. Thus the evidence presented by Sirotyuk does not rule out stabilization models based upon particulate contamination. However, his measurements do give evidence for the effect of surface-active agents on the stabilization mechanism. This model proposed by Sirotyuk has been supported and extended by Yount et al. [29, 30] with regard to measurements of bubble growth in supersaturated gelatin. In Yount's case, the skin covering the bubble is composed of surface-active agents that are initially permeable, but later become impermeable and stabilize the bubble at some critical radius.

2.4. *Crevice model*

One of the earliest models proposed for microbubble stabilization was that due to Harvey et al. [18] in which gas is trapped in a conical crevice in a solid inhomogeneity present in the liquid. This model has received considerable attention and has been examined by Strasberg [27], Apfel [3], Winterton [28] and Crum [8, 9], as well as several others. As this model has been successful in predicting many aspects of nucleation in boiling and cavitation, it will be examined in more detail here.

3. Application of the crevice model to nucleation in boiling and cavitation

Consider a particle of solid impurity containing a pocket of gas entrained in a crevice within the particle as in Figure 3. If this quantity of gas is to survive, the interface must be concave toward the liquid in order that the surface tension assist in preventing the gas from dissolving, and thus $\alpha_A > \pi/2 + \beta$. There also must be some hysteresis in the advancing and receding contact angles for water against the surface of the solid in order that the stabilization condition might be met. For the cavity shown in Figure 3 the condition for stability is

$$P_h = P_v + P_g + \frac{2\sigma}{R}, \tag{1}$$

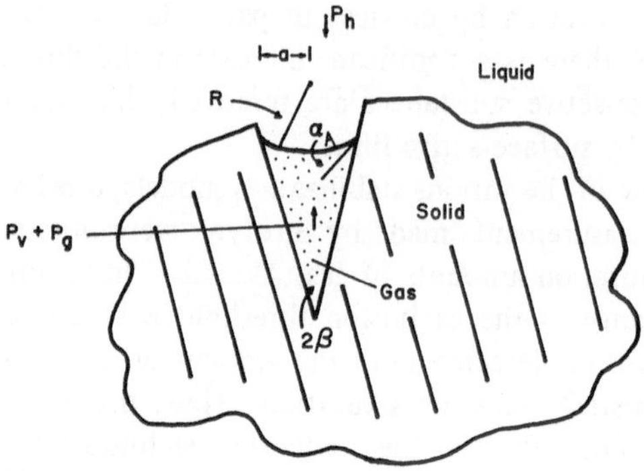

Figure 3. Mathematical model for stabilization of a gas pocket in a solid impurity.

where P_h is the hydrostatic pressure, P_v is the vapour pressure, P_g is the equilibrium gas pressure of the gas dissolved in the liquid, σ is the liquid surface tension and R is the radius of curvature of the interface.

Let us consider now some specific experiments.

3.1. Nucleation of boiling

Winterton [28] has applied the crevice model to the study of boiling in water. In a specific application, he has examined the dependence of a previously-imposed deactivation pressure on the vapor pressure required for boiling. Suppose a deactivation overpressure is applied to the cavity shown in Figure 3. The interface will advance, provided the advancing contact angle is reached, and take a position such that the half-width of the interface is given by

$$a = \frac{2\sigma}{P_h' - P_v' - P_g'} \, |\cos(\alpha_A - \beta)|, \tag{2}$$

since the radius and half-width are related in a crevice larger than critical size by

$$a/R = |\cos(\alpha_A - \beta)|. \tag{3}$$

The primes indicate that these quantities may change upon application of the pressure. Suppose next that the liquid is heated until boiling occurs. The nucleation of boiling will occur when the interface bows out and the receding contact angle is reached as shown in Figure 4. The condition for nucleation is then

$$P_v = P_h - P_g + \frac{2\sigma}{R'}, \tag{4}$$

where R' is the radius of curvature of the air water interface when the receding contact angle is reached. Combining equations (2), (3) and (4), we find that

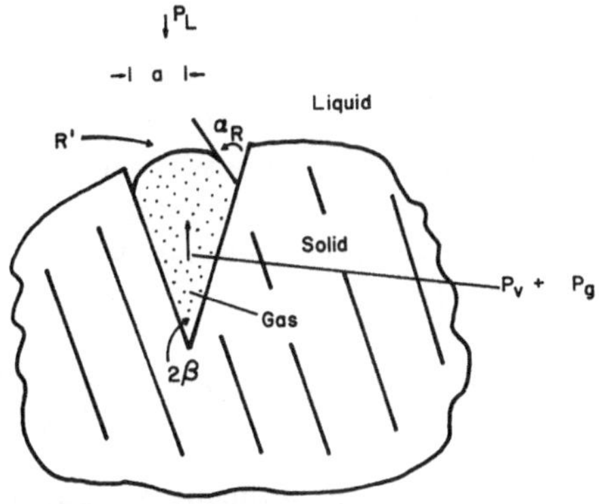

Figure 4. Mathematical model for nucleation of a free gas bubble from a gas pocket in a solid impurity.

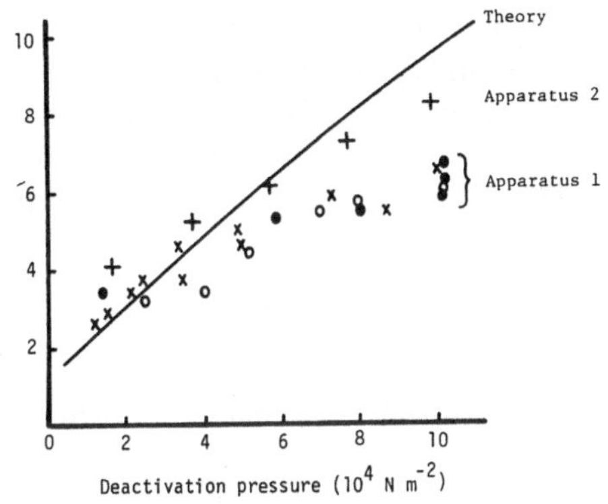

Figure 5. Dependence of the vapor pressure required for boiling of water as a function of a preimposed deactivation pressure. The symbols are for several measurements with two different sets of apparatus and the solid line is equation (5). After Winterton [28].

the magnitude of the vapor pressure required to achieve boiling is given by

$$P_v = P_h - P_g + \frac{|\cos(\alpha_R - \beta)|}{|\cos(\alpha_A - \beta)|}(P_h' - P_v' - P_g'), \tag{5}$$

provided $\alpha_R > \beta$. A similar expression is obtained for $\alpha_R < \beta$ [28].

Winterton has applied this equation to his measurements of the effect of a previously-applied deactivation pressure on the vapor pressure required for boiling. His results are shown in Figure 5 for two different experimental systems and for the condition that $\gamma = |\cos(\alpha_A - \beta)|/|\cos(\alpha_R - \beta)| = 1$. One sees that reasonable agreement is obtained between the measured and predicted results.

3.1. Nucleation of acoustic cavitation

Crum [8, 9] has applied this crevice theory to the study of the threshold for acoustic cavitation inception for a variety of parameters. Consider the inception of transient cavitation in degassed water. Equation (5) can be modified to give an equation for the acoustic pressure amplitude, P_A, required to cause the interface to advance, and thus nucleate a cavity that will grow unstably. This expression is given by [3]

$$P_A = (P_h - P_v - P_g) + (P_h - P_v - P_g)\left|\frac{\cos(\alpha_R - \beta)}{\cos(\alpha_A - \beta)}\right|. \tag{6}$$

Crum [8, 9] was able to remove the difficulty associated with the determination of the magnitudes of the contact angles and especially their variation with surface tension by use of the equation

$$\cos\alpha_E = -1 + C/\sigma, \tag{7}$$

where α_E is the equilibrium contact angle for water on nonpolar solids, C is a constant that depends upon the surface properties of the solids and is approximately 50 for paraffin and beeswax. This relationship has been shown to apply for a wide variety of materials [5] and surface tensions. Combining equations (6) and (7) gives an expression for the acoustic cavitation threshold that can be directly applied

$$P_A = (P_h - P_v - P_g) + \frac{P_h - P_v - P_g}{\delta} \times$$

$$\{\cos\phi(C/\sigma - 1) + \sin\phi\,[1 - (C/\sigma - 1)^2]^{1/2}\} \tag{8}$$

where $\phi = \alpha_H + \beta$, $\delta = |\cos(\alpha_A - \beta)|$, and α_H is the hysteresis angle. Likely candidates for particulate matter are the paraffins, beeswax, and a variety of organic nonpolar solids that are hydrophobic and possess significant hysteresis angles for water.

We would now like to apply this equation to measurements of the acoustic cavitation threshold of water. Specific details of the experimental apparatus are given elswhere [8, 9]. Figure 6 shows the variation of the cavitation threshold of distilled water with surface tension for two different gas concentrations. The results are somewhat remarkable in that lower surface tensions tend to give significantly higher thresholds! Included on Figure 6 is the theoretical prediction of the threshold as given by equation (8) and shows that the agreement is rather good. The theory also accounts for the variation of the threshold with temperature, dissolved gas concentration, and applied static pressure [8, 9].

In order to apply a further test of the theory, an attempt was made to explain the recent discovery that long chain polymers can significantly reduce the hydrodynamic cavitation index [12, 20]. Accordingly, measurements

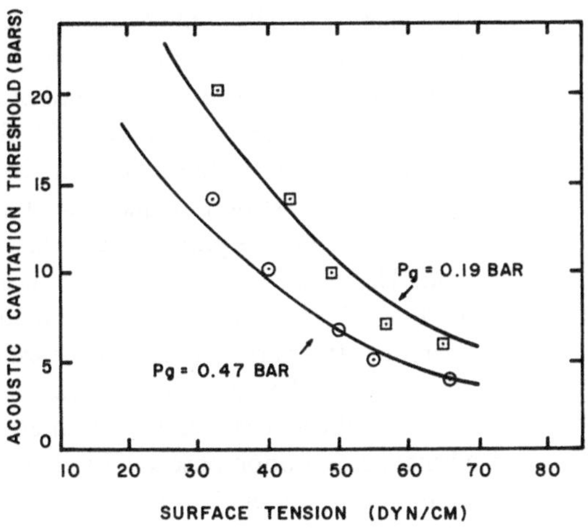

Figure 6. Variation of the acoustic cavitation threshold of water with liquid-vapor surface tension for two values of the dissolved gas content. The solid lines are for equation (8) with $\delta = 0.035$ and $\phi = 35°$. The frequency was 36 kHz and the temperature was 25°C.

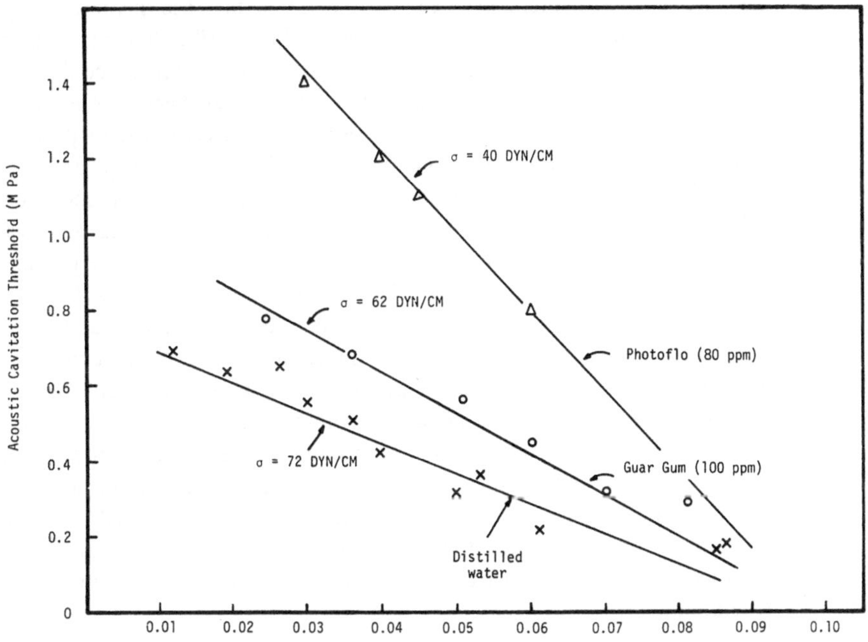

Figure 7. Variation of the acoustic cavitation threshold of water with dissolved gas content for distilled water, an aqueous solution of guar gum at a concentration of 100 ppm by weight, and for an aqueous solution of Photoflo at a concentration of 80 ppm by weight. The solid lines are for equation (8) with $\delta = 0.035$ and $\phi = 35°$. The frequency was 36 kHz and the temperature was 25°C.

were made of the effect of guar gum, a large molecular weight polymer, on the acoustic cavitation threshold. It has been thought, at least in the hydrodynamic sense, that these additives 'reduced the cavitation index by changed flow dynamics rather than fluid physical properties' [20]. Shown in Figure 7 is the variation in the acoustic cavitation threshold with dissolved gas content for distilled water, for an aqueous solution of guar gum at a concentration of 100 ppm, and for an aqueous solution of photoflo, a surface-active agent, at a concentration of 80 ppm. The surface tension of these liquids was found to be respectively 72 dyn/cm, 62 dyn/cm and 40 dyn/cm. With the identical set of constants used to predict the variation of the threshold with surface tension, temperature, etc. in our earlier work, we can quite closely predict the effect of polymer additives on the acoustic cavitation threshold of water. If it is true that the hydrodynamic cavitation observed by these workers [12, 20] is due to inhomogeneous nuclei, rather than free bubbles, say, then we can give a reasonable explanation of their results.

4. Nucleation in tissue

In the past few years there has been a tremendous increase in the development of ultrasonic systems for medical usage. The possible hazards of such systems are not yet fully known and it is of paramount importance that the safety of these systems be well established before they become too widespread. In that vein we examine in this section some cases where possible explanations can be given for some observed effects of damage of tissue by ultrasound. Because these systems operate primarily in the megaherz frequency domain, and because transient cavitation thresholds at these frequencies are relatively large, the primary cavitational effects are possibly due to growth of bubbles by rectified diffusion. We have modified the existing equations [10, 11, 21] that describe the growth of bubbles by rectified diffusion so that they are now applicable at megaherz frequencies. We have then examined the growth of bubbles in biological tissue for the cases reported of damage by applied ultrasound, and have tried to correlate the damage with the growth of bubbles through resonance size. For rectified diffusion to occur there must be some pre-existing nuclei. Rather than speculate at this point on the nature of these nuclei, let us initially assume the presence of free air bubbles that will grow by rectified diffusion according to our theory. We shall return to the explanation of how free air bubbles can be generated later.

4.1. The effect of pulse length on damage experienced by insonated pea roots

In an examination of the effect of ultrasound on pea roots, Child et al. [6] detected a dependence on the damage due to the roots on the length of the sound pulse. Their results are shown in Figure 8. Note that exposures less than approximately 1 msec had much less effect than exposures greater than

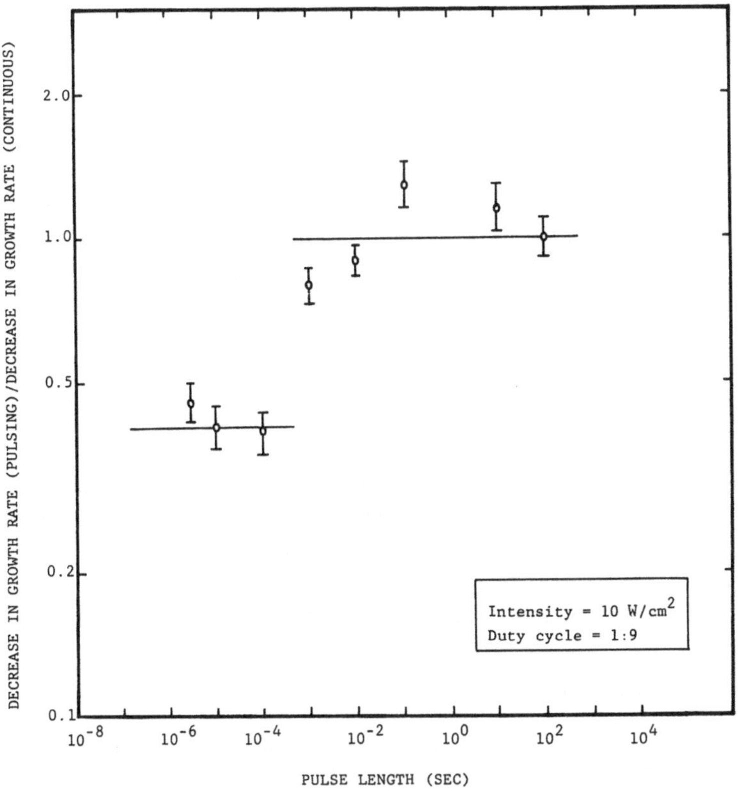

Figure 8. Variation in the relative growth rates of pea roots exposed to pulsed ultrasound as a function of pulse length. The intensity was 10 ± 2 W/cm²; the driving frequency was 2.3 ± 0.05 MHz; the temperature was 22 ± 2°C. After Child et al. [6].

1 msec. Further, note that exposures longer than 1 msec had approximately the same effect even though the pulse length was increased by 5 orders of magnitude. We think these results are consistent with the growth of air bubbles by rectified diffusion. Consider Figure 9, which shows the growth curves for a range of free air bubbles exposed to a particular level of pulsed ultrasound. Note that for times greater than about 10 msec all nuclei that could grow will have done so in the 10 msec length pulse. Longer pulse lengths have very little additional effect. This result seems consistent with the experimental results. Note also that the times required to grow through resonance are on the order of a msec, and are independent of the initial size of the nucleus. Thus, if damage is done by bubbles undergoing large pulsation amplitudes as they pulsate near and through resonance, then the rectified diffusion results can explain the behavior observed by Child et al. for pulse lengths less than 1 msec. That is, unless the ultrasound is on for this length of

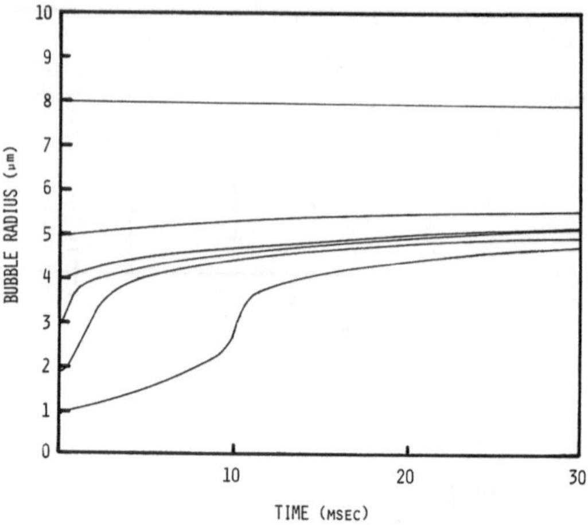

Figure 9. Growth times for air bubbles by rectified diffusion for various initial radii. The intensity was 30 W/cm²; the driving frequency was 1.0 MHz; the surface tension was 68 dyn/cm. These curves were obtained by numerical integration of the equations for rectified diffusion as extended to megaherz frequencies (see [10]).

time, the bubble has insufficient time to grow through resonance and cause damage.

4.2. Acoustic emissions from insonated tissue

Morris and Coakley [22] have observed the acoustic emissions from bean roots exposed to high intensity ultrasound. They irradiated the tips of the bean roots and measured the harmonic and subharmonic emissions from the roots during insonation. They observed delay times on the order of a few seconds, even at intensities of 50 W/cm^2, before they observed any significant emissions. After the emissions started, however, they observed distinct bursts of subharmonic and harmonic activity with characteristic growth times on the order of a few tenths of a millisecond. These emissions were similar to those observed by Neppiras and Coakley [23] for cavitation in water at megaherz frequencies.

Shown in Figure 10 are growth curves for air bubbles driven through resonance at an intensity of 30 W/cm^2 and at a frequency of 1.0 MHz. This intensity and frequency are applicable to Morris and Coakley's work. Note that the time required for the bubble to grow through resonance, increase dramatically in size and consequently undergo large pulsation amplitudes is on the order of a msec. These times are consistent with the acoustic emission bursts from the bean roots.

It is important to make some observations about the nuclei involved in these studies of irradiated plant tissue. First, we assume that the gas channels known to be present in these tissues provide the nuclei for the bubble growth and do not cause damage by their own stable pulsations. This seems possible

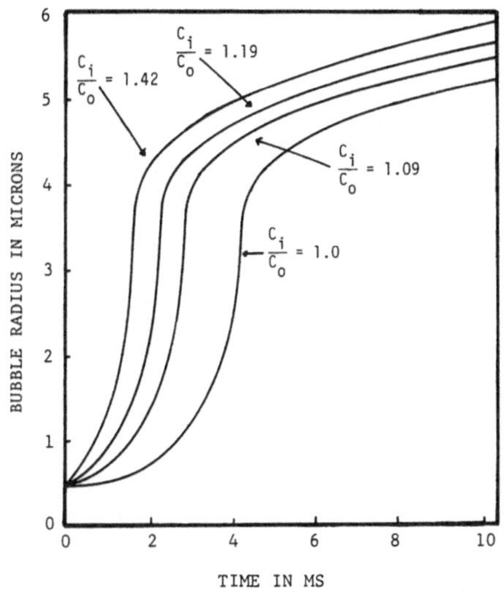

Figure 10. Growth of an air bubble by rectified diffusion from an initial radius of 0.48 μm for different levels of gas concentration ($C_i/C_0 = 1.0$ for saturation). The intensity was 30 W/cm^2; the driving frequency was 1.0 MHz; the surface tension was 68 dyn/cm. These curves were obtained by numerical integration of the equations for rectified diffusion as extended to megaherz frequencies (see [10]).

because there appears to be no effect of increasing the pulse length once damage has occurred. If the damage were caused by air channel vibration, then a longer pulse length ought to be associated with increased damage. Also, the acoustic emissions are similar to those observed for cavitation in water, which obviously does not possess gas channels. It is possible that the gas channels could be the source of the air that comprise the nuclei. If the channels of air were vibrated with sufficient amplitude, portions of gas could be broken off and form microbubbles that are smaller than resonance size. These microbubbles could then grow rapidly, causing damage and emitting subharmonics, break up and regrow in a continual process. Whatever the process, this area needs considerably more study in order to ascertain the mechanisms for damage.

4.3. Bubble growth in animal tissue

These have also been some recent indications of bubble growth in animal tissue. Child et al. [7] have reported damage to *Drosophila* larvae exposed to pulsed ultrasound and ter Haar and Daniels [17] have observed the presence of bubbles in guinea pigs exposed to therapeutic ultrasound. The role that nuclei play in these systems will be an exciting challenge for the future.

5. Summary and conclusions

We have examined various models for microbubble stabilization and how these bubbles may nucleate boiling and cavitation. There is evidence for stabilization by ionic charges, surface active agents, by crevices in particulate matter and by channels of air in plant tissues. We can explain some observed effects by each of these models and with varying degrees of success. The real challenge as it appears to the author, is not to seek reasons for rejecting this or that model, but to find a single model that can explain all the effects. It is entirely likely that on the air-water interface of a pocket of gas stabilized in a crevice in a solid particle there are surface active substances and ions that influence the behavior of that interface. Perhaps in order to ascertain the role that microbubbles play in nucleation, we shall need to create a grand unified model that combines the attributes of each of the special models.

Acknowledgement

The author wishes to acknowledge the assistance of his capable graduate student, Gary Hansen, in much of this work and the financial support of the Office of Naval Research and the National Science Foundation.

References

1. Akulichev VA (1966) Hydration of ions and the cavitation resistance of water. Sov Phys Acoustics 12: 144–149.
2. Alty T (1926) The origin of the electrical charge on small particles in water. Proc Roy Soc, Ser A. 112: 235–251.
3. Apfel RE (1970) The role of impurities in cavitation threshold determination. J Acoust Soc Amer 48: 1179–1189.
4. Apfel RE (1970) A novel technique for measuring the strength of liquids. J Acoust Soc Amer 49: 145–155.
5. Bargeman D and Van Voorst Vader F (1973) Effect of surfactants on contact angles at nonpolar solids. J Coll Inter Sci 42: 467–472.
6. Child SZ, Carstensen EL and Miller MW (1975) Growth of pea roots exposed to pulsed ultrasound. J Acoust Soc Amer 58: 1109–1110.
7. Child SZ, Carstensen EL and Lam SK (1981) Effects of Ultrasound on Drosophilia: III. Exposure of larvae to low-temporal average-intensity, pulsed irradiation. Ultrasound in Med and Biol 7: 167–173.
8. Crum LA (1979) Tensile strength of water. Nature 278: 148–149.
9. Crum LA (1980) Acoustic cavitation thresholds in water. In Lauterborn W (ed) Cavitation and Inhomogeneities in Underwater Acoustics. New York: Springer-Verlag.
10. Crum LA (1980) Measurements of the growth of air bubbles by rectified diffusion. J Acoust Soc Amer 68: 203–211.
11. Eller AI and Flynn HG (1965) Rectified diffusion during nonlinear pulsations of cavitation bubbles. J Acoust Soc Amer 38: 493–503.
12. Ellis AT, Waugh JG and Ting RY (1970) Cavitation suppression and stress effects in high-speed flows of water with dilute macromolecule additives. J Basic Engr Trans ASME 92: 459–466.
13. Epstein PS and Plesset MS (1950) On the stability of gas bubbles in liquid-gas solutions. J Chem Phys 18: 1505–1509.
14. Flynn HG (1964) Physics of acoustic cavitation in liquids. In Mason WP (ed) Physical Acoustics, Vol. I, Part B. New York: Academic.
15. Fox FE and Herzfeld KF (1954) Gas bubbles with organic skin as cavitation nuclei. J Acoust Soc Amer 26: 985–989.
16. Greenspan M and Tschiegg CE (1967) Radiation-induced cavitation; apparatus and some results. J Res Nat Bur Stds 71C: 299–312.

17. ter Haar G and Daniels S (1981) Evidence for ultrasonically induced cavitation *in vivo*. Physics in Med and Biol 26: 1145–1149.
18. Harvey EN, Barnes KK, McElroy WD, Whitely AH, Pease DC and Cooper KW (1944) Bubble Formation in Animals, I. Physical Factors. J Cell Comp Physiol 24: 1–22.
19. Herzfeld KF (1957) Comment. In Sherman FS (ed) Proc First Sympos Naval Hydrodynamics (Nat Acad Sci Wash DC) pp 319–320.
20. Hoyt JW (1976) Effect of polymer additives on jet cavitation. J Fluids Engr 98: 106–112.
21. Hsieh DH and Plesset MS (1961) Theory of rectified diffusion of mass into gas bubbles. J Acoust Soc Amer 33: 206–215.
22. Morris JW and Coakley WT (1980) The nonthermal inhibition of growth and the detection of acoustic emissions from bean roots exposed to 1 MHz ultrasound. Ultrasound in Med and Biol 6: 113–118.
23. Neppiras EA and Coakley WT (1970) Acoustic cavitation in a focused field in water at 1 MHz. J Sound Vib 45: 341–373.
24. Sirotyuk MG (1970) Stabilization of gas bubbles in water. Sov Phys Acoustics 16: 237–240.
25. Sirotyuk MG (1971) Experimental investigations of ultrasonic cavitation. In Rozenberg LD (ed) High Intensity Ultrasonic Fields. New York: Plenum.
26. Sirotyuk MG (1971) Elasticity and strength of stable gas bubbles in water. Sov Phys Acoustics 16: 482–484.
27. Strasberg M (1959) Onset of ultrasonic cavitation in tap water. J Acoust Soc Amer 31: 163–176.
28. Winterton RHS (1977) Nucleation of boiling and cavitation. J Phys D: Appl Phys 10: 2041–2056.
29. Yount DE, Hunkle TD, D'Arrigo JS, Ingle FW, Yeung CM and Beckman EL (1977) Stabilization of gas cavitation nuclei by surface-active compounds. Av Sp Env Med 48: 185–191.
30. Yount DE (1978) Skins of varying permeability: a stabilization mechanism for gas cavitation nuclei. J Acoust Soc Amer 65: 1429–1439.

Superheated drop nucleation for neutron detection

ROBERT E. APFEL, B.-T. CHU and JOHN MENGEL

Applied Mechanics, Yale University, P.O. Box 2159, New Haven, CT 06520, USA

Abstract. Moderately superheated drops of liquid in a gel will vaporize when exposed to neutrons. The basic characteristics of such detectors are described, with emphasis on a model for the dynamics of the process of vapor bubble nucleation.

Introduction

'It is well known that liquids can be superheated, but this metastable state is generally observed to be fragile and short-lived owing to the abundance of microscopic gas pockets and bubbles (heterogeneous nucleation sites) at liquid-solid interfaces either on particles or container surfaces. Only *one* of these sites needs to be present to act as the seed for the growth of a vapor bubble and, hence, the termination of the superheated state. Even in the absence of these sites, the natural radiation background will be responsible for the termination of the superheated state in a relatively short time' [1] as Glaser has demonstrated with the bubble chamber [3].

Unlike the bubble chamber, which undergoes only brief superheating and rapid pressure recycling, superheated drops introduced into a gel-like host are continuously sensitive, long-lived bubble chambers [1, 2]. 'By subdividing the sample into drops it has been assured that one nucleation event does not consume the sample, and, therefore, the repressurization procedure for traditional bubble chambers is avoided. Each drop is, itself, a miniature, neutron-sensitive bubble chamber. Such a detector needs no power source, because the drops themselves represent stored mechanical energy which is released when triggered by radiation.' Furthermore, the total amount of vapor evolved from the radiation-induced nucleation of drops can serve as a convenient integrated measure of the total exposure of the detector to a particular type of radiation.

'The theory of radiation-induced nucleation of superheated drops involves the nuclear physicist (if the interaction of an uncharged particle with a nucleus is being considered), the atomic physicist for the interaction of ions with matter, and the fluids physicist for describing the dynamic process resulting in a macroscopic vapor bubble. Since this "event" includes different physical phenomena occurring over a length scale covering up to twelve orders of magnitude, it should not be surprising that no existing theory has been capable of making accurate estimates of, for example, the threshold energy of neutrons required to nucleate bubbles in a given liquid superheated to a known degree' [1].

'There has been some controversy about mechanisms for vapor bubble

Applied Scientific Research 38: 117–122 (1982) 0003–6994/82/0382–0117 \$00.90.

nucleation, but it is generally agreed that Seitz's thermal spike theory is appropriate [4]. In this theory it is presumed that the atomic agitation along the path of the ion is equivalent to a hot spot which literally explodes, creating a sufficiently large vapor bubble' [1]. If the vapor bubble reaches a critical radius R_c defined by $2\gamma(T)/\Delta P$, then it will continue to grow and consume the liquid sample; here $\gamma(T)$ is the surface tension and $\Delta P = P_v(T) + P_g - P_o$, with P_v the pressure of the vapor in the cavity, P_g the partial pressure associated with non-condensible, dissolved gas, and P_o the externally applied pressure. ΔP is one measure of the degree of superheat of the liquid, and defines the critical size at a temperature, T.

In what follows, we shall outline the progress we have made in treating the theoretical problem which begins with the sudden deposition of a thermal spike in a superheated liquid and ends up with a vapor bubble which may or may not reach the critical size, as defined by the degree of superheat of the liquid.

Theoretical considerations

We first ask: What happens after an ion, resulting from a neutron-nucleus interaction, has imparted its energy to the molecules of the liquid? The highly agitated molecules around the ion track interact with their neighboring molecules, imparting to the latter some of their energy. Such interactions propagate radially outward from the ion path, resulting in the macroscopic observation of a rapidly expanding region of an extremely hot fluid. This is the thermal spike envisioned by Seitz [4]. The rapid expansion of this highly energized region is resisted by the inertia of the surrounding fluid so that an extremely high pressure is also produced in this region. Transfer of energy from this region to the surrounding medium is achieved through the dual process of shock propagation and heat conduction. The sudden expansion produces a very strong shock wave which propagates into the surrounding fluid. The shock is so intense in the initial stage that the temperature and pressure of the fluid within the shock enclosure far exceed the critical temperature and pressure of the fluid. Consequently, in the initial stage one cannot say whether the fluid within the shock enclosure is in the liquid state or the vapor state. As time proceeds, the expansion process slows down and the shock wave decays, with accompanying decreases in the temperature and pressure of the fluid in the shock enclosure. Before long, the pressure and temperature at some place within the enclosure reach the critical pressure and temperature when a vapor-liquid interface is formed. The location at which this occurs, relative to the critical radius R_c, is important. If the initial radius of the cavity is greater than the critical radius which balances surface tension with the excess pressure inside the cavity, the vapor cavity will grow indefinitely. On the other hand, if the initial radius of the cavity is sufficiently small, the cavity growth will slow down until its velocity becomes zero, after which it collapses under the action of surface tension.

A dynamic theory of radiation-induced cavitation should, therefore, take into account the compressibility effects, the conduction effect, and the influence of surface tension. The theory should tell us how energy is transmitted from a highly localized region to the more distant fluid, how temperature and pressure distributions in the fluid change with time, when and where a vapor-liquid interface is first formed, and whether the vapor cavity so formed will grow indefinitely or will eventually collapse. Above all, the theory should allow us to determine the threshold energy of the incident radiation which will result in cavitation.

To construct such a theory, we shall make two assumptions: First, the behavior of the medium can be described by the usual macroscopic fluid equations. Secondly, energy is deposited uniformly along an *infinite* line and the threshold energy producing a cylindrical cavity which will grow indefinitely is identified with the threshold energy for cavitation observed in experiments. Neither assumption is strictly true. The first assumption not only implies that the fluid may be treated as a continuum but also demands that all microscopic processes occur at a rate much faster than the fastest macroscopic process of interest. Obviously this is not the case at the earliest stage of the fluid expansion. The second assumption is certainly very far from being the case when the cavity has grown to macroscopically observable size (e.g., 0.1 mm). However, the assumption is probably acceptable up to the formation of a critical size cavity.

The mathematical problem is defined as follows: For the fluid we have five equations: the continuity expression, momentum equation, energy equation, equation of state (Horvath-Lin [5] for high temperatures and pressures), and a thermodynamic expression which allows the calculation of one thermodynamic variable if given two others. The five variables are density, fluid velocity, pressure, internal energy, and temperature, and the givens are initial and critical temperatures, pressures, and densities, as well as heat capacity and thermal conductivity.

Boundary conditions include mass, momentum, and energy conservation across the shock, and a global energy equation, given the energy deposited per unit length of the ion track. (This energy must be estimated from stopping power data for ions traversing a given material.)

Because we assume an infinite line source, the variables are solved for as a function of radial distance, r, and time, t. It is convenient, however, to normalize position to the shock radius $R(t)$, $[x = r/R$ and $0 \leqslant x \leqslant 1]$, and time through $\bar{R} = R/L$, where L is a size parameter related to the effective radius of a cylinder which contains as much energy as that deposited at $t = 0$.

The problem can be broken down into two major stages: In the first, the fluid inside the shock has pressures and temperatures above the critical parameters. The second stage begins when, at some position inside the shock, the critical parameters are reached and two phases, liquid and vapor, are identifiable. At this point, an interface with an appropriate temperature-dependent

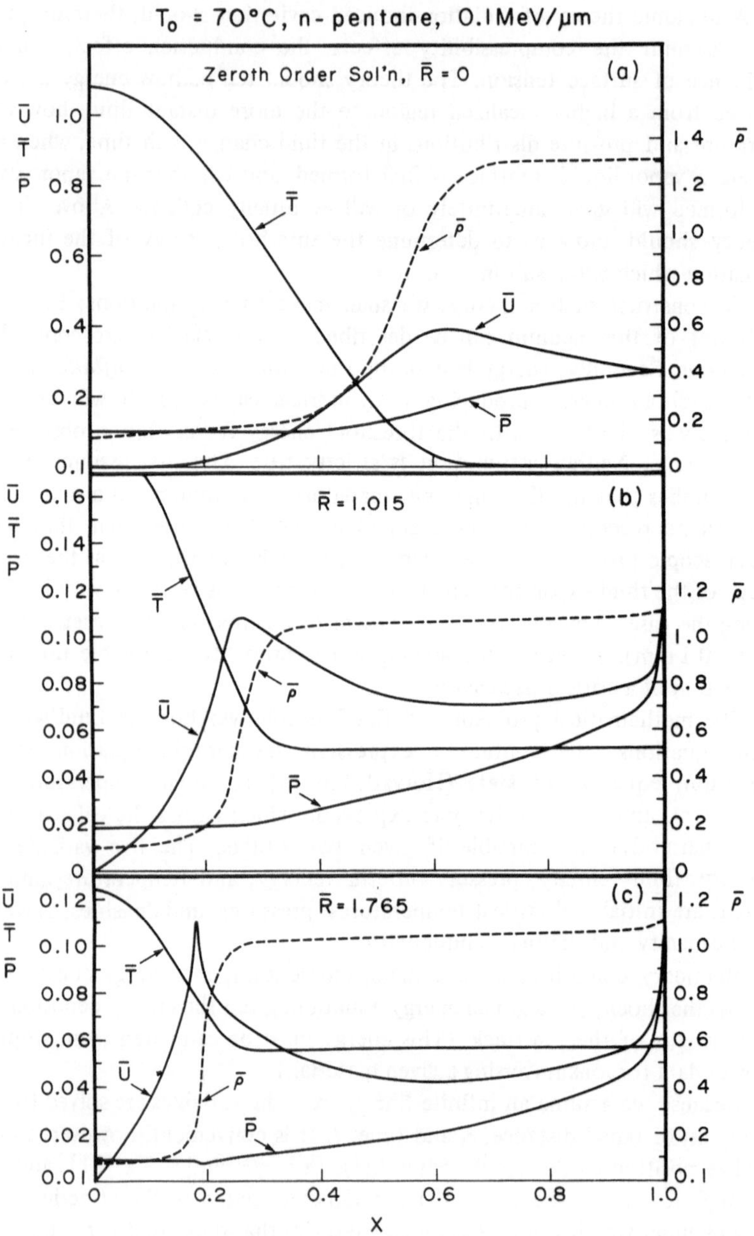

Figure 1. Nondimensional pressure $\bar{P}$, velocity $\bar{U}$, and temperature $\bar{T}$ (left ordinate), and $\bar{\rho}$ (right ordinate) versus nondimensional radius, x, for three times before a vapor bubble forms in n-pentane at 70°C (35°C superheat) for a deposited energy of 0.1 MeV/μm. (a) Very early time [note left ordinate scale differs from cases (b) and (c)]; (b) intermediate time, (c) time just before two-phase formation at $x \simeq 0.18$.

surface tension must be introduced. The second stage follows the dynamics of the vapor bubble, with two possible outcomes: either infinite growth, the critical size being surpassed, or bubble collapse and extinction, the zero velocity condition for the bubble occurring for a size less than the critical size.

Thus far, the first stage has been solved by employing a finite difference scheme after dealing with the singularity at $t = 0$. Figures 1(a) and 1(b) give examples of the normalized pressure $\bar{p}$, velocity $\bar{u}$, temperature $\bar{T}$, and density $\bar{\rho}$ distributions within the shock enclosure at two times before a vapor cavity is formed.

The particular case we are considering is n-pentane at 70°C (35° above its boiling point) for an energy deposited per unit length of 0.1 MeV/μm (= 0.0016 ergs/cm). Figure 1(c) shows the instant just before a vapor cavity is formed. It becomes clear that the density distribution becomes more and more like a step function as the temporal variable $\bar{R}$ increases. Indeed, the critical state is found to occur at $\bar{R} = 1.7786$ – that is moments after the distribution shown in Figure 1(c) for which $\bar{R}$ is 1.765. It occurs at an x equal to 0.18484, or just short of 20% of the distance out to the shock, pointing out that bubble formation via a thermal spike mechanism is a very inefficient process. The figures clearly show that the flow field can be separated into two regions: one with low density and high temperature, in which pressure is relatively uniform and flow velocity rises with increasing x or r; the other, or outer region, has a more or less constant temperature and density, a gradually increasing pressure (increasing with x) and a decreasing velocity. The vapor-liquid cavity first forms somewhere near the demarcation line (or, better, the transition zone) of these two regions. The pressure dip in this transition zone is real. It is the result of 'softening' of the medium near the critical point where the isothermal compressibility becomes very large. On the other hand, the sharp velocity peak is really not so 'sharp', because viscosity has been neglected in this computation. Likewise, the sharp velocity and pressure rise near the shock front will be wiped out when viscous effects are included.

Ongoing work

For the practical cases of interest to neutron detection (superheats of 20°–60°C and neutron energies up to 15 MeV), the initial vapor bubble size is smaller than the critical size as defined by the surface tension and degree of superheat. We are now, therefore, focussing on the dynamics of a vapor bubble including, in addition to the effects already incorporated into the model, both viscosity and surface tension. We should then have a complete dynamic theory from which thresholds for bubble production should be extractable. Moreover, with 'exact' numerical results we can have reference points for possible analytic models which employ significant simplifications.

Acknowledgments

Primary responsibility for the theoretical aspects of our program rests with one of us (Chu), while the other (Apfel) is responsible for the experimental program. This work has been supported by the Heat Transfer program of the National Science Foundation and is currently supported by the Office of Health and Environmental Research of the Department of Energy.

References

1. Apfel, RE (1979) The superheated drop detector. Nucl Instr and Meth 162: 603–608.
2. Apfel RE (1979) Detector for neutrons and other radiation. US Patent 4,143,274.
3. Glaser DA (1952) Some effects of ionizing radiation on the formation of bubbles in liquids. Phys Rev 87: 665.
4. Seitz F (1958) On the theory of the bubble chamber. Phys Fluids 1: 2–13.
5. Horvath C and Lin HJ (1977) A simple three-parameter equation of state with critical compressibility-factor correlation. Cana J Ch Eng 55: 450–456.

Observations of nuclei in cavitating flows

JOSEPH KATZ and ALLAN ACOSTA

California Institute of Technology, Pasadena, CA 91125, USA

Introduction

The present report focuses on the source of the large difference between the theoretical strength of pure liquid [12] and the actual tension that is required to initiate cavitation in technical fluids such as test facilities and natural waters. This discrepancy is commonly explained by the existence of nuclei, either solid particles or vapor and gas bubbles that permit phase transition to take place near equilibrium. The existence of these nuclei, their source and lifetimes have occupied much space in the technical literature for decades [1, 2, 3, 8]. Yet, direct observations of tests in applications to naval hydrodynamics and hydraulic machinery flows has not provided much information about these nuclei. Their existence, however, and their effect on cavitation is in no doubt as is demonstrated by the series of photographs of a propeller that were taken in the 'Vacu-Tank' of NSMB [9] shown in Figure 1. There, addition of 'nucleating' sources to the water by electrolysis clearly increases the number of visible cavitating bubbles on the blade surfaces.

The importance of these nuclei has led to the development of several detection and observation techniques [4, 6, 10, 11]. These include the microscopic observations of water samples, the 'coulter counter', 'single particle light scattering', the 'acoustic' techniques for detection of gas bubbles, holograms with microscopic observation of the reconstructed image, and finally the venturi 'liquid quality' meter [11].

Some results of the nuclei number density distribution function $N(R)$ that were accumulated from various sources [6] in different experimental facilities and some open sea measurements are presented in Figure 2. The results include holographic observations in the Caltech HSWT which are found to be mostly solid particles and the Caltech LTWT which are found to be mostly bubbles. This difference in the type of nuclei may explain why is it easier to cavitate the same body in the LTWT (provided that the flow is non-separating) while the liquid in the HSWT can support a certain tension [6]. Note the several orders of magnitude difference in the populations shown and the correspondence between the oceanic acoustic measurements of Medwin [10] that display the population of bubbles and the holographic observations of plankton [5, 13]. It is interesting that some of these facilities are well above the 'natural' levels and it may be inferred that the Vacu-Tank may be well below. Thus, significant 'scaling' errors may occur while applying laboratory results to field cavitation phenomena [1, 2, 3].

Applied Scientific Research 38: 123–132 (1982) 0003–6994/82/0382–0123 $01.50.

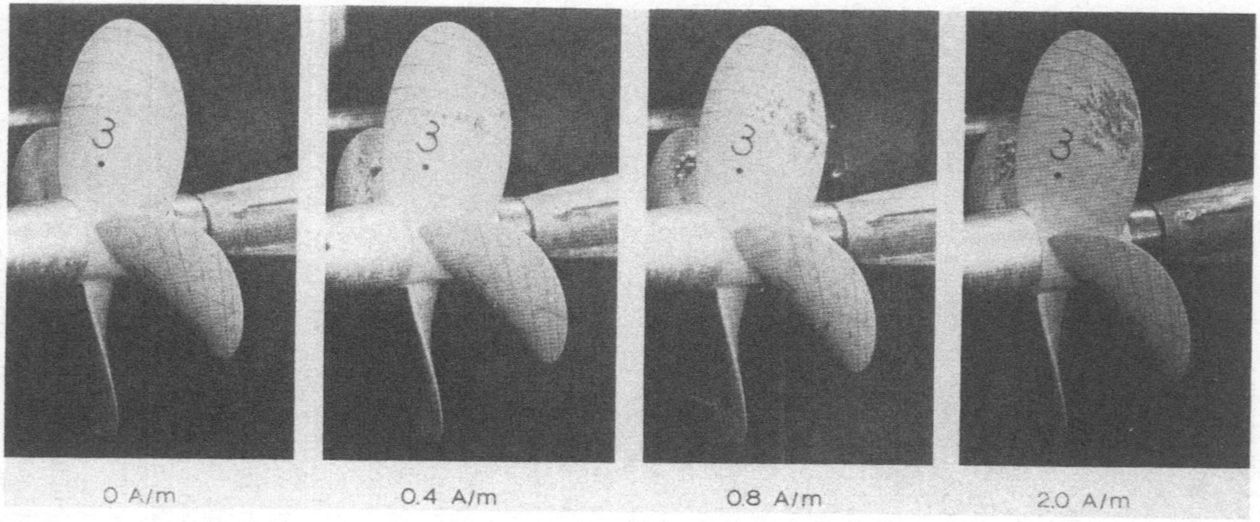

Figure 1. The dependence of cavitation on a propeller on the amount and size of bubbles that are added to water by electrolysis [2].

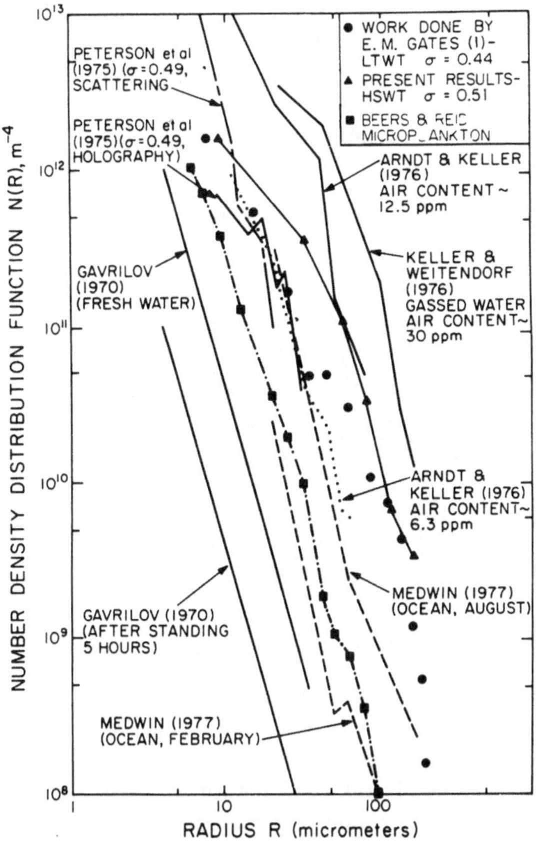

Figure 2. Nuclei distributions from various sources. The data of the HSWT and Beers et al. [5] are superposed on a graph presented in [6].

Nuclei and cavitation inception

Our own interest has been to identify the type, the location and the role of the nuclei in the cavitation process in steady flows past bodies. Following other investigators in this country and elsewhere (see Gates [6] for details) we have opted for the holographic system which provides one with the ability to record details to about 10 micrometers diameter. The details of the holocamera and the reconstruction system used herein are illustrated in Figures 3 and 4 respectively. This is an in-line system with a small modification in the reconstruction procedure that enables us to use it as a schlieren recording system as well [7].

Figure 5 is a photograph of cavitation inception on a hemispherical head form body with a downstream facing step (1 mm height) at the tangency point. The flow in this region is visualized by heating the body about one

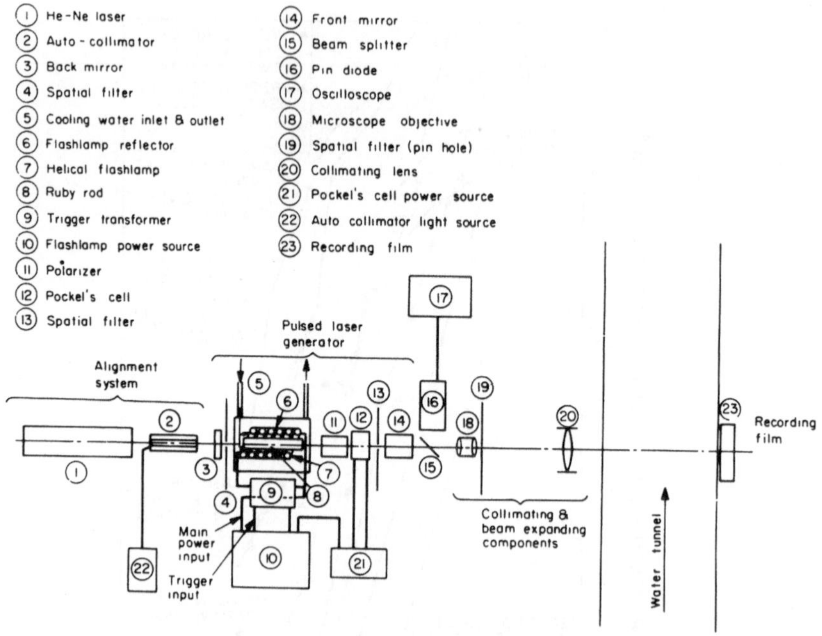

Figure 3. Diagram of the holocamera.

degree Centigrade and reconstructing the hologram of this flow as a schlieren picture (Figure 6) that shows clearly the laminar separation, the transition in the shear layer, and part of the turbulent mixing zone. Holograms that are recorded just prior to the formation of visible cavitation reveal the appearance of a large number of microbubbles at the unstable part of the shear layer upstream of the reattachment zone as is shown in Figure 7. A detailed map of the bubble population in different sections of the flow field including part of the background 'free stream', the separated zone and a small region upstream is demonstrated in Figure 8. This map provides us with the

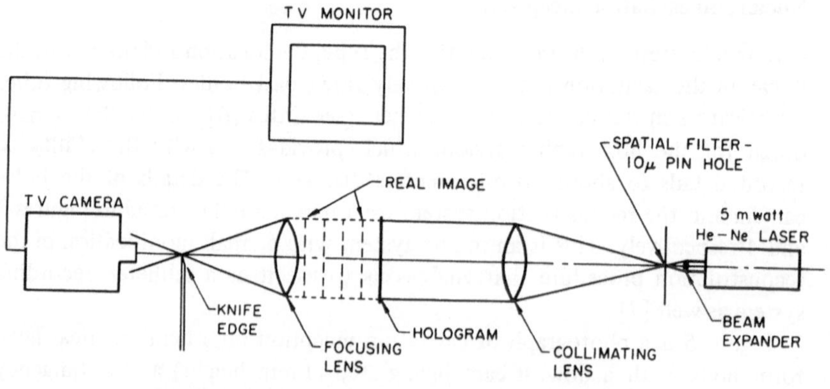

Figure 4. Diagram of the schlieren reconstruction system.

Figure 5. A photograph of cavitation inception on the hemispherical step body.

supply of free stream nuclei and the growing bubble population in the turbulent mixing zone during the transition from fully wetted flow to visible cavitation.

A schlieren-reconstructed hologram of the separated zone on the well-observed hemispherical head form body is presented in Figure 9. A hologram of the flow field (Figure 10) reveals the formation of a cluster of microbubbles upstream of and at the reattachment zone just prior to inception of macroscopic cavitation. None of these bubbles are attached to the surface of the body; their sizes vary from 10 to 150 micrometers diameter.

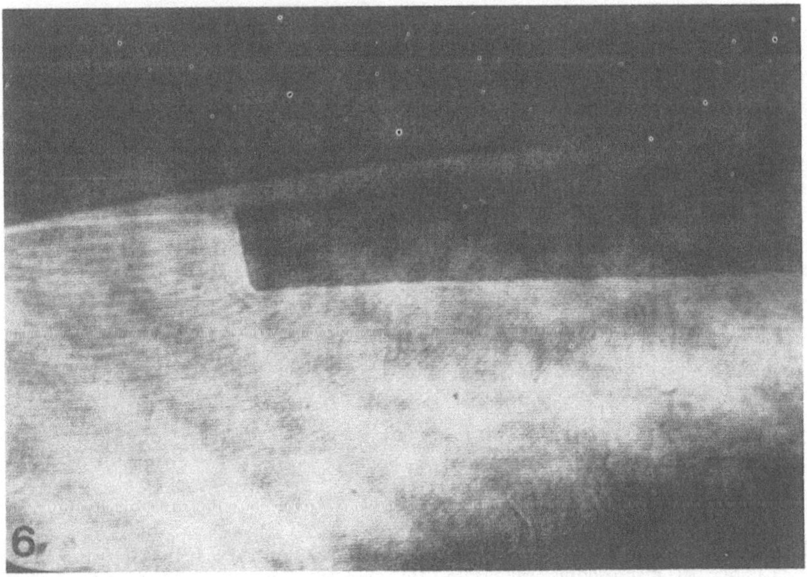

Figure 6. A schlieren reconstructed hologram on the surface of the hemispherical step body. $Re_D = 1.74 \times 10^5$.

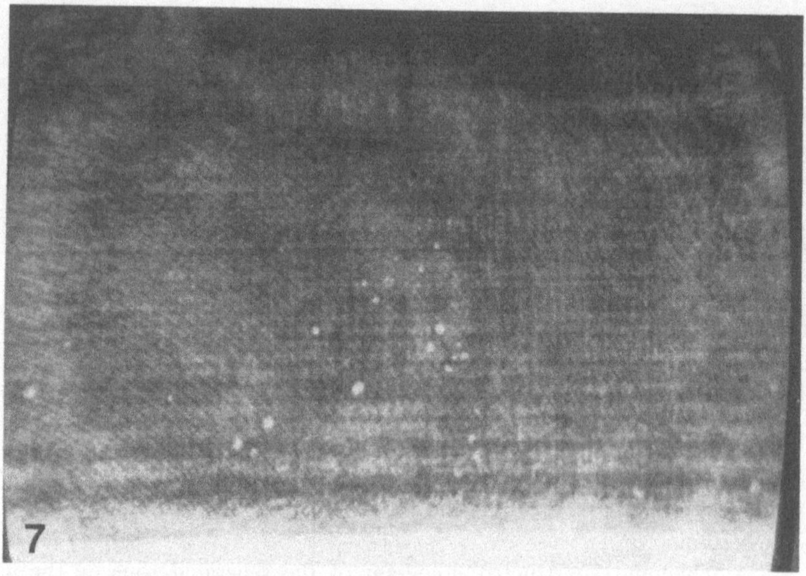

Figure 7. A hologram showing the appearance of microbubbles in the turbulent mixing zone extending from 3.5 to 10.5 mm behind the step just prior to the inception of visible cavitation. $Re_D = 2.78 \times 10^5$; $\sigma = 1.07$.

Step Body

$Re_D = 2.78 \times 10^5$

$\sigma = 1.11$

Bubble Diameter in microns

Section	0-20	20-50	50-75	75-100	100-150	150-200	200-250
a	11	1	2	–	–		
b	7	1	1				
c	10	–	1				
d	12	1	1	1			
e	10	2	1	–	–	–	1
f	4	1	–	–	–	2	
g	8	–	1				

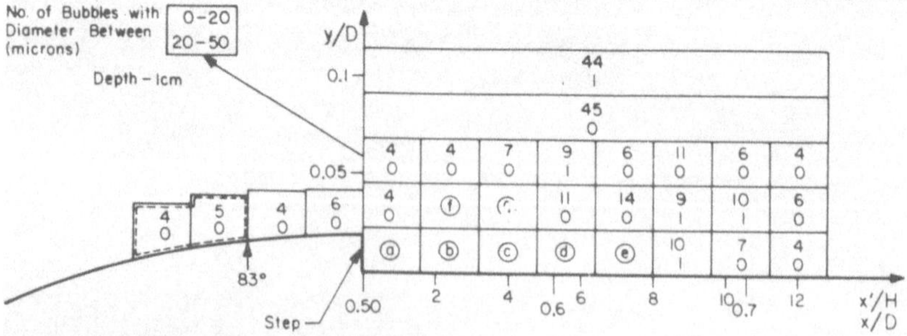

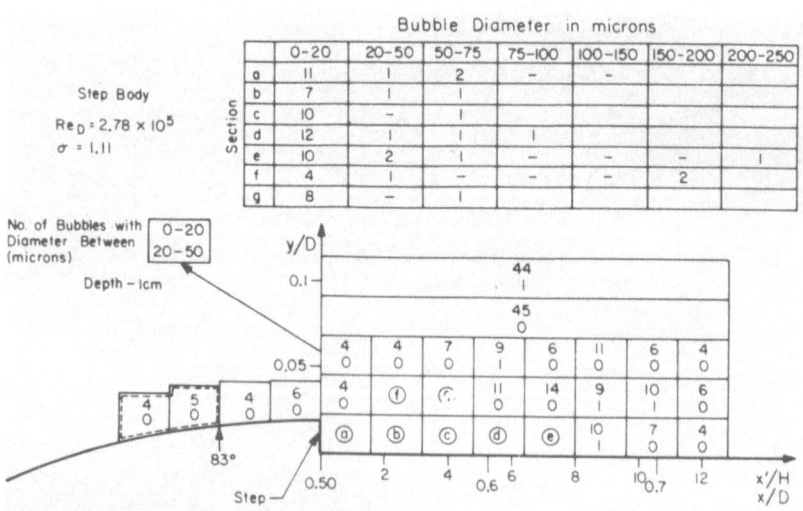

Figure 8. A map of the bubble population in the neighborhood of the step prior to the appearance of macroscopic cavitation. $Re_D = 2.78 \times 10^5$; $\sigma = 1.11$ ($\sigma = (p_\infty - p_v)/\frac{1}{2}\rho U^2$, p_∞ being the ambient pressure in the fluid, p_v the vapor pressure, ρ the density and U the flow speed).

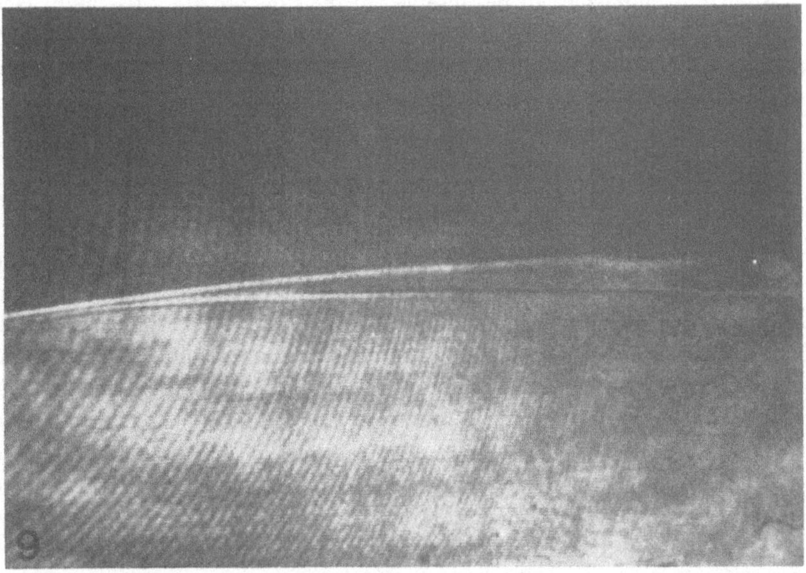

Figure 9. A schlieren-reconstructed hologram of the separation zone on the surface of the hemispherical body. $Re_D = 1.79 \times 10^5$.

Figure 10. Inception of cavitation on the surface of the hemispherical body. $Re_D = 3 \times 10^5$; $\sigma = 0.61$. The photograph covers the region between $x/D = 0.593$ and 0.607 (x being the axial distance and D is the body diameter).

Similar phenomena can be observed in the flow around a flat blunt face body [7], characterized by a much larger separation region; that is a large number of microscopic bubbles appear in turbulent mixing zone just prior to the formation of macroscopic cavities.

Nuclei control and future plans

Being an important scaling factor makes the control of free stream nuclei population an essential part of any cavitation test. Experiments in the Caltech LTWT shown in Figure 11 reveal, on the one hand, almost an order of magnitude difference in the bubble population at the same cavitation index. Whereas, on the other hand, even at sharply different cavitation indices, 0.61 vs 1.7 the bubble density is almost equal (the air content is kept constant, 10 ppm). These somewhat 'surprising' phenomena are due to the procedures that preceded the recording of the hologram. In one case, that is represented by the lower $N(R)$ distribution in Figure 11 the tunnel was run at $\sigma = 5.5$ for at least ten minutes before lowering the pressure whereas in the larger case the cavitation index was kept equal to 2.5. The final σ to which the tunnel is brought seems to be much less

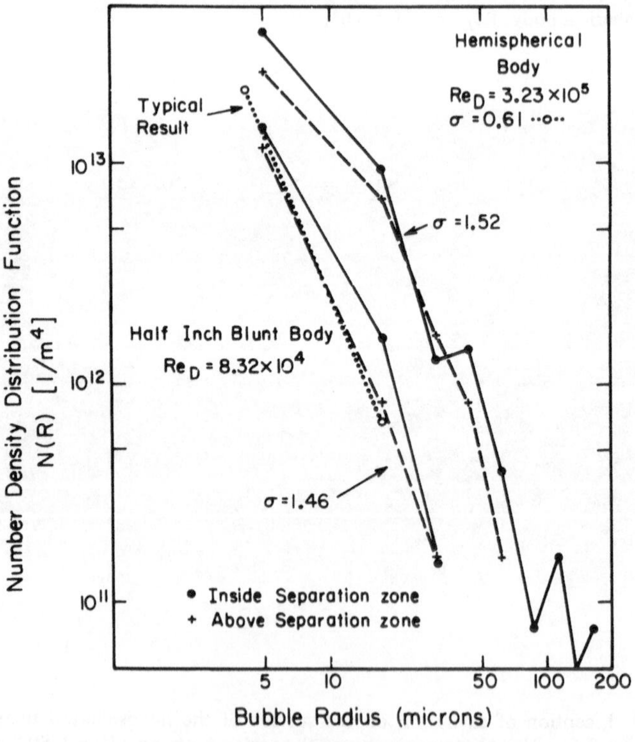

Figure 11. Bubble distribution in the Caltech LTWT determined by holography.

important than the steady state conditions prior to pressure reduction. These results (and they are representative of detailed experiments that lead to the same conclusion [7]) suggest the proper procedure for a repeatable cavitation test when the nuclei population is an important parameter. Note that the lowest $N(R)$ found in the LTWT is still considerably larger than the oceanic values that are presented in Figure 1.

There are very few undisturbed measurements of $N(R)$ in natural waters or even in the circuits of large hydraulic structures. Due to the long-standing interest in our marine applications of cavitation we think that it is of some value to pursue the idea of increasing our overall knowledge of the natural science of particulates (debris, planktonic and microbubbles) in natural waters. To this end we have adapted the components shown in Figure 3 into an underwater holocamera. Our goal is to make an additional contribution to the available data of the nuclei distribution in the ocean that will be used in the future for modeling of the natural waters. We also plan to coordinate these observations while the simpler liquid quality meter designed in the Delft Hydraulic Laboratory [11]. These results will assist in judging whether different oceanic particulates are a good source of cavitation nuclei.

Acknowledgment

This work was supported by Naval Sea Systems Command General Hydromechanics Research Program administered by the David Taylor Naval Research and Development Center under Contract No. N00014-75-C-0378. We would also like to thank Barbara Katz (Sea Grant, Cal State University of Long Beach) for helpful information.

References

1. Acosta AJ and Parkin BR (1975) Cavitation inception – A selective review. J Ship Res 19 (4): 193–205.
2. Acosta AJ and Parkin AR (1980) Report of the ATTC Cavitation Inception Committee Proc 19th Am Towing Tank Conf 2: 829–858.
3. Arndt REA (1981) Cavitation in fluid machinery and hydraulic structures. Am Rev Fluid Mech 13: 273–328.
4. Billet ML and Gates EM (1979) A comparison of two optical techniques for measuring cavitation nuclei. International Symposium on Cavitation Inception, ASME. New York.
5. Beers JR, Reid FMH and Steward GL (1973) Microplankton of the North Pacific Central Gyre. Population Structure and Abundance, June 1973. Int Revue ges Hydrobiol 60 (5): 607–638.
6. Gates EM and Acosta AJ (1978) Some effects of several free-stream factors on cavitation inception of axisymmetric bodies. 12th Symp Naval Hydrodynamics. Washington DC: 86–108.
7. Katz J (1981) Some cavitation phenomena associated with separated flows. Ph D Thesis, California Institute of Technology, Pasadena, CA (in preparation).
8. Knapp RT, Daily JW and Hammitt FG (1970) Cavitation. New York: McGraw-Hill Book Company.

9. Kuiper G (1981) Cavitation Inception on Ship Propeller Models. Ph D Thesis, Technische Hogeschool Delft. Wageningen: Veenman.
10. Medwin H (1977) In situ acoustic measurements of microbubbles at sea. J Geophys Res 82 (6): 921–976.
11. Oldenziel DM (1979) Bubble cavitation in relation to liquid quality. Ph D Thesis Technical University Twente. Delft: Delft Hydraulic Laboratory, Publication No. 211.
12. Plesset MS (1969) The tensile strength of liquids. Cavitation State of Knowledge. ASME: 15–25.
13. Stewart GL and Bears JR (1973) Proc Soc Photo-Opt Inst Engrs 41: 183–188.

On the stability of gas bubbles in liquid-gas solutions

MILTON S. PLESSET* and SATWINDAR S. SADHAL**

*California Institute of Technology, Pasadena, CA 91125, USA;
University of California, Los Angeles, CA 90024, USA;
**University of Southern California, Los Angeles, CA 90007, USA

Abstract. It was shown some time ago by use of diffusion theory that a gas bubble in a liquid-gas solution was unstable. This problem has been reconsidered recently in two papers both of which propose to develop a stability analysis solely from thermodynamic considerations. The first of these studies purports to find stability for a gas bubble in a liquid-gas solution. Some possible sources of error in this analysis are mentioned here. The second study considers a particular system of a bubble in a liquid drop immersed in a second liquid in which the gas is insoluble. A condition of stability is then found. This system is reconsidered here simply in terms of the ideas of diffusion theory. The stability conditions may then be stated in simple physical terms.

1. Introduction

The question of the stability of gas bubbles in a liquid-gas solution is of concern in several problems of physical interest. Among these may be mentioned the extinction of sound or light in water by air bubbles and the diffusion of air into bubbles formed in cavitating flow of a liquid [5]. Microbubbles of air in saturated or supersaturated solutions of air in water have also been invoked to explain the observed modest tensile strength of water; theory on the other hand predicts a tensile strength much larger than that observed. It was shown, however, some time ago [2] that gas bubbles in liquid-gas solutions of any concentration are unstable.

The stability of a gas bubble in a liquid-gas solution has been reconsidered recently in two papers [1, 4] both of which develop a stability analysis solely from thermodynamic considerations. The earlier analysis [2] was based entirely on the diffusion process as governed by the diffusion equation. In the first paper to which we have referred [1], the author proposes to find stability conditions by examination of the Helmholtz free energy, F. The familiar thermodynamic condition for equilibrium in terms of this function is $dF = 0$, and the condition for a stable equilibrium is $d^2F > 0$. The author finds from his thermodynamic analysis that there are stable gas bubbles in liquid-gas solutions. The thermodynamic analysis is incorrect since the Helmholtz free energy is a minimum for stability only when the system has constant total volume. The analysis is made on the basis of a fixed mass of solvent (the liquid) and of a fixed mass of gas (in solution and in the bubble). If now one considers a variation in the gas bubble volume, say δV, then the volume of the solution essentially does not change so that the total system undergoes the same volume change δV. Whether this is the important error in the paper is not clear.

Applied Scientific Research 38: 133–141 (1982) 0003–6994/82/0382–0133 $01.35.

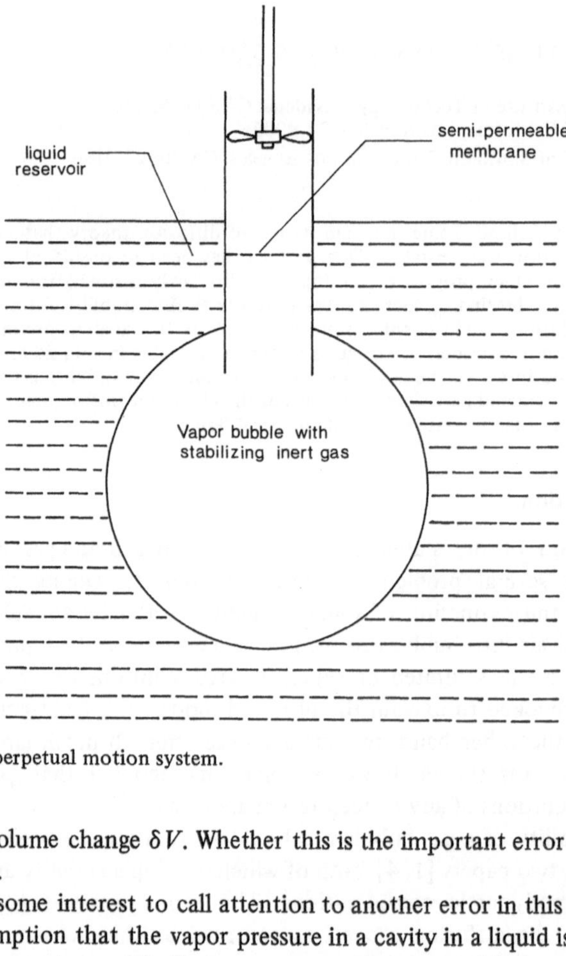

Figure 1. A perpetual motion system.

the same volume change δV. Whether this is the important error in the paper is not clear.

It is of some interest to call attention to another error in this paper which is the assumption that the vapor pressure in a cavity in a liquid is affected by the cavity curvature with the consequence that it would then be different from the vapor pressure over a liquid with a flat surface. It must be said that this supposition has appeared in many places in the literature and in some textbooks [3], yet it is easy to show that this supposition cannot be correct by invoking the second law of thermodynamics. To see this result we consider a body of liquid all at the temperature T with a flat surface over which the vapor pressure is $p_v(T)$, Figure 1. We now consider a bubble of any radius, R, within the liquid. The bubble can be stabilized with an inert gas which is insoluble in the liquid, and perfect gas behavior may be assumed both for this inert gas and for the vapor so that there will be no interaction between them. Let the vapor pressure in the bubble be $p_v(R, T)$. The bubble may now be connected to the space above the flat surface with a tube that has a semipermeable plug, permeable to the vapor and impermeable to the inert gas. If $p_v(R, T) \neq p_v(T)$, one can run an engine indefinitely, and the

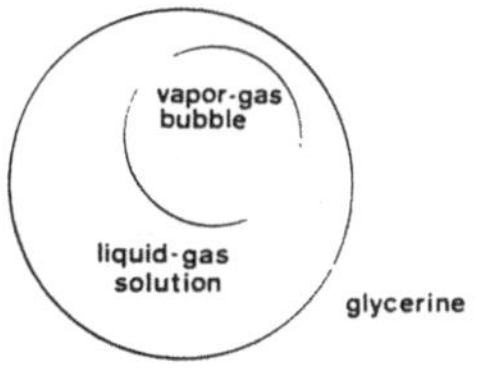

Figure 2. A finite solute-solvent system.

contradiction with the second law is evident. The conclusion must be that $p_v(R, T) = p_v(T)$ for any cavity radius R.

The analysis of Cha's paper [1] appears to be limited to small, or certainly finite, systems, and this important limitation is not made explicit. For a bounded system, it would seem most necessary to specify explicitly conditions at the boundary of the system, but these are not considered.

The flaws just described do not apply to the very interesting paper by Mori et al. [4]. These authors have explicitly considered a finite system which consisted of a bubble of gas (and vapor) in a liquid drop (Freon) in which the gas is soluble; this system was suspended in another liquid (glycerine) of relatively large volume (Figure 2). The solubility of the gases considered while significant in liquid Freon was negligible in glycerine. Experimental observations with several gases demonstrated that there were stable bubbles. The authors analyzed the stability of a bubble for this configuration from thermodynamic considerations. The thermodynamic stability criterion was properly based on the Gibbs thermodynamic potential, G, with equilibrium given by $dG = 0$ and stability by $d^2G > 0$. The experimental measurements were in agreement with the theoretical stability analysis. These authors also point out that the possibility of the existence of a stable bubble disappears with an increase in the volume of the liquid-gas solution. This behavior agrees, as it should, with the results of diffusion theory [2].

The bubble stability which was observed experimentally and which was determined analytically from thermodynamic arguments by Mori et al. [4] is very readily understood on physical grounds. There can be no transport of dissolved gas across the outer boundary of the liquid Freon drop because it is immersed in another liquid, glycerine, in which the gas is essentially insoluble. The stability of the gas bubble arises since the inventory of dissolved gas in the liquid is not only finite but is relatively small. Thus, for example, if the liquid drop is supersaturated relative to the gas pressure in the bubble, the bubble will grow by diffusion of dissolved gas from the liquid into the bubble. The dissolved inventory in the liquid will be depleted with a decrease in the supersaturated condition. Similarly, if the liquid is undersaturated, gas will diffuse from the bubble into the liquid and increase the liquid inventory. Clearly, the magnitude of the total dissolved inventory is an essential parameter for a stable bubble. If the magnitude of this dissolved inventory relative to the gas mass in the bubble is too large, there cannot be a stable bubble in the drop.

The specific conditions for stability can be obtained from the notions of diffusion theory as may be shown in the next section.

2. Analysis

Let us consider an isothermal system at a temperature T consisting of a liquid drop of Freon-21 surrounded by glycerine as shown in Figure 2. Within this drop we suppose that there is a gas bubble. The drop size is of the order of 100 μm. The surface tension effects are sufficient to assure that both the drop and the bubble are spherical in shape.

In the analysis that follows we use the notation of Mori et al. [4] in which the subscript 1 refers to Freon, the subscript 2 refers to the solute gas, the prime denotes the gas phase, the double prime denotes the liquid phase, and the triple prime refers to the surrounding impenetrable liquid (glycerine).

The pressure in the bubble is given by

$$p' = p''' + \frac{2\sigma}{r} + \frac{2\sigma_{ll}}{r_{ll}}, \tag{1}$$

where p''' is the pressure in the glycerine, σ is the interfacial tension of the liquid-gas interface, and σ_{ll} is the glycerine-Freon interfacial tension. The drop radius is denoted by r_{ll} and the bubble radius by r.

The partial pressure of the solute gas in the bubble is given by

$$p_2' = p' - p_1',$$

or

$$p_2' = p''' - p_1' + \frac{2\sigma}{r} + \frac{2\sigma_{ll}}{r_{ll}}, \tag{2}$$

where p_1' is the vapor pressure of the Freon. From the ideal gas equation the number of moles of the solute in the gas phase is given by

$$n_2' = \frac{p_2' V'}{RT},$$

or

$$n_2' = \left[(p''' - p_1') + \frac{2\sigma}{r} + \frac{2\sigma_{ll}}{r_{ll}}\right] \frac{4\pi r^3}{3RT} \tag{3}$$

where $V' = (4/3)\pi r^3$ is the bubble volume and R is the universal gas constant.

The number of moles of solute in solution is

$$n_2'' = n_{20} - n_2' = n_{20} - \frac{p_2' V'}{3RT}, \tag{4}$$

where n_{20} is the total number of moles of the solute in the entire system. In the same way, the number of moles of Freon in the liquid phase is

$$n_1'' = n_{10} - \frac{p_1' V'}{RT}, \tag{5}$$

where n_{10} is the total number of moles of Freon in the system. Here n_{10} and n_{20} are constants.

The solute mole fraction in the liquid phase is therefore given by

$$x_2'' = \frac{n_2''}{n_1'' + n_2''},$$

or

$$x_2'' = \frac{n_{20} - p_2' V'/RT}{n_{10} + n_{20} - p' V'/RT}. \tag{6}$$

By assuming that the liquid is incompressible we may take its volume V'' to be fixed. Therefore, we have

$$r_{ll} = (r_d^3 + r^3)^{1/3}, \tag{7}$$

where

$$V'' = (4/3)\pi r_d^3 = \text{constant}, \tag{8}$$

and r_d is the drop radius for a completely collapsed bubble. With the use of eqations (1), (2) and (7) in equation (6) we obtain

$$x_2'' = \frac{n_{20}RT - \left[p''' - p_1' + \dfrac{2\sigma}{r} + \dfrac{2\sigma_{ll}}{(r^3 + r_d^3)^{1/3}}\right]\dfrac{4}{3}\pi r^3}{(n_{10} + n_{20})RT - \left[p''' + \dfrac{2\sigma}{r} + \dfrac{2\sigma_{ll}}{(r^3 + r_d^3)^{1/3}}\right]\dfrac{4}{3}\pi r^3}. \tag{9}$$

Equation (9) represents the solute mole fraction in the liquid phase, averaged over the liquid volume. In a nonequilibrium state we would expect a non-uniform distribution over the liquid volume. In particular, the mole fraction at the liquid gas interface in a state of nonequilibrium would be different from that given by equation (9). It would be given instead by Henry's law

$$x_{2,s}'' = Kp_2' = K\left[p''' - p_1' + \frac{2\sigma}{r} + \frac{2\sigma_{ll}}{(r^3 + r_d^3)^{1/3}}\right], \tag{10}$$

where $1/K$ is the Henry's law constant.

In a state of equilibrium we would have

$$x_2'' = x_{2,s}''. \tag{11}$$

Since the solute mole fraction is very small, we may approximate the denominator on the r.h.s. of equation (9) as follows

$$(n_{10} + n_{20})RT - \left[p''' + \frac{2\sigma}{r} + \frac{2\sigma_{ll}}{r_{ll}}\right]\frac{4\pi r^3}{3} \simeq n_{10}RT. \qquad (12)$$

With the equilibrium condition (11) and, after some algebra, one finds the relation between the equilibrium bubble radius and the outside pressure:

$$p''' - p_1' = -\frac{2\sigma}{r} - \frac{2\sigma_{ll}}{(r^3 + r_d^3)^{1/3}} + \frac{n_{20}}{n_{10}K + (4\pi/3RT)r^3}, \qquad (13)$$

which was also obtained by Mori et al. [4]. The important purpose of our discussion, however, is to examine equations (9) and (10) and deduce the stability limits on the basis of diffusion theory. We have plotted x_2'' and $x_{2,s}''$ as functions of r, both on the same graph. The stability analysis is discussed below.

3. Results and discussion

In Figure 3(i–iii) for $(p''' - p_1')$ fixed at 1×10^5 N/m^2, the dissolved solute (oxygen in this example) mole fractions, n_2'', are plotted as functions of r for three different values of the total solute content, n_{20}. We have taken $\sigma = 1.8 \times 10^{-2}$ N/m, $\sigma_{ll} = 6.34 \times 10^{-2}$ N/m, $K = 1.97 \times 10^{-8}$ m^2/N and $T = 25°$C. For the case 3(i) we have $n_{20} = 1.2 \times 10^{-10}$ mole. We observe that the curves intersect at two points, $r = R_A$ and $r = R_B$. These are the two states of equilibrium. We notice that for $r < R_A$, we have $x_2'' < x_{2,s}''$. Since the surface concentration is greater than the average concentration elsewhere in the liquid, we have the solute diffusing away from the bubble, thereby causing it to diminish. For $R_A < r < R_B$, we have $x_2'' > x_{2,s}''$ and we would have diffusion into the bubble, causing it to grow. Finally, for $r > R_B$, we find $x_2'' < x_{2,s}''$ and the bubble would diminish.

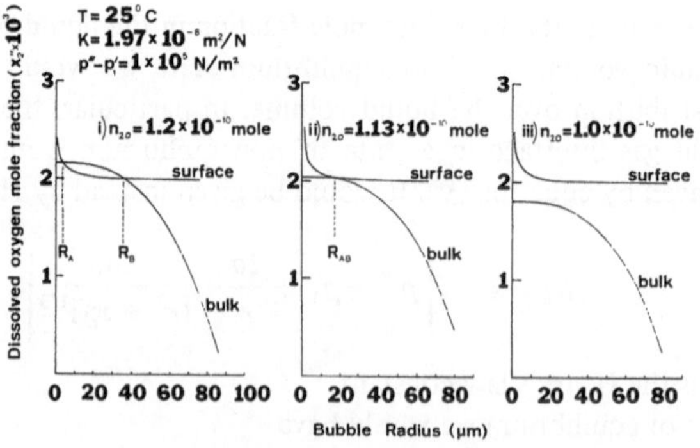

Figure 3. Variation of solute mole fraction with bubble radius at bubble surface and bulk of drop for fixed $(p''' - p_1')$ and various values of n_{20}.

We deduce from this analysis that for any radius r near R_B with $r \gtrless R_B$, we would have r tending towards R_B. The state $r = R_B$ is clearly a state of stable equilibrium. For any disturbance from this state, the system would return to this state.

The equilibrium state $r = R_A$, however, is unstable. A disturbance would cause the bubble either to collapse or grow to the radius $r = R_B$.

For the case 3(ii), the total oxygen content is $n_{20} = 1.13 \times 10^{-10}$ mole, with all other conditions being the same as 3(i). Here the two equilibrium points merge into one. We see that at nonequilibrium states the surface concentration is higher than the average concentration within the drop. The bubble in this case would always tend to diminish. The equilibrium state may be referred to as 'just unstable'. A disturbance that would reduce r from $r = R_{AB}$ would cause the bubble to collapse. With an opposite disturbance the equilibrium would be restored at $r = R_{AB}$.

In the case 3(iii) the solute content is small enough so that no equilibrium state is found. We always have $x''_{2,s} > x''_2$ and the bubble is unconditionally unstable at the given values of p''' and T.

It is quite clear from Figure 3 that as we reduce the oxygen content in the system, keeping p''' and T fixed, the stable radius $r = R_B$ reduces up to a minimum $r = R_{AB}$. At such a state the solute content is the minimum possible for a stable bubble to exist. By locating the maximum of n_{20} in equation (13) for fixed p''' and T, we can indeed find the minimum stable radius, R_{AB}. In Figure 4 we have plotted r versus n_{20} for various values of $(p''' - p'_1)$. The locus of the minima for n_{20} is the stability line.

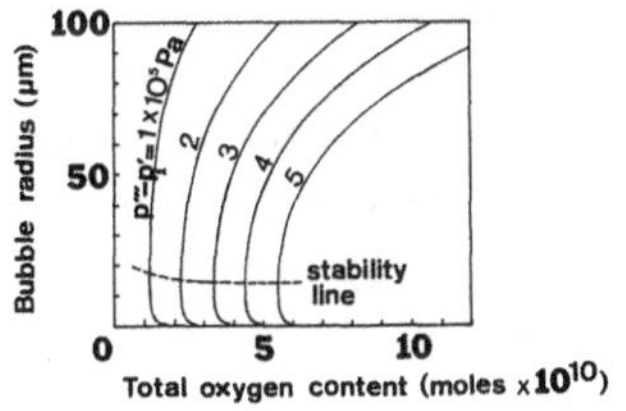

Figure 4. Plots of r vs n_{20} for various values of $(p''' - p'_1)$.

In Figure 5, plots of x''_2 and $x''_{2,s}$ versus r are given for $n_{20} = 2 \times 10^{-10}$ mole, $T = 25°C$ and three different values of $(p''' - p'_1)$. As in Figure 3, we find that, at a sufficiently low pressure, a stable bubble can exist. As the pressure is raised, the stable radius gets smaller to a minimum value. At higher pressures a stable bubble cannot be found. The locus of the maxima of $(p''' - p'_1)$ in equation (13) would therefore define a stability line. Such a stability line was found by Mori et al. [4]. These authors plotted a family of curves for $(p''' - p'_1)$ versus r for several values of n_{20}, and the locus of the maxima is the stability line. They also showed that the stability line given by the locus of $(p''' - p'_1)_{max}$ agrees with that obtained by thermodynamic analysis,

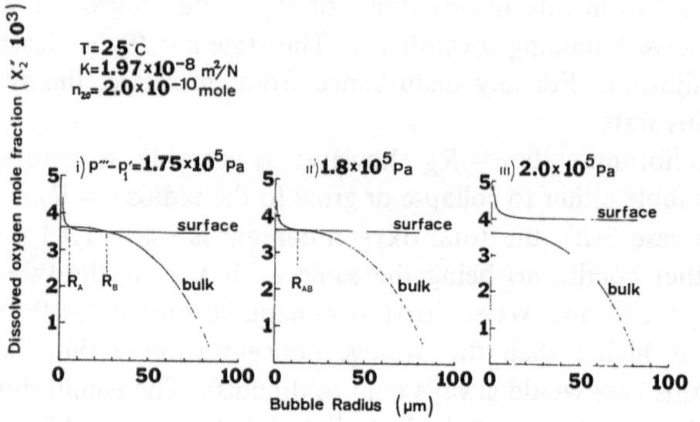

Figure 5. Variation of solute mole fraction with bubble radius at bubble surface and bulk of drop for fixed n_{20} and various values of $(p''' - p'_1)$.

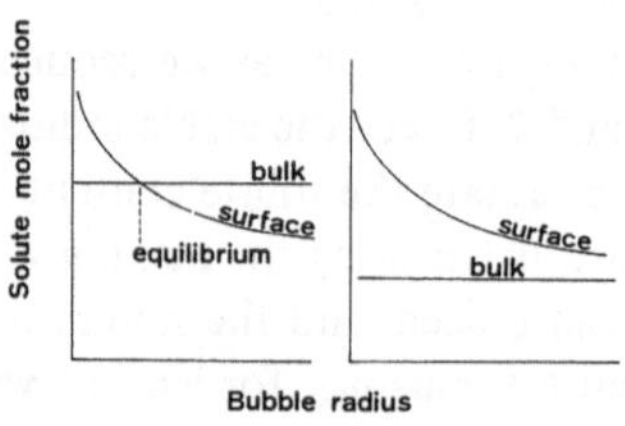

Figure 6. The limiting case of the drop being infinitely large.

except for high solubility gases such as CO_2. They state that the reason for this agreement is probably because of the inapplicability of Henry's law for CO_2 in Freon-21. The question, as it stands, is still not answered.

It is also of interest to observe the behavior of the system as we increase its size; that is, we increase the drop size and the total solute content, while keeping $n_{20}/(n_{10} + n_{20})$ constant. In equation (9) if we let n_{10} and n_{20} become very large, we find

$$x''_2 = \frac{n_{20}}{n_{10} + n_{20}} = \text{constant}. \tag{14}$$

In Figures 3 and 5, the graph for x''_2 would be represented by a straight line. This is shown schematically in Figure 6. We see that when we do have equilibrium, it exists as just one point. We also note from our diffusion approach that this point is unstable, as expected [2].

4. Conclusion

The present analysis shows that the conditions for stable equilibrium in a 'small' isothermal system can be established on the basis of the diffusion theory. The essential feature is the simple notion that mass flows from a higher to a lower concentration.

The results show that two states of equilibrium may exist for a finite solute-solvent system. The equilibrium state at the lower bubble radius is unstable, while the one at the higher radius is stable. The status of the stability in each case is determined solely from consideration of the concentration at the bubble surface and that within the drop. We also find that we can recover the previously known result of instability [2] by allowing the system to become large.

We have also commented on a fundamental error regarding the vapor pressure in a bubble. It is often stated that the vapor pressure within a bubble is different from that at a flat surface having the same temperature [1, 3]. This view can be shown to be incorrect by invoking the second law of thermodynamics. It is of course correct that in the case of a droplet in its own vapor we have an increased vapor pressure. This result was first given by William Thompson.

References

1. Cha YS (1981) On the equilibrium of cavitation nuclei in liquid-gas solutions. J Fluids Eng (ASME), September.
2. Epstein PS and Plesset MS (1950) On the stability of gas bubbles in liquid-gas solutions. J Chem Phys 18: 1505.
3. See, for example, Landau L and Lifshitz E (1958) Statistical Physics, p 465. Pergamin Press, Ltd.
4. Mori Y, Hijikata K and Nagatani T (1977) Fundamental study of bubble dissolution in liquid. Int J Heat Mass Transfer 20: 41.
5. Plesset MS (1949) The dynamics of cavitation bubbles. J Appl Mech 16: 277.

PART FOUR

Bubble dynamics

Bubble dynamics: a review and some recent results

ANDREA PROSPERETTI

Istituto di Fisica, Universitá degli Studi, 20133 Milano, Italy

Abstract. Several aspects of small-amplitude oscillations of bubbles containing gas, vapor, or a gas-vapor mixture are discussed. An application to pressure-wave propagation in a bubbly liquid is described. Nonlinear forced oscillations are considered in the light of recent research on forced oscillations of nonlinear systems. The growth of vapor bubbles, an extension of the Rayleigh-Plesset equation to non-Newtonian liquids and appreciable mass transfer at the interface, and a boundary integral numerical method for non-spherical cavitation bubble dynamics are also briefly discussed.

1. The growth of gas bubbles

The gradual growth of a gas bubble from a nucleus (which must be distinguished from the explosive growth characteristic of cavitation) is strongly favoured by the presence of an oscillating pressure field, such as an imposed sound field or pressure fluctuations caused by mechanical vibrations and turbulence. In the absence of such pressure oscillations growth is only possible if the liquid is supersaturated with gas, and even then it is a slow process due to the small diffusion coefficient of gases in liquids, which has the typical order of magnitude of 10^{-5} cm^2/sec [7].

It appears that the mechanisms by which pressure fluctuations affect the growth process are essentially two, rectified diffusion and coalescence. Crum [5] has recently carried out an experimental study of the growth by rectified diffusion of air bubbles of radius of the order of 10^{-3} cm in water in a 22.1 kHz sound field. His data are in general agreement with a modification of the theory of Eller [6], although certain important aspects do not appear to be satisfactorily resolved. In the first place, while the data on threshold conditions agree very well with theory for saturation conditions, discrepancies of the order of 30% appear for a 5% undersaturation. The data for 4% supersaturated water are few and it is difficult to decide whether a systematic discrepancy between theory and experiment exists in this case, although it cannot be ruled out. Another point concerns the measured negative growth (i.e. dissolution) rates at low pressure amplitude (below 0.2 bar) which in some cases appear to be underpredicted and in other cases overpredicted by the theory. It is my general impression that, although a gratifying agreement between theory and experiment emerges from Crum's study, more work on both fronts should be carried out. Experiments with different liquids, in different saturation conditions, and at different temperatures would give very valuable information. On the theoretical side I would like to see the several available pieces of information (detailed knowledge of damping mechanisms

Applied Scientific Research 38: 145–164 (1982) 0003–6994/82/0382–0145 $03.00.

and thermal behavior, nonlinear aspects, etc.) brought together into a coherent picture. In addition to threshold conditions a special consideration should be given to the growth and dissolution processes in which a thick boundary layer caused by the net mass diffusion into or out of the bubble is superimposed on the thin oscillatory one caused by the radial pulsations. It appears that the method of multiple scales used by Skinner [28, 29] could be improved upon to deal with this situation.

Two other points emerge from Crum's work which are important, but on which very little is known. The first one is the large effect of surface oscillations on the growth rate, already pointed out by other investigators and notably by Gould [12], on which little more than a qualitative explanation is available. The second one concerns the important anomalies caused by the addition of surfactants to the liquid. This brings us to the problem of surface adsorption at the bubble interface, a question of the utmost importance also in nucleation [32] and in the translatory motion of bubbles in liquids [13], which is still mysterious.

The second growth mechanism mentioned above, coalescence, is very effective once bubbles have grown somewhat beyond the size of a nucleus. Its importance stems from the fact that attractive forces, the so-called Bjerknes forces, exist between pulsating bubbles. Their physical origin can be explained observing that a bubble tends to move against a pressure gradient with an acceleration proportional to its volume. In an oscillating pressure field the gradient reverses periodically and the bubble will exhibit a net drift in the direction which the pressure gradient has when the bubble volume is greatest.

A quantitative description of the phenomenon can readily be obtained if the pressure field varies only slightly over a distance comparable with the bubble size, for in this case the resultant of the pressure forces acting on the bubble surface S

$$\mathbf{F} = -\oint_S p(\mathbf{x}, t)\,\mathbf{n}\,\mathrm{d}S, \tag{1}$$

can be approximated by

$$\begin{aligned}\mathbf{F} &\simeq -\oint_S [p(\mathbf{x}_0, t) + (\mathbf{x} - \mathbf{x}_0)\cdot\nabla p + \ldots]\,\mathbf{n}\,\mathrm{d}S \\ &= -V\nabla p + \ldots,\end{aligned}$$

where $\mathbf{n}$ is the outwardly directly unit normal, $\mathbf{x}_0$ is a point interior to the bubble, and V is its volume. Averaging over a cycle we then have

$$\langle\mathbf{F}\rangle \simeq -\langle V\nabla p\rangle. \tag{2}$$

In the linearized (acoustic) approximation we may distinguish between a primary Bjerknes force, due to the pressure field driving the oscillations, and a secondary Bjerknes force, due to the pressure gradient produced by a neighbouring bubble. In the first case, if we set

$$p(\mathbf{x}, t) = p_0 - p'(\mathbf{x}) \cos \omega t, \tag{3}$$

$$V(t) = V_0 [1 + \delta \cos (\omega t + \varphi)], \tag{4}$$

where p_0, V_0 are the average values of p and V, we find from (2)

$$\langle \mathbf{F}_1 \rangle = \tfrac{1}{2} V_0 \delta \cos \varphi \boldsymbol{\nabla} p', \tag{5}$$

which shows that a bubble driven below resonance ($\cos \varphi > 0$) will be attracted by the pressure antinodes, while the reverse will occur for a bubble driven above resonance, for which $\cos \varphi < 0$.

To treat the bubble-bubble interaction we set

$$\boldsymbol{\nabla} p = \boldsymbol{\nabla} [\rho R' (R' \ddot{R}' + \ldots) |\mathbf{x} - \mathbf{x}'|^{-1}],$$

to find, with $V' = V_0' [1 + \delta' \cos (\omega t + \varphi')]$,

$$\langle \mathbf{F}_2 \rangle = -\frac{1}{8\pi} \rho V_0 V_0' \delta \delta' \omega^2 \cos (\varphi - \varphi') \frac{\mathbf{x} - \mathbf{x}'}{|\mathbf{x} - \mathbf{x}'|^3} . \tag{6}$$

Here primed quantities refer to the second bubble, and $\mathbf{x} - \mathbf{x}'$ denotes the distance between the two bubbles. It can be seen that this force is attractive if both bubbles are driven below or above resonance (because then $\cos (\varphi - \varphi') > 0$), and repulsive otherwise. Since in a sound field of frequency up to a few tens of kHz the radius of a resonant bubble is greater than 10^{-2} cm, in most cases the secondary Bjerknes force will be attractive and it will promote coalescence. The importance of this process is of course enhanced by the fact that in these conditions most bubbles would collect near the pressure antinodes where the oscillation amplitude, and hence the attractive force, is large, and the average separation small.

We can estimate the time necessary for coalescence by balancing the secondary Bjerknes force with the Stokes viscous drag acting on a bubble. If the two bubbles involved are equal we find that each one of them has the velocity

$$v = \frac{\rho V_0^2 \delta^2 \omega^2}{48\pi^2 \mu R_0 r^2} \tag{7}$$

where μ is the liquid viscosity, R_0 is the average radius, and $r = |\mathbf{x} - \mathbf{x}'|$ is the instantaneous separation. Since the two bubbles move towards each other the above expression equals $-\frac{1}{2} \mathrm{d}r/\mathrm{d}t$. Integrating from $r = r(0)$ to $r \simeq 0$ we readily have the following estimate of the coalescence time

$$t_{\text{coalescence}} = \frac{8\pi^2 \mu R_0}{\rho V_0^2 \delta^2 \omega^2} r^3(0). \tag{8}$$

To give an idea of the orders of magnitude involved we may take the liquid to be water with a typical value of 0.2 cm for the average initial separation $r(0)$

(corresponding to an average nuclei population of about 100 per cc [15]). Setting $\omega = 2\pi \times 20\,\text{kHz}$ and $\delta = 0.1$ we find estimates of the coalescence time of 2.3×10^8 sec, 2280 sec, 2.3×10^{-2} sec respectively for $R_0 = 10^{-4}$ cm, 10^{-3} cm, and 10^{-2} cm. The very rapid decrease with increasing R_0 (actually as R_0^{-5}) makes therefore this mechanism an effective one for bubbles somewhat larger than the nucleus size.

2. Small-amplitude radial oscillations of gas bubbles

It is clear from the previous considerations that the response of bubbles to variable pressure fields plays a central role in several important processes. Unfortunately, even in the small-amplitude (linearized) approximation, the detailed analysis of the phenomenon is very involved if a serious attempt at completeness is made [8, 9].

In general, if the bubble is immersed in a pressure field of the form

$$p(t) = p_\infty(1 - \epsilon e^{i\omega t}), \tag{9}$$

with $|\epsilon| \ll 1$, and if one sets $R(t) = R_0[1 + X(t)]$, where R_0 is the equilibrium radius, in the linearized approximation it is found that $X(t)$ can be taken to satisfy an equation of the damped oscillator type

$$\ddot{X} + 2\beta\dot{X} + \omega_0^2 X = \frac{p_\infty}{\rho R_0^2}\, e^{i\omega t}, \tag{10}$$

where dots denote time differentiation. The 'effective' damping constant β and natural frequency ω_0 however are found to depend in a complicated way on the value of the radius, of the driving frequency, and of the physical parameters.

The most significant factors which contribute to the damping β are the thermal energy loss from the bubble to the liquid, the acoustic scattering by the bubble, the liquid viscosity, the phase change at the bubble boundary, and the exchange of the dissolved gas between the liquid and the bubble. The relative importance of these mechanisms depends on several factors. For instance, in a 'cold' liquid such that the vapor concentration in the bubble is small phase changes will not contribute appreciably. If the bubble radius is sufficiently large (say, greater than 10^{-2} cm) viscous damping in liquids like water will be small too. At very low frequency gas diffusion across the bubble boundary may be important, especially at low pressure, but most often it will be overshadowed by thermal processes (the compression work done on the bubble is transformed into heat which is lost to the liquid). Finally, at large frequency, the dominant mechanism is the energy loss by sound radiation. We show in Figures 1 and 2 some examples of the variation of β with ω for an air bubble in water. Other results of this type can be found in [23]. Figure 1, taken from [9], refers to a temperature of 20°C and a pressure of 1 atm. The results with and without gas diffusion are represented by the solid and dashed

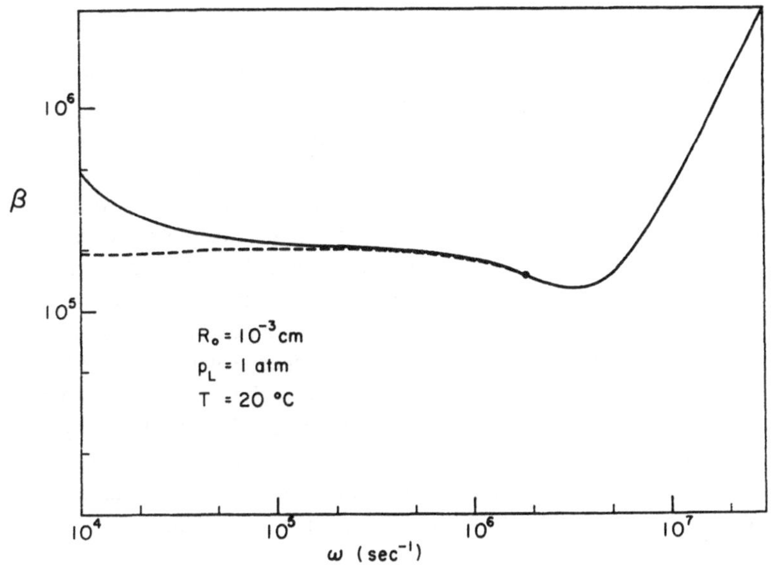

Figure 1. Effective damping coefficient for forced oscillations of a gas-vapor bubble in water at 20°C and 1 atm pressure. The results with and without gas diffusion are represented by the solid and the dashed lines respectively. The open circle indicates the position of the true resonance frequency (from [9]).

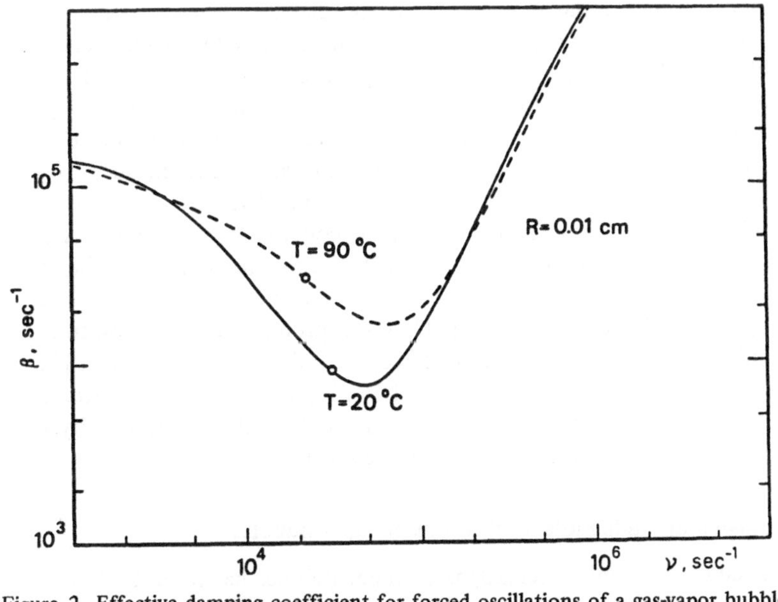

Figure 2. Effective damping coefficient for forced oscillations of a gas-vapor bubble in water at 20°C (solid line) and at 90°C (dashed line) at 1 atm pressure. The open circles indicate the position of the true resonance frequency.

lines respectively. In Figure 2 we compare the values of β for 1 atm pre and temperatures of 20°C and 90°C, both for a gas concentration c sponding to saturation at the lower temperature. In these figures the circles mark the position of the true resonance frequency.

The denomination 'effective' natural frequency for the parameter ($f(\omega)$ appearing in (10) is of course somewhat improper. Physically quantity is related to the magnitude of the restoring force, while m matically it specifies (with β) the position of the pole of the response X i complex plane. The true resonant frequency must be obtained from the dition $\omega_0 = f(\omega_0)$ an equation which it is too complicated to write (explicitly but which can be solved numerically. Some results for this qua with the neglect of gas diffusion and phase change will be found in A very simple and widely used way to represent the response of a bubl an imposed pressure oscillation is a polytropic relation of the type $p_{i0}(R_0/R)^{3K}$. The question is then how to select the polytropic expon(appropriately. In the framework of the linearized theory this quantity c related to the computed ω_0 by the well-known result of Minnaert

$$\omega_0 = \frac{1}{R_0}\left[\frac{3K}{\rho}\left(p_\infty + \frac{2\sigma}{R_0}\right) - \frac{2\sigma}{\rho R_0}\right]^{1/2},$$

and thus it is found to depend on the various parameters as ω_0 itself In general it can be said that the bubble will essentially behave isother (i.e., $K \simeq 1$) if the thermal diffusion length in the gas $(\chi_G/\omega)^{1/2}$ is gr than the radius, while it will tend to an adiabatic behavior (i.e., $K \simeq$ $R_0/(\chi_G/\omega)^{1/2} \gg 1$ *and* R_0 is much less than the wavelength of sound i bubble λ_G. When $\lambda_G \lesssim R_0$ pressure nonuniformities in the bubble be important and K can actually take values outside the range $1 \leqslant K \leqslant \gamma$ The situation in the large-amplitude oscillation case would of course be more complex, with K depending also on the oscillation amplitude.

An important application of the concepts discussed so far is t(problem of wave propagation in a liquid containing gas bubbles. A cı point here is the connection of the single bubble response with the av behavior of the composite medium. We have applied the approach of I [11] and Waterman and Truell [31] to this problem, and we show in Fi 3 and 4 a comparison of our results [10] with the data of Silberman [27 gas void fractions of 5.8×10^{-2}% and 1%. Figure 3 refers to the attenu coefficient, expressed in dB/cm, and Figure 4 to the actual celerity c waves. The agreement appears to be quite satisfactory.

3. Small-amplitude radial oscillations of vapor bubbles

The subject of radial oscillations of vapor bubbles has recently drawn attention [30, 19, 14] especially because of the occurrence of two (resonance frequencies for a given radius.

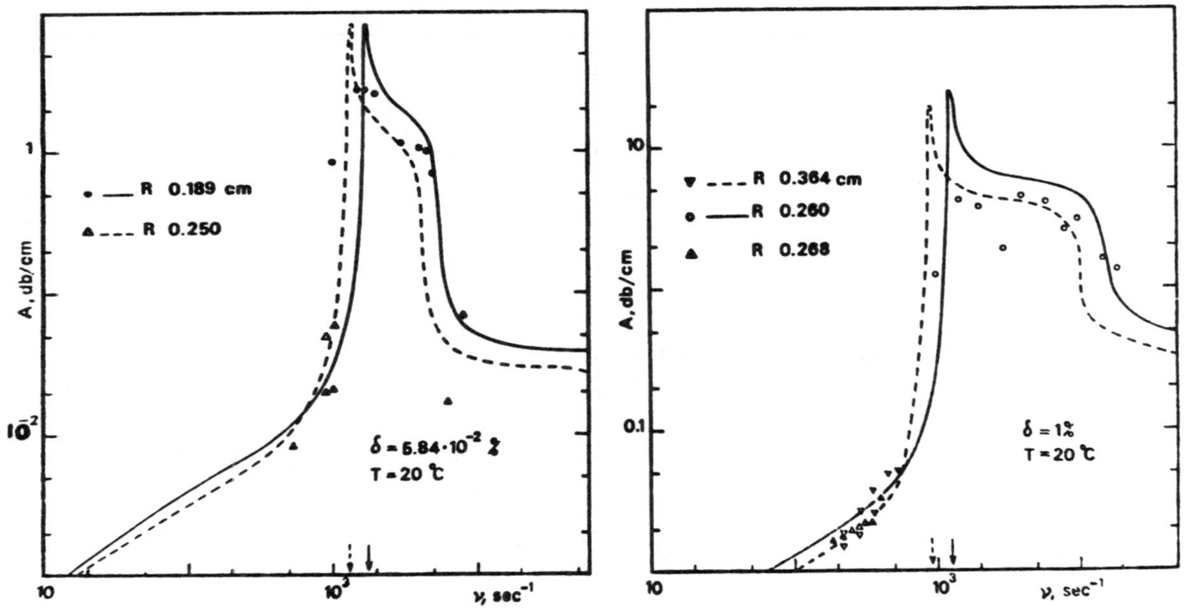

Figure 3. Attenuation coefficient for linear pressure wave propagation in a liquid containing gas bubbles (from [10]).

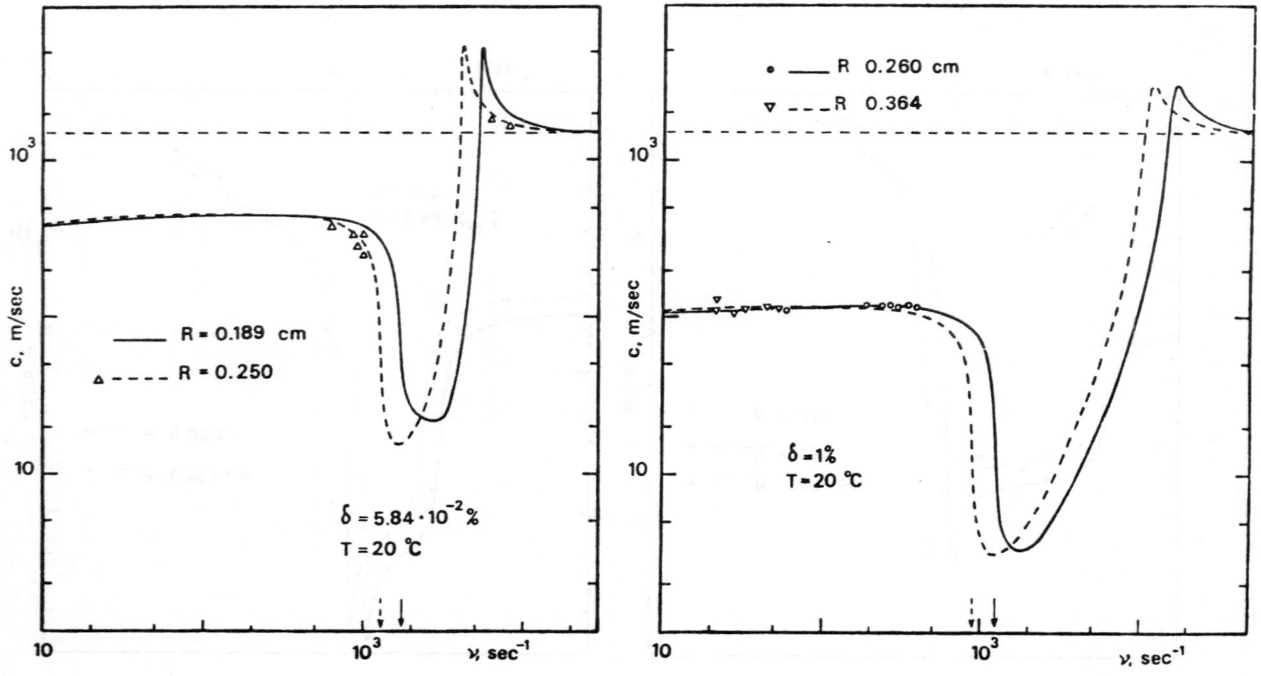

Figure 4. Celerity of linear pressure waves propaging in a liquid containing gas bubbles (from [10]).

To explain how a vapor bubble can exhibit a positive stiffness at all we may imagine a slight radius increase ΔR, which will cause the vaporization of a mass of liquid given by $\Delta m = 4\pi R^2 \Delta R \rho_v$ approximately, where ρ_v denotes the vapor density. The latent heat necessary for this change of phase is $L\Delta m$, and it will cause a temperature drop of the bubble surface by an amount ΔT such that $4\pi R^2 (\chi_L/\omega)^{1/2} \rho_L c_L \Delta T = L\Delta m$, where χ_L and c_L denote the liquid thermal diffusivity and specific heat respectively. (The above estimate is correct only if $(\chi_L/\omega)^{1/2} \ll R_0$; otherwise an additional term is present, see [14].) Finally, to the surface temperature drop ΔT corresponds a pressure drop $\Delta p = \mathrm{d}p^e/\mathrm{d}T\, \Delta T$, where the derivative is taken along the saturation line. If we introduce a 'stiffness' parameter k in terms of which to write this restoring force as $F = -k\Delta R$, the above argument leads to the following expression

$$k = 4\pi R^2 \left(\frac{\omega}{\chi_L}\right)^{1/2} \frac{L\rho_v}{c_L \rho_L} \frac{\mathrm{d}p^e}{\mathrm{d}T}. \tag{12}$$

We can now estimate the first resonance frequency recalling that the kinetic energy of a liquid in which a bubble is in radial motion is given by $\frac{1}{2} 4\pi R^3 \rho_L \dot{R}^2$, which shows that the 'equivalent' mass of the dynamical system under consideration is $M_{eq} = 4\pi R^3 \rho_L$, so that $\omega_0^2 = k/M_{eq}$ or, by (12),

$$\omega_0^3 R_0^2 = \frac{1}{\chi_L} \left(\frac{L\rho_v}{c_L \rho_L^2} \frac{\mathrm{d}p^e}{\mathrm{d}T}\right)^2, \tag{13}$$

a result which, up to a numerical factor, agrees with that of a more detailed analysis [19].

The origin of the second resonance is quite different. At very low frequency inertial effects are unimportant, and a balance is possible between the restoring force previously described and the surface tension force

$$-k\Delta R \simeq 4\pi R_0^2\, \Delta \left(\frac{2\sigma}{R}\right)$$

or, again by (12),

$$\omega_0 R_0^4 = \chi_L \left(\frac{2\sigma c_L \rho_L}{L\rho_v} \frac{\mathrm{d}T}{\mathrm{d}p^e}\right)^2 \tag{14}$$

which again agrees with the result of more detailed analyses [19]. This condition corresponds to a resonance situation as would happen for the mechanical oscillator $\delta \ddot{x} + (k_1 - k_2)x = \epsilon e^{i\omega t}$ in the limit $\delta \to 0$, $k_1 \to k_2$. Since very little momentum is stored in the motion ($\delta \to 0$) the equivalent mass is negligible. Furthermore also the net restoring force vanishes, and hence the amplitude of the oscillation can be arbitrarily large. In practice of course the residual inertia and, more importantly, the damping mechanisms, will limit the oscillation amplitude. An interesting feature of the results (13) and (14) is the functional dependence of ω_0 on the equilibrium radius R_0 which is quite different in the two cases and from the gas bubble result (11).

In analyzing the problem of forced oscillations by means of the general conservation equations of continuum mechanics sometimes the quantity ω_0^2 appearing in (10) is found to be negative. Physically this behavior corresponds to a 'restoring' force which actually is destabilizing. This phenomenon is found especially in the case of vapor bubbles in cold liquids. For instance, a vapor bubble of radius 0.1 cm in water at 20°C does not exhibit a positive stiffness below $\omega \sim 4 \times 10^3$ sec^{-1} and does not have a resonance frequency at all, while for a temperature of 50°C the stiffness is positive already for $\omega \sim 100$ sec^{-1}.

The considerations of this and of the previous section show how much physics goes into determining the response of a bubble in a sound field. Conversely, measurements of this response can provide a powerful tool to investigate these phenomena. In order for this approach to be viable, however, it appears that a qualitative jump in the experimental approach is required. For instance, in addition to the usual acoustic measurement techniques, optical methods based on the use of the laser will be very useful. In this way some very important quantities such as accommodation coefficients for evaporation and condensation, the dynamical behavior of surface tension, and many others may become accessible. The required purity of the liquid sample may be difficult to obtain with water, but not as nearly with other liquids which would be just as interesting.

4. Nonlinear oscillations of gas bubbles

Our theoretical understanding of nonlinear radial oscillations of gas bubbles is incomparably more limited than for the linear case. All existing studies have essentially been based on the numerical solution of the Rayleigh-Plesset equation with a polytropic approximation for the internal pressure

$$R\ddot{R} + \tfrac{3}{2}\dot{R}^2 = \frac{1}{\rho}\left[p_{i0}\left(\frac{R_0}{R}\right)^{3K} - p_\infty(1 - \epsilon \cos \omega t) - \frac{2\sigma}{R} - 4\mu\frac{\dot{R}}{R}\right], \tag{15}$$

or extensions thereof to account for compressibility effects in the liquid (see e.g. [17, 4, 26, 16]). Thus, the various exchange processes of heat and mass between the liquid and the bubble described in the previous sections have never been considered in the nonlinear domain. An approximate procedure which might be valid for small-amplitude, nonlinear oscillations would be the use of an effective liquid viscosity obtained from the damping parameter

β of the linear case and of a similarly obtained value of K. The limits of validity of this procedure, however, are unknown. Since the solution of (15) depends critically on these quantities, experimental observation of various aspects of nonlinear oscillations can be a useful diagnostic technique. Although a full exploitation of this approach must await further theoretical progress, some interesting possibilities in this direction exist already as indicated in [22].

In his classic study of the steady solutions of (15) Lauterborn [17] found that when the driving amplitude ϵ exceeds a certain value (e.g., about 0.8 for $R_0 = 10^{-3}$ cm in water at 1 atm) no steady solution could be computed. This finding ties in with recent research on the behavior of forced nonlinear dynamical systems which has shown that, when the driving is sufficiently strong, the motion ceases to be periodic and, although perfectly deterministic, exhibits in a sense a chaotic behavior. Interestingly enough, the route to this chaotic behavior goes through a successive series of bifurcations to subharmonic oscillations of increasingly longer period. The first step of this process, from a bubble cycle of the duration of $2\pi/\omega$ to one of the duration $4\pi/\omega$, can perhaps be seen in the numerical results of Apfel [1] for equation (15). Furthermore, near the value of the forcing where chaotic behavior sets in, subharmonics are found to appear in bursts separated by non-subharmonic oscillations. These observations are perhaps relevant to explain the observed connection between acoustic cavitation and the presence of a subharmonic signal. One may speculate that 'chaos' for equation (15) entails the possibility of violent collapses characteristic of strong cavitation, and this large amplitude motion, either for the individual collapsing bubble or for neighbouring ones, can be intimately connected with subharmonic emission as indicated. (The other mechanism suggested in [21] is not in conflict with the previous one and could still be present, particularly at low values of the forcing.) The situation here is very complex, and it is impossible to draw definite conclusions at this time. However a general comment suggested by the preceding considerations is that at large driving amplitudes the search for periodic solutions of (15) may not be possible or fruitful. Transient behavior, as already remarked in [21], may be important. Very interesting numerical results on transient oscillations can be found in [2].

In the light of the very complicated structure of the solutions of (15) for large ϵ, one should also realize that the bubble response will become markedly different upon even slight changes of the values of the parameters, initial conditions, and so on. This remark should induce to some caution in the interpretation of large effects caused by apparently small causes (e.g. the consideration of the shock waves in the bubble [3]). The large amplitude motion is, in a sense, unstable and therefore can be changed markedly even by slight modifications of the conditions, without the result having any fundamental physical significance.

Finally, we wish to present a result which may be useful in numerical work.

We start from (15) with $\mu = 0$ writing $p_\infty(t)$ in place of $p_\infty(1-\epsilon\cos\omega t)$, where $p_\infty(t)$ is taken to be periodic. Upon integration from the end of the nth cycle to that of the $(n+1)$st cycle we easily obtain

$$R_{n+1}^3\dot{R}_{n+1}^2 - R_n^3\dot{R}_n^2 = \frac{2}{3(K-1)}\frac{p_{i0}}{\rho}R_0^{3K}\left[R_n^{3(1-K)} - R_{n+1}^{3(1-K)}\right]$$

$$+\frac{2\sigma}{\rho}(R_n^2 - R_{n+1}^2) + \frac{2}{3}\frac{p_\infty(t_n)}{\rho}(R_n^3 - R_{n+1}^3)$$

$$+\frac{2}{3\rho}\int_{t_n}^{t_{n+1}} R^3\dot{p}_\infty\,\mathrm{d}t, \tag{16}$$

where indices refer to conditions at the end of the corresponding cycle and the fact has been used that $p_\infty(t_n) = p_\infty(t_{n+1})$. Consider now a steady, periodic oscillatory solution $S(t)$ for which $S_{n+1} = S_n$, $\dot{S}_{n+1} = \dot{S}_n$, and a perturbation thereof

$$R = S(t)[1 + A(t)], \tag{17}$$

where A is taken to be small and possibly slowly varying with time. Taking the instants t_n such that $\dot{S}(t_n) = 0$, upon substitution of (17) into (16) and linearization in A we readily find

$$(A_{n+1} - A_n)S_n^3\left[p_\infty(t_n) - p_{i0}\left(\frac{R_0}{S_n}\right)^{3K} + \frac{4\sigma}{S_n}\right] = \int_{t_n}^{t_{n+1}} AS^3\dot{p}_\infty \mathrm{d}t. \tag{18}$$

But since S satisfies (15) we can simplify this relation to read

$$A_{n+1} - A_n = -\frac{1}{S_n^4\ddot{S}_n}\int_{t_n}^{t_{n+1}} AS^3\frac{\dot{p}_\infty}{\rho}\,\mathrm{d}t. \tag{19}$$

If t_n is taken in correspondence of (relative) maxima of $S(t)$, then $\ddot{S}_n < 0$ and this relation shows that the periodic solution S is stable if the integral has opposite sign to A_n. Conversely, if t_n is taken in correspondence of (relative) minima of S. This result may be used, for instance, in numerical work to monitor convergence towards a stable periodic oscillation.

5. The dynamics of vapor bubbles

The growth process of a vapor bubble under a constant ambient pressure can be subdivided essentially into three phases [25]. During the first one, of very short duration, surface tension forces are the dominant restraining mechanism of the growth. As soon as the radius has grown by approximately one order of magnitude liquid inertia becomes important. The growth velocity in this inertial range can be easily obtained by equating the kinetic energy of the liquid, $2\pi R^3 \rho \dot{R}^2$, to the work done by pressure forces, $\frac{4}{3}\pi R^3 (p_v - p_\infty)$, with the result

$$\dot{R} = \left(\frac{2}{3}\frac{p_v - p_\infty}{\rho}\right)^{1/2}. \tag{20}$$

Here the vapor pressure of the liquid p_v is taken to be constant, which can be shown to be a good approximation for large superheats, at least initially. As the bubble grows, more and more latent heat is required to keep it filled up with vapor. Eventually therefore energy transport to the bubble boundary becomes the major restraining mechanism. Since this transport occurs at a rate given approximately by

$$4\pi R^2 K \frac{T_\infty - T_b}{(\chi_L t)^{1/2}}, \tag{21}$$

(where K is the thermal conductivity of the liquid, T_∞ is the liquid temperature far from the bubble, and T_b is the boiling temperature at the pressure p_∞) and the latent heat requirement is

$$L \frac{\mathrm{d}}{\mathrm{d}t}\left(\tfrac{4}{3}\pi R^3 \rho_v\right) \simeq 4\pi R^2 \rho_v L \dot{R}, \tag{22}$$

we have, equating (21) and (22) and inserting a numerical factor coming from the detailed analysis [25],

$$\dot{R} = \left(\frac{3}{\pi}\right)^{1/2} \frac{K}{L\rho_v} \frac{T_\infty - T_b}{(\chi_L t)^{1/2}}. \tag{23}$$

The estimate (21) for the heat supply to the bubble wall is based on the assumption of a thin thermal boundary layer. It can be shown that this assumption is justified provided that the Jacob number

$$\mathrm{Ja} = \frac{\rho L}{\rho_v c_L (T_\infty - T_b)} \tag{24}$$

is greater than about 3 [25].

The three different stages indicated above in principle exist in any conditions of temperature and pressure. However, the practical situation may be such that only one of the limiting processes is of importance. To clarify this point we may inquire for what value of the radius the inertial and thermal growth velocities (20) and (23) coincide. Since for smaller radii the limiting mechanism is the liquid inertia, while for greater values of the radius thermal energy transport dominates the growth, this value of the radius essentially identifies the switchover point between the two limiting processes. By an obvious procedure we find

$$R = \frac{6}{\pi}\frac{1}{\chi_L}\left[\frac{K(T_\infty - T_b)}{L\rho_v}\right]^2\left(\frac{3}{2}\frac{\rho}{p_v - p_\infty}\right)^{1/2}. \tag{25}$$

(The initial, surface-tension dominated stage has been disregarded here. Hence these considerations apply only to conditions of sufficiently high superheat, even though qualitatively they apply also to the general case.) We show in the table the values of R predicted by this equation for the case of water at different temperatures T_b. We have assumed $T_\infty = T_b + 10°C$, and $p_v = p^e(T_\infty)$, $p_\infty = p^e(T_b)$. The table also shows the corresponding values of $\rho_v(T_b)$, which is the quantity from which the result most strongly depends. It can be seen that in a cold liquid the thermal constraint would become dominant only so late in the bubble history that it ceases to be of any practical importance. One usually refers to these 'cold' bubbles as cavitation bubbles, and the approximation that their internal pressure remains essentially constant and equal to p_v during their entire life is quite justified.

T_b (°C)	R (cm)	$\rho_v \times 10^{-3}$ (g/cm³)
20	23.8	0.0173
40	1.6	0.0512
60	0.16	0.130
80	2.2×10^{-2}	0.293
100	3.8×10^{-3}	0.598

The preceding considerations lead one to suspect that the growth behavior of vapor bubbles, at least approximately, may exhibit some universal character when expressed in terms of suitably scaled variables. Indeed this is the case, as shown in [25].

6. The Rayleigh-Plesset equation for mass transfer and non-Newtonian liquids

The Rayleigh-Plesset equation of bubble dynamics

$$R\ddot{R} + \tfrac{3}{2}\dot{R}^2 = \frac{1}{\rho}\left(p_i - p_\infty - \frac{2\sigma}{R} - 4\mu\frac{\dot{R}}{R}\right), \tag{26}$$

is rigorously valid only for a Newtonian liquid under conditions of negligible mass exchange across the bubble boundary. The equation can be modified so that it is possible to lift these two restrictions with the result [24]

$$R\dot{U} + \tfrac{3}{2}U^2 - \frac{J}{\rho_L}\left[2U + J\left(\frac{1}{\rho_B} - \frac{1}{\rho_L}\right)\right]$$

$$= \frac{1}{\rho_L}\left[p_i - p_\infty - \frac{2\sigma}{R} + 3\int_R^\infty r^{-1}\tau_{rr}\,dr\right], \tag{27}$$

where U is the liquid velocity at the bubble wall, ρ_B is the density of the bubble contents at the wall τ_{rr} is the (r, r) component of the stress tensor, and J, the mass flux at the interface, is given by

$$J = \rho_L(U - \dot{R}). \tag{28}$$

For a Newtonian liquid $\tau_{rr} = 2\mu\, \partial u/\partial r$ and (27) becomes

$$R\dot{U} + \tfrac{3}{2}U^2 - \frac{J}{\rho_L}\left[2U + J\left(\frac{1}{\rho_B} - \frac{1}{\rho_L}\right)\right]$$

$$= \frac{1}{\rho_L}\left(p_i - p_\infty - \frac{2\sigma}{R} - 4\mu\frac{U}{R}\right). \tag{29}$$

The mass flux correction appearing here is usually small. For example, for a growing vapor bubble,

$$4\pi R^2 J = \frac{\mathrm{d}}{\mathrm{d}t}\left(\tfrac{4}{3}\pi R^3 \rho_B\right) \simeq 4\pi R^2 \rho_B \dot{R},$$

so that $J \simeq \dot{R}\rho_B$. Then the correction induced by J in the equation is of the order of

$$\frac{J^2/\rho_L\rho_B}{U^2} \simeq \frac{\rho_B/\rho_L}{(1 + \rho_B/\rho_L)^2},$$

and hence small far from the critical point of the liquid. For growth in the inertial range, for example, in place of (20) equation (29) gives, with the neglect of surface tension and viscosity,

$$\dot{R} = \left(\frac{2}{3}\frac{p_v - p_\infty}{\rho_L}\right)^{1/2}\left[1 + \frac{4}{3}\frac{\rho_B}{\rho_L} - \left(\frac{\rho_B}{\rho_L}\right)^2\right]^{-1/2}.$$

7. A numerical method for non-spherical cavitation bubble dynamics

Under conditions of potential flow the standard mathematical formulation of free surface flows is (see e.g. [20])

$$\mathbf{u} = \nabla\varphi, \qquad \nabla^2\varphi = 0, \tag{30}$$

$$\frac{\partial\varphi}{\partial t} + \tfrac{1}{2}u^2 = \frac{p_\infty - p_i}{\rho} \quad \text{on the free surface}, \tag{31}$$

$$\frac{\partial\varphi}{\partial n} = 0 \quad \text{on rigid surfaces},$$

$$\varphi \to 0 \quad \text{as} \quad |\mathbf{x}| \to \infty. \tag{32}$$

Surface tension effects have been neglected in (31) and, as already remarked,

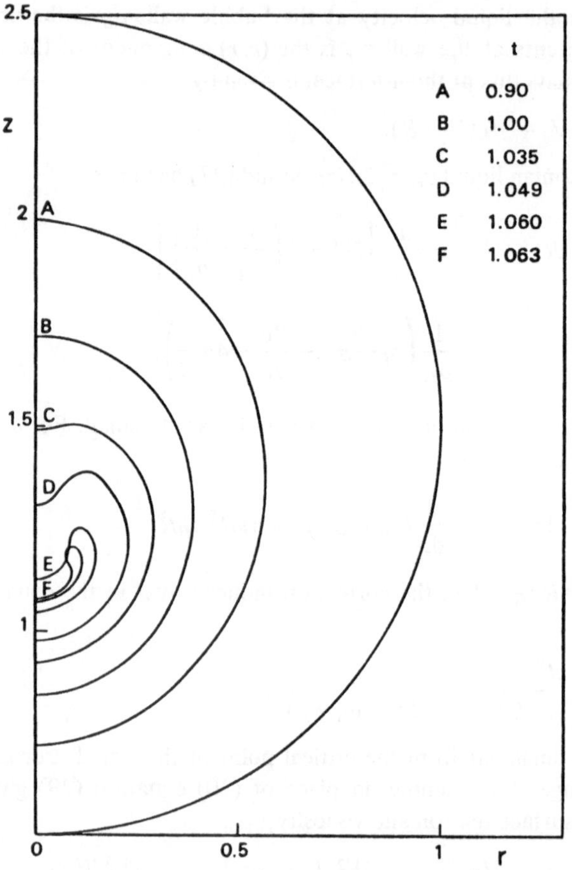

Figure 5. Successive shapes of an initially spherical bubble collapsing in the neighborhood of a rigid plane wall located at 1.5 times the initial radius from the bubble center.

in the application to bubble dynamics p_i can be considered constant in a cold liquid. Equation (31) can be converted to a Lagrangian equation for the surface potential, which can therefore be considered prescribed on the bubble surface at any given time step, provided that the velocity is known [20]. We are currently developing a boundary integral approach to the solution of this problem. The starting point is Green's identity applied to a point $\mathbf{x}$ on the boundary

$$\varphi(\mathbf{x}) = \frac{1}{2\pi}\int_S \left(|\mathbf{x}-\mathbf{x}'|^{-1} \frac{\partial\varphi}{\partial n'} - \varphi \frac{\partial}{\partial n'} |\mathbf{x}-\mathbf{x}'|^{-1} \right) dS'. \tag{33}$$

Since φ can be considered known on the bubble, this equation may be viewed as an integral equation for $\partial\varphi/\partial n$, which can be solved by standard methods. The tangential derivative of φ on the surface can also be readily computed

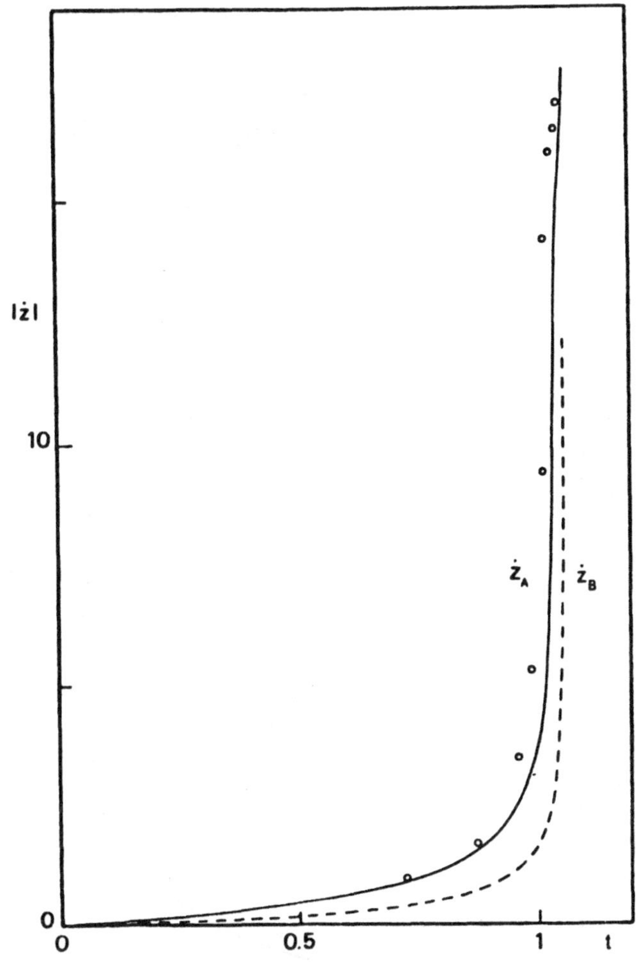

Figure 6. Comparison between the velocities on the axis of symmetry computed by the method of Section 7 (solid line, 'north' pole; dashed line, 'south' pole) and the results of Chapman and Plesset [20] (open circles).

from the knowledge of φ. In this way the bubble boundary can be displaced to generate the new bubble shape at time $t + \Delta t$. This method has the great advantage that one does not need to solve for the potential in the liquid at all, with large savings in computer time and coding.

We show in Figure 5 a succession of bubble shapes generated by a preliminary version of our program. The initial bubble shape is a sphere placed at a distance of 1.5 times the radius from a plane rigid boundary (not shown). Figure 6 shows a comparison of the computed velocities on the axis of symmetry (solid line) with the results of Chapman and Plesset [20] (open circles). It can be seen that our code predicts lower velocities than those found by

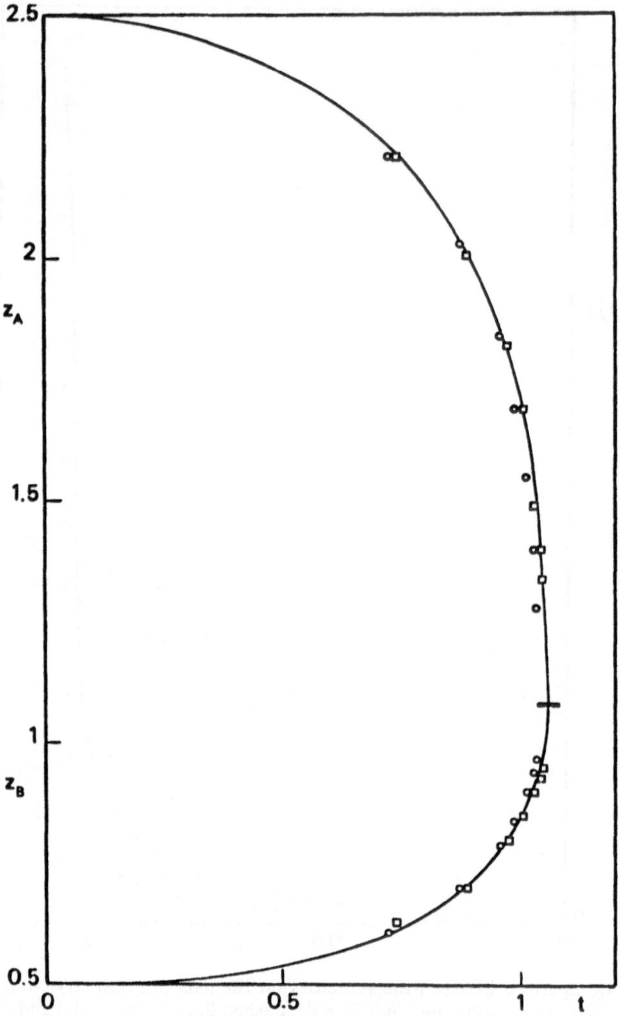

Figure 7. Comparison between the positions of the 'north' and 'south' poles of the collapsing bubble of Figure 5 (solid lines) with the results of [20] (open circles) and the experimental ones of [18] translated forward in time (open squares). The horizontal bar marks the completion of the collapse.

these authors. Since the experimental results of Lauterborn and Bolle [18] would seem to support the numerical results of [20], this difference may be due to inaccuracies in our computations. However experimentally the estimate of the initial instant $t = 0$ for the collapse is inherently rather imprecise, and it may be that the results of Lauterborn and Bolle actually refer to times a little later than those indicated. To test this hypothesis we show in Figure 7 the time evolution of the positions of the 'north' and 'south' poles of the bubble as computed by us (solid line) and by Plesset and Chapman (open circles). The open squares are Lauterborn and Bolle's experimental results translated forward by a small amount. The agreement with our results is then excellent, which may lend support to the previous conjecture and hence to our computations. We propose to refine our numerical procedure to examine this matter in a more satisfactory detail.

Note added in proof: Our most recent computations give a situation opposite to that depicted in Figure 6, with the results of Chapman and Plesset slightly below ours. We have not yet resolved the matter to our satisfaction.

Acknowledgments

Some of the results described in Section 2 have been obtained in collaboration with M. Fanelli and M. Reali of Ente Nazionale per l'Energia Elettrica (National Power Agency of Italy).

The application of the boundary integral method described in Section 7 was suggested by L. Guerri of the University of Pavia, who also wrote the computer program. To these colleagues and friends the author wishes to express his gratitude.

This study has been supported by Ministero della Pubblica Istruzione of Italy.

References

1. Apfel RE (1980) Some new results on cavitation threshold prediction and bubble dynamics. In Lauterborn W (ed) Cavitation and inhomogeneities in underwater acoustics, pp 79–83. New York: Springer.
2. Borotnikova MI and Soloukhin EI (1964) A calculation of the pulsations of gas bubbles in an incompressible liquid subject to a periodically varying pressure. Sov Phys Acoust 10: 28–32.
3. Bykovtsev GI and Rozarenov GS (1975) Pulsation of a spherical bubble in an incompressible fluid. Fluid Dyn Sov Res 10: 322–324.
4. Cramer E (1980) The dynamics and acoustic emission of bubbles driven by a sound field. In Lauterborn W (ed) Cavitation and inhomogeneities in underwater acoustics, pp 54–63. New York: Springer.
5. Crum LA (1980) Measurements of the growth of air bubbles by rectified diffusion. J Acoust Soc Am 68: 203–211.
6. Eller AI (1972) Bubble growth by gas diffusion in an 11-kHz sound field. J Acoust Soc Am 52: 1447–1449.
7. Epstein PS and Plesset MS (1950) On the stability of gas bubbles in liquid-gas solutions. J Chem Phys 18: 1505–1509. Erratum, ibid. 19: 256.
8. Fanelli M, Prosperetti A and Reali M (1981) Radial oscillations of gas-vapor bubbles in liquids. Part I: Mathematical formulation. Acustica 47: 253–265.
9. Fanelli M, Prosperetti A and Reali M (1981) Radial oscillations of gas-vapor bubbles in liquids. Part II: Numerical examples. Acustica 49 (2).
10. Fanelli M, Prosperetti A and Reali M (1981) The propagation of linear pressure waves in a bubbly liquid. J Acoust Soc Am (submitted).
11. Foldy LL (1945) The multiple scattering of waves. Phys Rev 67: 107–119.
12. Gould RK (1974) Rectified diffusion in the presence of, and absence of, acoustic streaming. J Acoust Soc Am 56: 1740–1746.

13. Harper JF (1981) Surface activity and bubble motion (these Proceedings).
14. Hsieh DY (1979) On oscillating vapor bubbles. J. Acoust Soc Am 66: 1514–1515.
15. Keller A (1972) The influence of cavitation nucleus spectrum on cavitation inception, investigated with a scattered light counting method. J Basic Eng 94: 917–925.
16. Lastman GH and Wentzell RA (1981) Comparison of five models of spherical bubble response in an inviscid compressible liquid. J Acoust Soc Am 69: 638–642.
17. Lauterborn W (1976) Numerical investigation of nonlinear oscillations of gas bubbles in liquids. J Acoust Soc Am 59: 283–293.
18. Lauterborn W and Bolle H (1975) Experimental investigation of cavitation bubble collapse in the neighbourhood of a solid boundary. J. Fluid Mech 72: 391–399.
19. Marston PL (1979) Evaporation-condensation resonance frequency of oscillating vapor bubbles. J Acoust Soc Am 66: 1516–1521.
20. Plesset MS and Chapman RB (1971) Collapse of an initially spherical vapor cavity in the neighborhood of a solid boundary. J. Fluid Mech 47: 283–290.
21. Prosperetti A (1975) Nonlinear oscillations of gas bubbles in liquids: transient solutions and the connection between subharmonic signal and cavitation. J Acoust Soc Am 57: 810–821.
22. Prosperetti A (1977) Application of the subharmonic threshold to the measurement of the damping of oscillating gas bubbles. J Acoust Soc Am 61: 11–16.
23. Prosperetti A (1977) Thermal effects and damping mechanisms in the forced radial oscillations of gas bubbles in liquids. J Acoust Soc Am 61: 17–27.
24. Prosperetti A (1981) A generalization of the Rayleigh-Plesset equation of bubble dynamics. Phys Fluids (in press).
25. Prosperetti A and Plesset MS (1978) Vapor bubble growth in a superheated liquid. J Fluid Mech 85: 349–368.
26. Rath HJ (1980) Free and forced oscillations of soherical gas bubbles and their translational motion in a compressible fluid. In Lauterborn W (ed) Cavitation and inhomogeneities in underwater acoustics, pp 64–71. New York: Springer.
27. Silberman E (1957) Sound velocity and attenuation in bubbly mixtures measured in standing wave tubes. J Acoust Soc Am 29: 925–933.
28. Skinner LA (1970) Pressure threshold for acoustic cavitation. J Acoust Soc Am 47: 327–331.
29. Skinner LA (1972) Acoustically induced gas bubble growth. J Acoust Soc Am 51: 378–382.
30. Wang T (1974) Effects of evaporation and diffusion on an oscillating bubble. Phys Fluids 17: 1121–1126.
31. Waterman PC and Truell R (1961) Multiple scattering of waves. J Math Phys 2: 512–537.
32. Yount DE and Yeung CM (1981) Bubble formation in supersaturated gelatin: a further investigation of cavitation nuclei. J Acoust Soc Am 69: 702–708.

Cavitation bubble dynamics – new tools for an intricate problem

W. LAUTERBORN

Drittes Phsyikalisches Institut, Universität Göttingen, Bürgerstrasse 42–44,
D-3400 Göttingen, Fed. Rep. of Germany

Abstract. With the help of laser produced bubbles in water and high speed photography and holography sophisticated experiments on cavitation bubble dynamics can be conducted. The observation of a bubble vortex ring after jet formation upon collapse of a spherical bubble in front of a plane solid boundary is reported. The vortex ring may expand and contract several times until it disintegrates into a ring of bubbles by some instability finally taking over. A critical discussion of our qualitative understanding of jet formation is included. In a second part the problem of the acoustic cavitation noise spectrum is discussed. Numerically obtained 'visible cavitation noise' plots from a single bubble already resemble those obtained experimentally from acoustic cavitation. A discussion shows that the theory should be extended to self-consistency.

1. Introduction

Cavitation, first observed on blades of ship propellers at the turn of the century (for retracing the history see [3]), in the course of time has become manifest in various branches of physics. It now plays a role in such diverse areas as fluid mechanics, acoustics, optics, and nuclear physics [8, 18, 13] and has connections to problems in biology and medicine [19, 1]. In elementary particle physics small-scale venturi nozzle bubble chambers are under investigation to study extremely short-lived particles via cavitation in the search for the top quark and a confirmation of the theory of quantum chromodynamics.

When looking for the common foundation of all cavitation problems it turns out that in the majority of cases cavitation bubble dynamics is involved. At first sight this problem does not look unsurmountably difficult but a closer occupation with it reveals a challenging (and sometimes frustrating) subject. In its complexity it approaches problems of stellar evolution and laser fusion but with much less effort spent on it and with accordingly much slower progress made. This paper is an account of some work done in the last years at the Third Physical Institute, University of Göttingen, in the attempt to push ahead our knowledge of dynamics of cavitation bubbles. Thereby an eye has been kept on two main problems in cavitation physics, i.e. cavitation damage and noise emission.

2. Liquid jets and bubble vortex rings

The problem of cavitation damage has kept busy many an investigator. A cornerstone has been set in 1966 by Benjamin and Ellis [2] in that they gave a convincing demonstration of jet formation for bubbles in a nonspherically

Applied Scientific Research 38: 165–178 (1982) 0003–6994/82/0382–0165 $02.10.

symmetric situation (influence of gravity, influence of a solid boundary nearby). Their findings strongly supported the view that jets may be the cause of damage encountered when a solid is subject to cavitation. In the theoretical part of the paper Benjamin tried to give arguments in favour of jet formation through a discussion of the Kelvin impulse. He started with a translating bubble which undergoes a collapse and could show that the principle of impulse conservation leads to an increase in translational velocity. This is very much like a spherical cavity must increase its radial velocity upon contraction. The argument is difficult to transfer to a bubble originally stationary in front of a solid boundary without initial impulse. But the bubble seems to acquire an impulse towards the boundary via the collapse and thus the argument of Benjamin may be applied in the later stages of collapse. In any case an increasing velocity of the bubble centroid towards the solid boundary can be measured during first collapse of an initially spherical bubble [9, 10]. It is found to decrease during rebound and regrowth of the bubble. The impulse conservation law may also be responsible for the erratic dancing motion of bubbles in a sound field in that low translating velocities due to acoustic pressure forces on the bubble are augmented during collapse. Slight unsymmetries in the pressure field due to the presence of other bubbles and shock waves then may alter the translational course of the bubble resulting in the observed zig-zag motions.

Whereas the self-propelling effect of a translating collapsing bubble seems reasonably explained via impulse conservation the additional feature of jet formation as observed experimentally seems to upset the whole analysis. In the analysis of Benjamin only integral values assigned to the *total* bubble like the kinetic energy or the Kelvin impulse are used. Impulse conservation alone then cannot supply enough information how to split the total impulse into centroid motion and jet motion because this would mean that from $A + B = C =$ const A and B could be determined separately. Therefore hints from experiments must be included.

The situation even gets worse in the case of an initially spherical bubble in front of a solid boundary. The initial impulse is zero and neither the self-propelling of the bubble towards the boundary nor jet formation can be predicted from impulse conservation. The adverse statement of Benjamin [21] is highly objectionable. It seems that even a qualitative theoretical discussion must resort to the special case and non-integral values must be used, i.e. different portions of the bubble (liquid) must be looked at. This has been done in [9, 10], and the conclusion was reached that it is the local curvature of the bubble wall which is of utmost importance. The development of the curvature must be followed as guessed from the flow field. It is then found for initially stationary bubbles that more highly curved parts of a bubble collapse faster than less curved parts. This relation can be derived from the Rayleigh collapse of spherical bubbles [20] in the following sense. Take an elongated bubble with a maximum (minimum) radius of curvature

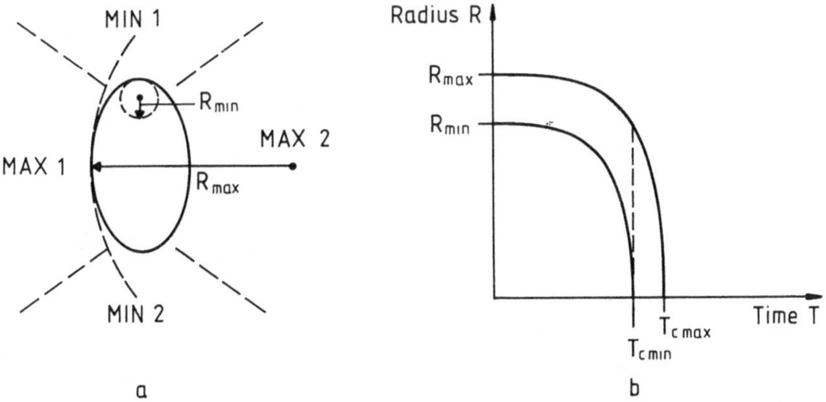

Figure 1. Qualitative theoretical discussion to understand jet formation; (a) elongated bubble with different curvatures of bubble wall; (b) Rayleigh collapse curves of spherical bubbles with radii of R_{min} and R_{max}, respectively.

R_{max} (R_{min}) in an extented liquid at initially constant pressure. We now divide the surrounding liquid into four volumes as a first approximation to look at different portions of the bubble or liquid as found necessary from the above discussion. The four volumes are depicted in Figure 1a and called MIN1, MAX1, MIN2, MAX2. For the exemplification of the idea it is sufficient to just compare the volumes MIN1 and MAX1. We now assume MIN1 and MAX1 as independent to a first approximation. Then the part of the bubble wall with the radius of curvature R_{min} will collapse much like a spherical bubble with radius R_{min}. The same holds true for R_{max}. As the Rayleigh collapse time T_c is proportional to the radius of the spherical bubble we get

$$\frac{T_{c\,\mathrm{min}}}{T_{c\,\mathrm{max}}} = \frac{R_{\mathrm{min}}}{R_{\mathrm{max}}},$$

i.e. the more highly curved part has collapsed first. Now why may this give rise to jet formation? It has to do with the extreme speeding up of the collapse towards its end as depicted in Figure 1b. In Figure 1b the total collapse curves are given for a bubble with a radius R_{min} and for one with R_{max}. It is easily seen that at the time $T_{c\,\mathrm{min}}$ when the bubble of radius R_{min} has collapsed the other bubble is still large. Thus a high speed liquid inflow from the highly curved part can be expected which cannot be stopped until it hits the opposite side of the bubble wall. In this view jet formation is not an impulse-conserving mechanism (in the sense of Benjamin) but a shaped charge effect. It derives its violence from essentially the same reason as spherical collapse. (The reason is the concentration of a finite amount of energy down to a singularity (a 'point'). This always leads to difficult problems like the collapse of stars (gravitating bodies) or laser fusion through the implosion of pellets or how to distribute the electron charge.)

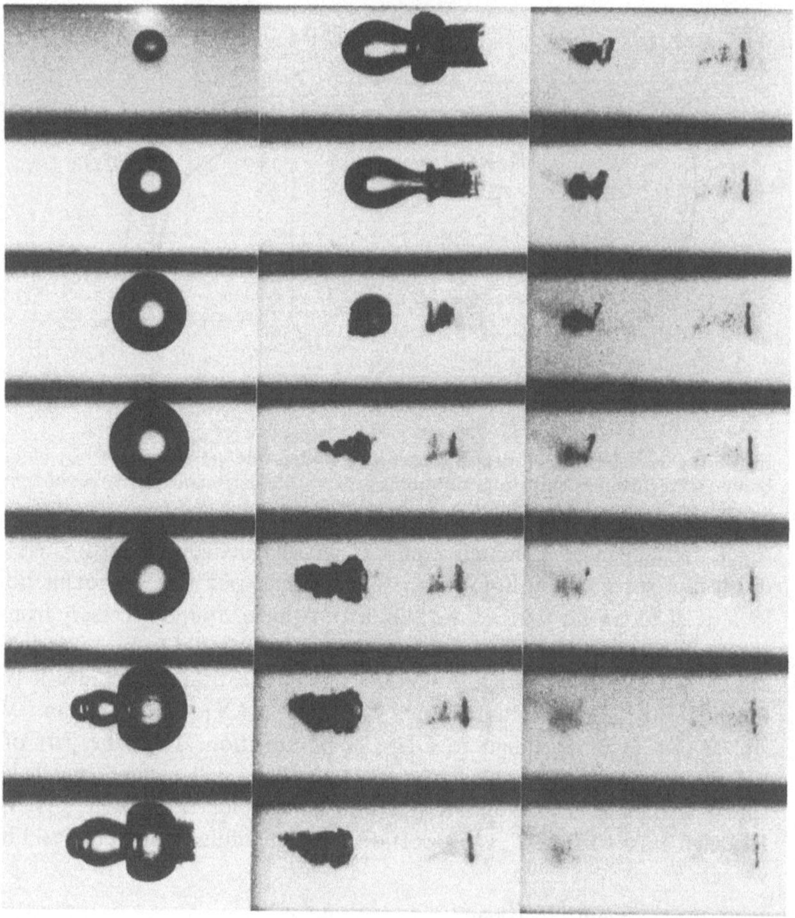

Figure 2. Appearance of stable structures (interpreted as vortex rings) during interaction of two subsequently produced bubbles in water (20 000 frames/s, series taken by W. Hentschel).

The paper of Benjamin and Ellis [2] also discusses the question of what may happen after jet formation has occurred. It is clear that when the jet strikes the opposite wall the bubble attains a torus-like shape. Now Benjamin predicts that this configuration may develop into a vortex ring to carry the original Kelvin impulse. As jet formation may occur without original impulse a slight modification seems in order in that the original impulse is modified to some impulse the jet has acquired during collapse.

It seems that the prediction of vortex ring formation (which implies a certain stability and flow field configuration) has never convincingly been demonstrated. A torus-like shape of the bubble through jet formation may not be a sufficient criterion. Now, our method of optic cavitation [9, 10, 13]

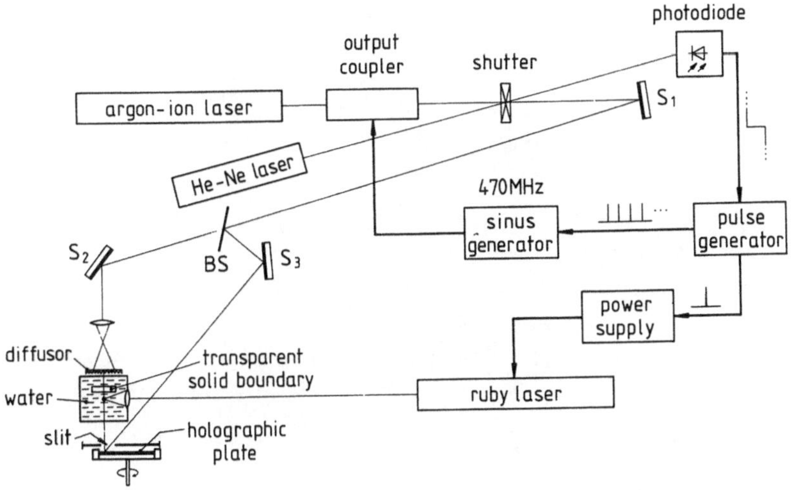

Figure 3. Experimental set-up for high-speed holocinematography of laser-induced cavitation bubbles at rates up to 25 000 holograms per second. S_1, S_2, S_3 highly reflecting mirrors, BS beam splitter.

developed to a high degree of sophistication gave us first hints that vortex ring formation may indeed play an essential part in the life of a cavitation bubble. We had built an apparatus to take high speed photographic pictures of bubble collapse at up to a million frames per second by triggering on its collapse [14]. The aim was to measure the jet velocities occurring on collapse of spherical bubbles in front of solid boundaries. But the result was a peculiar one. The collapsed state seemed to be not sharply defined (as e.g. a spherical collapse should be). Instead a bar was observed on the film upon collapse which persisted for several microseconds until the characteristic shape of a bubble with a jet appeared relatively slowly.

A more convincing example of vortex ring formation in cavitation bubble dynamics came from experiments where two bubbles had been produced with a time delay. Figure 2 gives an example. The second bubble is created about 220 μs after the first in its immediate vicinity. Several vortex rings seem to be generated with an especially stable one moving to the right at a velocity of about 4 m/s. The framing rate is 20 000 frame/s, and thus this vortex ring lasts for more than half a millisecond. It should be remarked that the two nearby created bubbles seem to repel each other after interaction.

The next confirmation for vortex ring formation in cavitation bubble dynamics came to us from experiments with transparent plane solid boundaries which allows looking through the bubble from the top (or bottom). The experiments were done in the course of our development of high speed holocinematography for application to cavitation problems [see 12, 6, 7]. A new set-up to reach higher framing rates is given in Figure 3. It is essentially one of the set-ups given by Ebeling [7] but with a little bit more sophistication

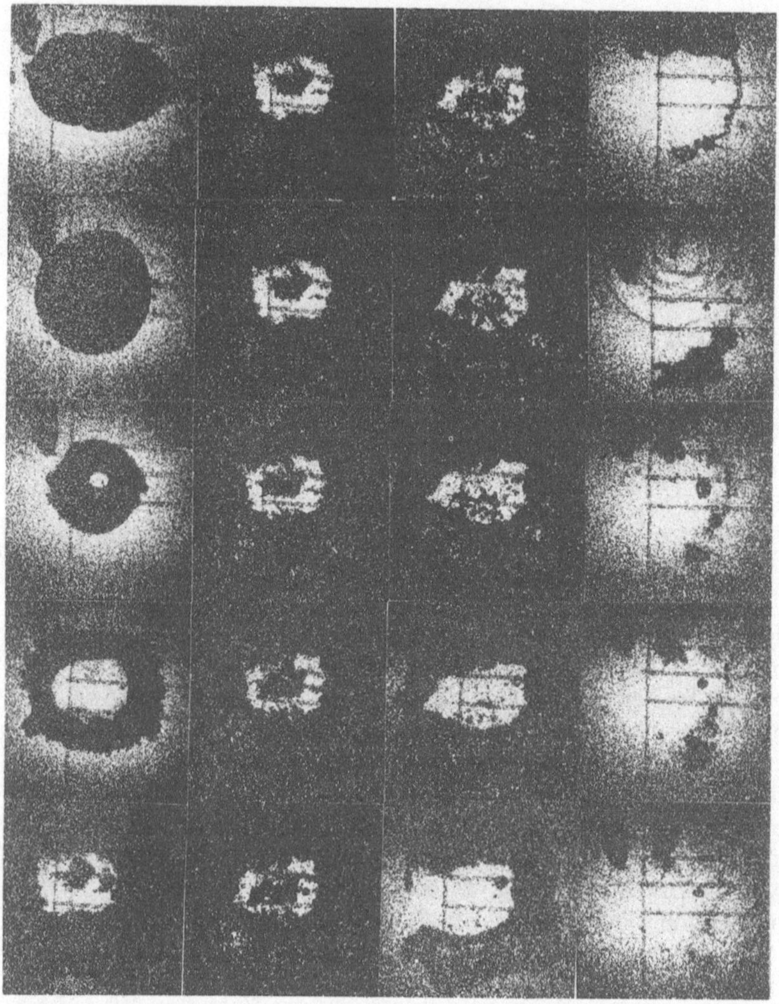

Figure 4. Reconstructed image series of a laser produced bubble in water taken at a rate of 25 000 holograms/s. The first frame (upper left) gives the bubble after breakdown, the next one below 19 frames later before collapse. The bubble is near a plane solid transparent boundary (a mm-scale on glass) located behind the bubble. A ring-shaped bubble is formed after first collapse. It forms a ring of bubbles upon second collapse (upper right frame) and emits several shock waves (next lower frame) (series taken by K.D. Merboldt).

to reach holographic framing rates of 25 000 holograms/s. A cavity-dumped argon ion laser delivers a series of pulses as determined by the pulse generator via the output coupler. Green light pulses ($\lambda = 514$ nm) of about 30 ns duration with a typical energy of 1.2 μJ at rates from 0–1 MHz are obtainable. The series is started by opening the shutter which is necessary for suppression of light leaking out from the output coupler. The photodiode then gets light and triggers the pulse generator and the ruby laser. The light pulses are split by the beam splitter BS to make up a usual holographic set-up with an off-axis reference beam. The holographic plate is rotated so fast behind a stationary slit that from one pulse to the next a different portion of the plate is illuminated. Some care is needed that the interference fringes will not blur due to the motion of the holographic plate. Details can be found in [16].

Figure 4 shows a sequence of pictures reconstructed from a holographic series taken at 25 000 holograms/s with the set-up of Figure 3. The first picture in the upper left corner gives the bubble shortly after formation. The second picture below is the nineteenth frame (760 μs later), and from now on every frame of the sequence is given with a time interval between frames of 40 μs. The transparent solid boundary is located behind the bubble (see Figure 3) several mm apart. It is immediately noticed that between the third and fourth frame given something peculiar must have happened. From other experiments [2, 9, 10, 13] it is known that inevitably a jet is generated towards the solid boundary. In this case we are looking from above on the bubble along the jet going down to the boundary. After first collapse the bubble essentially looks like a dark ring whereas a side view (from other experiments) gives a dark bar. Both views together give a torus as form for the bubble. In the next frame, 40 μs later, a dark spot appears in the middle. It seems that the jet has taken some gas or vapour with it out of the bubble. This gas has become unstable in its form around the jet and deflects the light out of its path. A pure jet of water in water can hardly be imagined to deflect light totally away. Obviously the torus like configuration is rather stable, and the gas or vapour taken along with the jet rearranges in the form of tiny bubbles which are viewed through the interior of the torus. The interpretation that a vortex ring (a 'bubble vortex ring') has been created seems straightforward. Interesting is the subsequent behaviour of the bubble vortex ring. The torus seems to open up on one side (third row, bottom frame) and collapses in parts to a ring of bubbles (upper left corner and next frame). Several shock waves are emitted from different portions of the ring. The ring structure now comprised of individual bubbles gets more and more diffuse and turns into a cloud of tiny bubbles. This seems to be the essential mechanism for the proliferation of bubbles from a single collapsing cavitation bubble.

As the quality of the holograms is not yet comparable to ordinary high speed cinematography we decided to have a closer look at the phenomenon with our rotating drum camera at 20 000 frames/s. In Figure 5 two examples of a bubble collapsing near a plane solid boundary are given. In Figure 5a the boundary is located in the upper part of each frame, and the jet accordingly points in that direction. The distance of the point of breakdown to the boundary is 6.25 mm, the height of each frame 9 mm. It can be seen that after the second collapse a short (not quite straight) black line is left which

a b

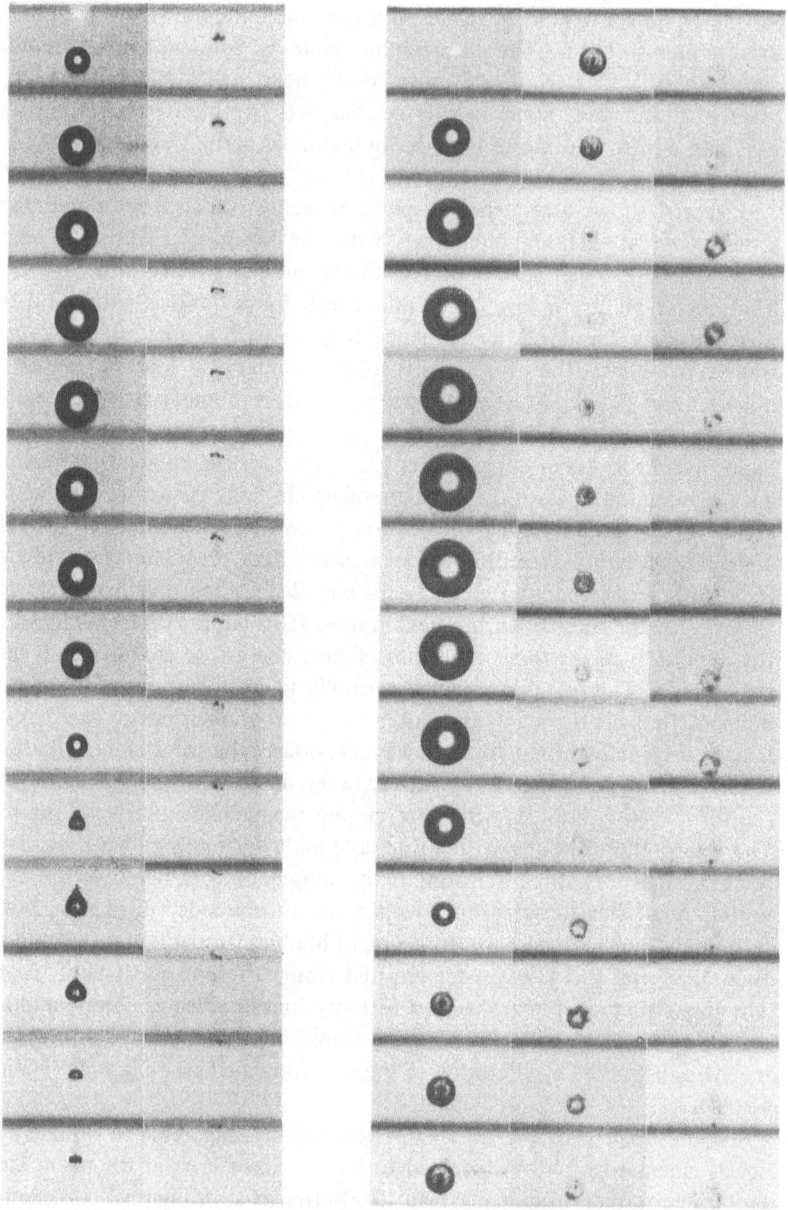

Figure 5. Dynamics of two different laser produced cavitation bubbles in water near a plane solid boundary at a distance of 6.25 mm from the site of breakdown, (a) side view with boundary above the bubble; (b) top view with boundary behind the bubble. Both series have been taken at 20 000 frames/s with a rotating drum camera. An oscillating ring structure (interpreted as vortex ring) is formed after first breakdown with jet formation towards the boundary (series taken by A. Vogel).

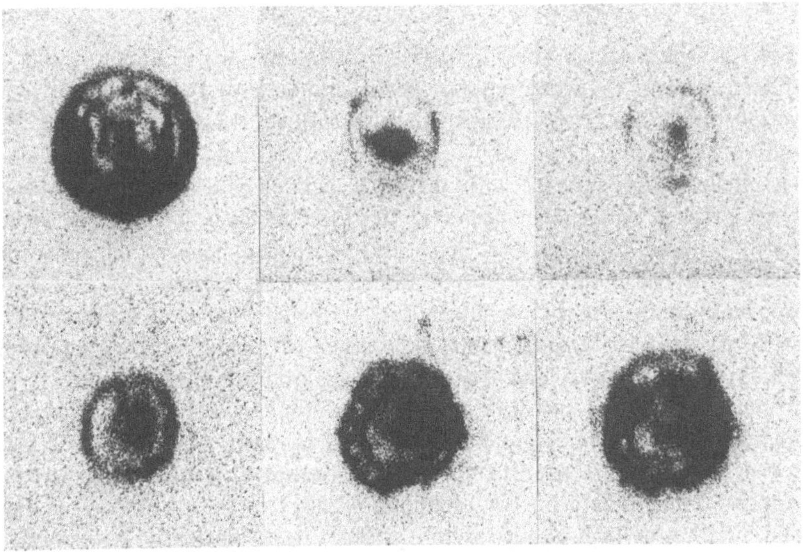

Figure 6. Enlargement of Figure 5(b), frame numbers 16–21, to show the ring structure during second collapse and rebound.

does not alter its appearance significantly for several 100 μs (last eight frames of Figure 5a) except for a motion towards the boundary. Figure 5b shows a top view of another slightly bigger bubble with the same distance of 6.25 mm from the boundary which this time is located behind the bubble. After the first collapse a ring-shaped structure emerges. This ring oscillates and collapses several times until it disintegrates into many bubbles but without even then losing its ring-like shape. The interpretation again is in terms of a bubble vortex ring. Figure 6 gives an enlarged view of the second collapse of the cavitation bubble and the first collapse of the vortex ring. The dark spot in the middle is due to bubbles the jet has left in the line of sight during its disintegration. This is suggested by the frames of Figure 5a. Remarkable is the regrowth of the torus from a really thin ring indicating an extreme stability.

With the above experiments bubble vortex ring formation is firmly established, and the prediction of Benjamin [2] originally given for a translating collapsing bubble is proven right also for the case of a bubble in front of a plane solid boundary. As vortex ring formation seems to be intimately connected with some form of jet formation and jet formation has been found to be a common phenomenon in cavitation bubble dynamics [10] it is conjectured that also bubble vortex ring formation is a common phenomenon in cavitation bubble dynamics.

3. Acoustic cavitation noise

The basic physical question in acoustic cavitation noise problems is what is the mechanism by which a sinusoidal sound wave of frequency f_0 is converted into a broad band noise spectrum with superimposed lines at $n \cdot f_0$, $n/2 \cdot f_0$, $n/3 \cdot f_0, \ldots$ ($n = 1, 2, \ldots$), i.e. a spectrum of 'periodic chaos'? Despite many attempts to clarify the mechanism no conclusive answer has yet been given. It seems clear, however, that the bubbles occurring in an acoustically stressed liquid play an active role via their dynamics [17]. Thus the question is one of the response of a nonlinear dynamical system to a single frequency input. This view will be refined in the later stages of the discussion.

As cavitation bubble dynamics is governed by highly nonlinear equations which do not seem to lend themselves to analytic solutions one must resort to numerical procedures. An extensive study along these lines of the response of single spherical bubbles to a sound field has been done [11]. The main results have been (i) that also bubbles driven below their main resonance may respond subharmonically (or ultraharmonically) to the driving sound field and (ii) that at high enough sound pressure levels obviously no stationary solution can be obtained indicating a transition to chaotic behaviour. The observed spectrum may readily be 'explained' by these findings but in fact they are just not in contradiction. The explanation gets more convincing by a few additional facts. As in the calculations oscillations with a one-half subharmonic component in the spectrum are obtained in by far the most cases the experimentally observed predominance of the $f_0/2$ line (together with its harmonics) in the noise spectrum as compared to the very seldom occurrence of the $f_0/3$ or $f_0/4$ lines (together with their harmonics) finds a natural explanation.

In the meantime we have conducted sophisticated experiments with the help of a minicomputer [15] and also developed refined albeit yet crude models for the noise production process [4]. Here I just want to describe another model approximating the usual experiments where the sound field amplitude is raised above the cavitation threshold and the noise is monitored. In this model the response of a single spherical bubble due to a linearly increasing sound field amplitude of fixed frequency is calculated. The oscillation of the bubble is described by a set of equations which include sound radiation (see [5] for the equations and a description). From overlapping segments of the radius-time curve thus obtained short time spectra are calculated and plotted in the manner of visible speech, in this context called 'visible cavitation noise'. Figure 7 gives an example of a radius-time curve for a bubble of 60 μm radius at rest driven by a sound field of 23.56 kHz (this peculiar value is due to an actual value used in experiments) the amplitude of which is raised at a rate of 0.2356 bar/ms (= 0.01 bar × 23.56 kHz = 100 oscillations per 1 bar increase). Only part of the total curve from 0 bar to 9 bar sound amplitude is shown, i.e. the part where the oscillations change to

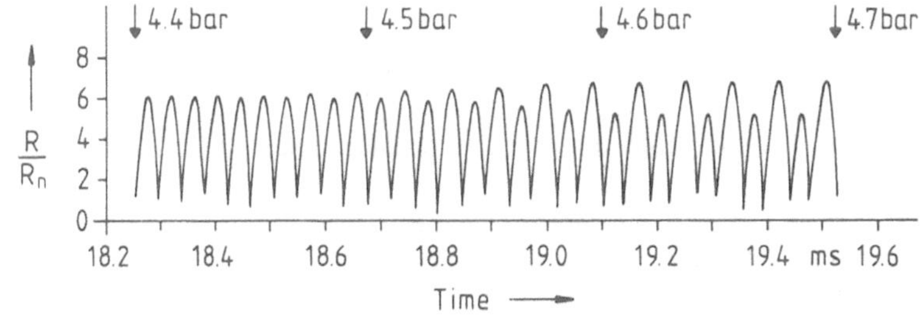

Figure 7. Radius-time curve of a spherical bubble oscillating in a sound field of frequency $f_0 = 23.56$ kHz with a sound field amplitude increasing linearly at a rate of 0.2356 bar/ms. The radius of the bubble at rest is 60 μm. Only part of the total curve calculated is shown (from 4.4 to 4.7 bar). (Calculations done by E. Cramer.)

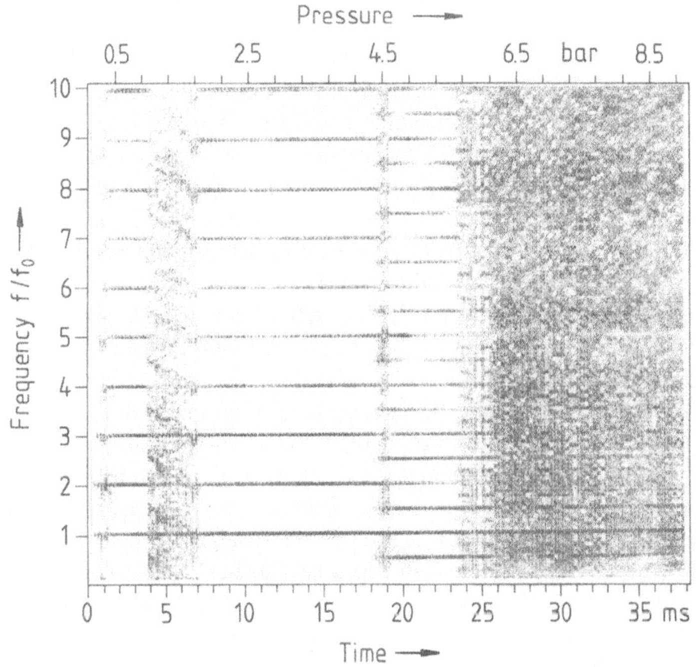

Figure 8. 'Visible noise' calculated from the bubble wall motion part of which is shown in Figure 7. The amplitude spectrum is plotted versus time with the spectral amplitude encoded logarithmically as gray level. (Calculations done by E. Cramer.)

become subharmonic. Figure 8 shows the corresponding visible cavitation noise plot. Gray scale plots (the spectral amplitude is encoded as gray level) always present some difficulties. We have used an electrostatic plotter with only black and white plotting capabilities and taken a 3 x 3 matrix which is filled with dots according to the amplitude in the spectrum on a logarithmic scale. Some characteristic patterns are readily observed.

(i) There is a sharp onset for the first subharmonic (at about 4.5 bar in this case).

(ii) Below the threshold for the first subharmonic spurious noise bands occur.

(iii) There is a second threshold for almost steady state noise throughout the spectrum. It is higher than the threshold for the first subharmonic.

Just all these features are also found in our computer controlled experiments. Thus we are left with the astonishing result that the calculated visible noise of a single bubble already matches the total noise spectrum of a whole cavitation bubble field although only qualitatively, of course. A critical discussion seems in order to see how this may come about. In an acoustic cavitation bubble field bubbles of different sizes are present. According to our calculations differently sized bubbles have different thresholds for the onset of the first subharmonic. An ensemble of bubbles should therefore smear out a sharp threshold. This is not observed experimentally. Two effects may be the reason. Firstly, the bubble with the lowest threshold will determine the onset and secondly, a smeared-out threshold can only be expected when all bubbles behave independently. This surely is not the case. There is a strong coupling between bubbles for geometrical reasons as in the dense clouds observed experimentally the bubbles often touch and coalesce as well as disintegrate into many bubbles according to the mechanism described in Section 2. Also a strong coupling between all bubbles due to the strong shock waves from collapsing bubbles must be assumed. This suggests a very complex behaviour. There seems to be only one way out of the difficulty. The statistics introduced by the large number of bubbles may lead to a largely similar behaviour of different cavitation bubble fields regardless of special interior variations. This behaviour then, on the average, may resemble that of a single bubble.

To be sure the above discussion is highly speculative. But it has shown how a rigorous theory of acoustic cavitation noise may be approached. The main aspect is that the theory so far is not self-consistent. A single bubble in a driving sound field is considered. But in a bubble field the acoustic emission of this bubble acts on the other bubbles and that of the other bubbles on the one considered. The driving acoustic field of any one bubble in the field is thus not the sound field applied but the acoustic cavitation noise field itself. We are thinking about a formulation of a theory which incorporates this feedback, i.e. a self-consistent theory of acoustic cavitation noise.

4. Conclusion

Model bubbles produced with the aid of a ruby laser have revealed new features in cavitation bubble dynamics, i.e. oscillating torus-like bubbles with a remarkable stability which strongly suggests that they are bubble vortex rings. They seem to appear in conjunction with jet formation, a common feature in bubble dynamics and therefore must be a common feature, too.

Concerning the problem of acoustic cavitation noise spectra a new line of investigations has been opened with visible cavitation noise and the notion of self-consistency. Theory seems to have a fair chance to explain at least qualitatively the transformation of a single frequency via nonlinear bubble dynamics to a complicated spectrum.

Acknowledgement

This work has been sponsored by the Fraunhofer-Gesellschaft, Munich, and the German Science Foundation. Also, I want to thank my numerous co-workers especially E. Cramer, W. Hentschel, R. Timm, A. Vogel, K.D. Merboldt and K.J. Ebeling without whose work manu et mente the results reported in this paper would not have come into existence.

References

1. Benedict JV, Harris EH and von Rosenberg DU (1970) An analytical investigation of the cavitation hypothesis of brain damage. Trans Amer Soc Mech Eng, J Basic Eng D92: 597.
2. Benjamin TB and Ellis AT (1966) The collapse of cavitation bubbles and the pressures thereby produced against solid boundaries. Philosophical Trans Royal Soc London A260: 221–240.
3. Burrill LC (1951) Sir Charles Parsons and cavitation, 1950 Parsons memorial lecture. Trans Inst Marine Engineers 63: 149–167.
4. Cramer E (1978) Experimentelle und numerische Untersuchungen zur Schallabstrahlung von Kavitationsblasenfeldern. PhD dissertation Göttingen (see also this volume).
5. Cramer E (1980) The dynamics and acoustic emission of bubbles driven by a sound field. See [13], pp 54–63.
6. Ebeling KJ and Lauterborn W (1978) Acousto-optic beam deflection for spatial frequency multiplexing in high speed holocinematography. Appl Optics 17: 2071–2076.
7. Ebeling KJ (1980) Application of high speed holocinematographical methods in cavitation research. See [13], pp 35–41.
8. Knapp RT, Daily JW and Hammitt FG (1970) Cavitation. New York: McGraw-Hill.
9. Lauterborn W (1974) Kavitation durch Laserlicht. Acustica 31: 51–78.
10. Lauterborn W and Bolle H (1975) Experimental investigations of cavitation-bubble collapse in the neighbourhood of a solid boundary. J Fluid Mech 72: 391–399.
11. Lauterborn W (1976) Numerical investigation of nonlinear oscillations of gas bubbles in liquids. J Acoust Soc Amer 59: 283–293.
12. Lauterborn W and Ebeling KJ (1977) High-speed holography of laser-induced breakdown in liquids. Appl Phys Letters 31: 663–664.
13. Lauterborn W (ed) (1980) Cavitation and inhomogeneities in underwater acoustics. Berlin: Springer.
14. Lauterborn W and Timm R (1980) Bubble collapse studies at a million frames per second. See [13], pp 42–46.
15. Lauterborn W and Cramer E (1981) On the dynamics of acoustic cavitation noise spectra. Acustica 50 (in press).
16. Merboldt KD (1981) Hochfrequenzholografie mit dem auskoppelmodulierten Argon-Ionen Laser. Diplomarbeit Göttingen.

17. Neppiras EA (1969) Subharmonic and other low frequency emission from bubbles in sound-irradiated liquids. J Acoust Soc Amer 46: 587–601.
18. Neppiras EA (1980) Acoustic cavitation. Phys Reports 61: 159–251.
19. Nyborg WL (1974) Cavitation in biological systems: In: Bjørnø L (ed) (1974) Finite-amplitude wave effects in fluids. Guildford, England: IPC Science and Technology Press.
20. Lord Rayleigh JW (1917) On the pressure developed in a liquid during the collapse of a spherical void. Phil Mag (Ser 6) 34: 94–98.
21. See [2], p. 233.

Bubble migration inside a liquid drop in a space laboratory

P. ANNAMALAI, N. SHANKAR, R. COLE and R.S. SUBRAMANIAN

Department of Chemical Engineering, Clarkson College of Technology, Potsdam, NY 13676, USA

Abstract. Commercial production of glasses for advanced applications often requires processing techniques substantially different from those in common use. In particular, containerless processing is desirable where melt temperatures are sufficiently high that the container wall reacts chemically with the melt and/or promotes crystallization. An ideal environment for containerless processing is provided by the NASA Space Shuttle program because in orbit, near free fall conditions prevail and little levitation is necessary. In such an environment, however, there are serious problems associated with convective mixing and buoyant fining (bubble removal) of glass melts. Alternate techniques for the promotion of mixing and for managing bubbles in space have been proposed by Subramanian and Cole and include thermocapillarity, rotation, oscillation, etc. This paper will describe these experiments and discuss two of a number of ongoing ground-based projects in support of the flight experiments.

1. Introduction

The recent successful flight of the Space Shuttle heralds the advent of a new age in the processing of materials. With the availability of long periods of near free fall conditions, materials may be processed in space without the need for a container. In high temperature advanced applications such as the commercial production of Nd^{+3} doped laser glass, containerless processing eliminates chemical reactions and/or crystallization promoted by the presence of the wall [4, 10]. Although it is theoretically possible for containerless processing to be carried out on earth using acoustic forces, the mass which can be levitated is limited. In orbit, however, little levitation is necessary and acoustic forces may be used to position, rotate, oscillate, and even shape, glass melts.

Homogenization and fining (bubble removal) are phenomena of great concern in glass processing. On earth, buoyant convection currents which arise in molten glass help to make the melt homogeneous, and a combination of buoyant rise and dissolution eliminate the many bubbles resulting from entrapment of air in the raw materials and from gases released in the glass forming reactions.

In orbit, where near free fall conditions prevail, it is anticipated that buoyant forces and therefore buoyant convection will be substantially diminished. Thus other means must be found to promote mixing of the melt and to manage the removal of gas bubbles. Accordingly, Subramanian and Cole [9] have proposed Space Shuttle experiments designed to develop an understanding of fluid flows and bubble motion in drops (both model fluids and molten glass) due to thermocapillarity, rotation, oscillation, expansion/

Applied Scientific Research 38: 179–186 (1982) 0003–6994/82/0382–0179 $01.20.

contraction, and other possible mechanisms available in a free fall environment.

An additional objective of our Space Shuttle experiments is an improved understanding of the centering mechanisms active in hollow shell formation processes. Glass shells of approximately 100 μm diameter with 1 μm thick walls (microballoons) are currently produced exclusively for the laser fusion process by KMS Fusion Inc. [5] and by the Lawrence Livermore Laboratory [6]. Although the two processes differ in many respects, they both utilize a drop furnace in which a falling drop of molten glass containing a gas bubble is transformed into a small hollow glass shell of very uniform wall thickness. Because buoyancy will force a bubble to rise within a molten drop, causing the wall to be thin at the top and thick at the bottom, the relatively large percentage of shells with acceptable wall thickness uniformity (200 nm [11]) is quite surprising. Clearly, a better understanding of the bubble centering mechanisms are needed to develop future processes for production of the larger shells (1–10 mm diameter) required for efficient fusion.

2. Space shuttle flight experiments

As illustrated in Figure 1, experiments are to be performed aboard the Space Shuttle to demonstate the effectiveness of thermocapillary flows to promote mixing (a), to demonstrate fining of bubbles by thermocapillary migration (b), and by rotation followed by thermocapillary migration (d), and to demonstrate the use of rotation (and other mechanisms) as a means of bubble centering (c).

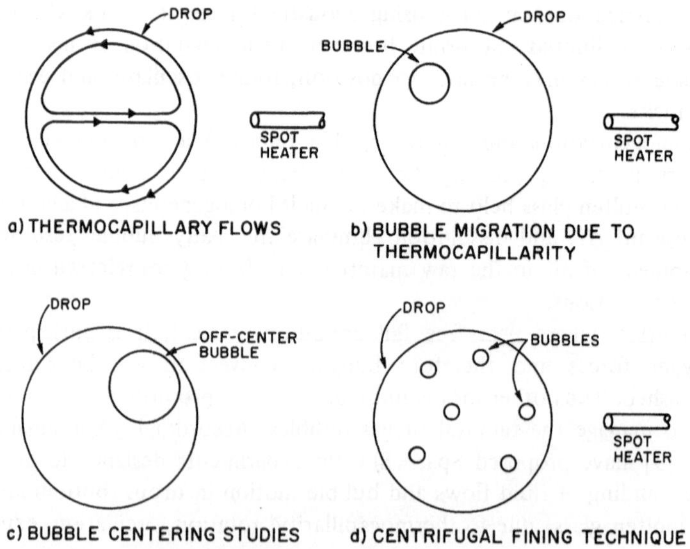

Figure 1. Space Shuttle flight experiments [9].

Homogenization (mixing). Thermocapillary flows, as illustrated in Figure 1a, will be generated by spot heating the surface of a levitated molten glass sphere. As the surface tension is a function of temperature (decreasing with increasing temperature), surface flows away from the hot spot will be initiated, which because of viscous drag will induce internal flows and mixing within the molten sphere.

Fining (thermocapillary migration). The experiment, as depicted in Figure 1b is identical to the previously described mixing experiment except for the presence of one or more gas bubbles within the drop. The internal convective motion induced by surface flows should sweep the bubble toward the drop surface in the region of the hot spot. Further, the thermal gradient within the drop will cause surface tension driven flows at the bubble-liquid interface which propel the bubble toward the hot spot, independent of the induced convection. Thus the two mechanisms are reinforcing.

Fining (centrifugal migration). The experiment, as illustrated in Figure 1d, is identical to the previously described fining experiment except that the drop is now rotated acoustically, causing the bubbles to migrate toward one or more axes of rotation and coalesce. The drop is again spot heated and the large bubble is removed by thermocapillary migration more rapidly than would the many smaller bubbles originally present.

Bubble centering. The drop, containing an off-center gas bubble, as shown in Figure 1c, is rotated acoustically about one or more axes to determine the degree to which the bubble can be centered. Additionally, other possible centering techniques such as oscillation and expansion/contraction will be investigated.

It is intended that all of the glass experiments be preceded by model fluid experiments either on the Space Shuttle or in the NASA SPAR (Space Processing Applications Rocket) program or both. Further, in preparation for the flight experiments and post-flight analysis, a substantial ground-based program is currently ongoing to aid in design and interpretation of the flight experiments. The program includes mathematical modeling of the physical processes involved, and experiments both with model fluids and low melting glasses ($< 1000°C$).

3. Ground-based studies

The following ground-based programs are currently active: (a) thermocapillary convection in liquid drops; (b) bubble migration in a rotating liquid; (c) bubble migration in a temperature gradient inside a liquid drop; (d) thermocapillary convection in molten glasses; (e) bubble rise in glass melts; (f) volatilization kinetics of a molten glass drop. Two of the studies ((b) and (c)) will be briefly discussed in this paper.

3.1. Bubble migration in a rotating liquid

As earlier indicated, the objective of the rotation study is to obtain a better understanding of the centering mechanisms responsible for the formation of uniform wall thickness glass microballoons currently used as targets in the laser fusion program. As these experiments and analysis represent the first phase of the investigation, it was not felt necessary to employ a rotating drop system. Instead, the experimentally simpler system of a liquid-filled rotating glass shell has been used.

Spherical glass shells (18.8, 24, and 33 mm radius) were blown from Pyrex glass tubing and coupled to a variable speed d.c. motor through a glass-to-metal joint. A typical experiment consisted of spinning the glass shell, filled with the Dow Corning 200 series silicone oil and containing a small gas bubble, about an axis maintained parallel to the ground. High speed motion picture photography was used to record the rotation of the liquid-filled glass shell and the trajectory of the bubble motion. A schematic of the experimental apparatus is shown in Figure 2. Analysis of the film was facilitated by a motion analyzer which provided a 25X magnification and read-out onto computer cards. The apparent position of the bubble was analytically corrected for optical distortion.

In order to interpret the experimental data, it is helpful to have some theoretical understanding of the migration process. However, as the system was spun-up from rest, the resulting flows are expected to be complex. Thus for the purposes of this initial investigation only a simplified, semiempirical approach is used in which the liquid is assumed to be in rigid body rotation at every instant with the rotation rate given by the instantaneous experimentally measured value. Secondary flows are completely ignored. The bubble experiences the following forces: (a) an upward buoyant force due to earth's gravity; (b) a radially inward buoyant force due to the acceleration experienced during rotation; (c) drag due to relative motion between the bubble and the neighboring liquid. The consequence of

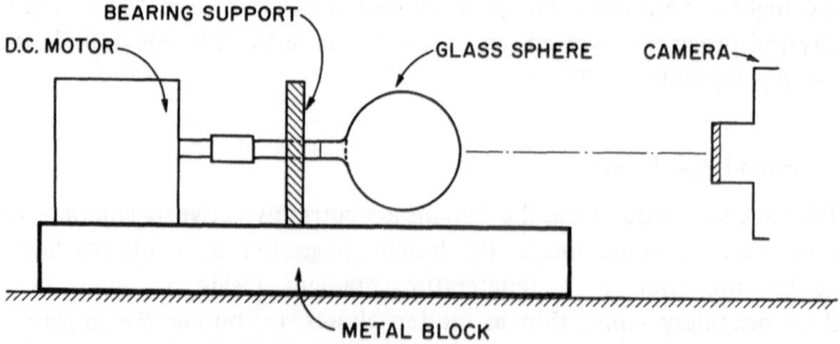

Figure 2. Schematic of experimental rotation apparatus.

the net resultant of these forces is the acceleration of the bubble and the liquid displaced by the bubble (according to the virtual mass concept). The value for the virtual mass coefficient is taken as 0.5. Expressions for the forces (a) and (b) are obtained from the hydrostatic equations. For the drag, the Rybczynski-Hadamard expression was used for Reynolds numbers less than 0.1 and a polynomial fitted to numerical results in the literature [1] for Reynolds numbers between 0.5 and 5. The vector equation was integrated using a fifth order Runge-Kutta numerical integration technique to obtain bubble position as a function of time.

Comparison with experiment for a typical run is shown in Figure 3. The circles represent the experimental data and the solid line corresponds to the numerical solution. The lower scale along the abscissa gives the number of sphere revolutions that have taken place at any time t corresponding to the upper scale. The scaled radial position 1.0 corresponds to the initial position

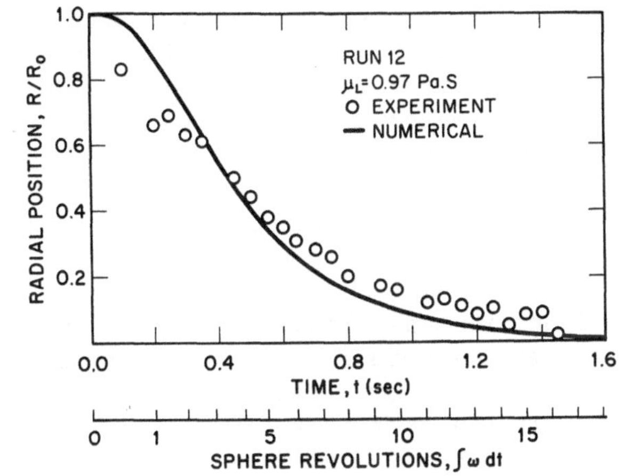

Figure 3. Bubble migration towards the rotation axis.

of the bubble. The slight differences between analysis and experiment are thought to be due to secondary flows [2] resulting from spinning the system up from rest and ignored in the analysis. Both experiment and theory, however, show conclusively that rotation is a mechanism which can aid the bubble centering process. A correlation of the experimentally determined migration times, suggested by a much simplified analysis leading to an analytical solution, indicates the migration time to decrease with increasing rotation rate and bubble size, and to increase with increasing viscosity.

3.2. Bubble migration in a temperature gradient inside a liquid drop

The objective of the migration study is to provide a predictive capability for the motion of a bubble in a liquid drop whose surface is subject to some prescribed arbitrary temperature distribution. For glass melts the Reynolds and Peclet numbers are expected to be small. Hence in this initial phase of the modelling process the accumulation and convective transport terms of the momentum and energy equations will be neglected. Even so, the full three-dimensional problem is quite formidable. Therefore the additional assumption is made here that the prescribed temperature field on the drop surface is symmetric about the axis joining the bubble and drop centers. Under this condition, migration of the bubble will occur only along this axis. Also, the bubble and drop are assumed to retain their spherical shapes, and the balance of normal stresses will be ignored.

When the bubble is so located (eccentrically within the drop), general solutions of the conservation equations for an incompressible, Newtonian fluid, are available in bipolar coordinates [3, 8]. The general solutions may be applied to the present problem by means of appropriate boundary conditions at the drop surface: (a) temperature variation on the surface is prescribed; (b) normal component of the liquid velocity vanishes; (c) tangential component of the stress is balanced by the gradient of interfacial tension. And at the bubble surface: (d) negligible heat flux into the bubble; (e) the normal component of the liquid velocity equals the normal component of the migration velocity of the bubble; (f) tangential component of the stress is balanced by the gradient of interfacial tension. By setting the net force on the bubble equal to zero, the migration velocity is obtained.

The temperature prescribed on the outer surface of the drop is expressed in a series of spherical harmonics. Because of the linearity of the problem, the effect of each of the pure modes of the prescribed temperature field may be considered independently. Isotherms for a P_2 mode prescribed temperature field are shown in Figure 4 (P_n is the Legendre polynomial of order n). T, K, and D represent scaled temperature, bubble radius, and bubble displacement from the drop center. The isotherms are normal to the bubble because of condition (d). Streamlines for the P_2 mode are shown in Figure 5. As expected, the bubble migrates toward the nearest warm pole. Similar results are obtained for higher modes of the prescribed surface temperature. In general, the migration velocity increases from a value of zero for $D = 0$ (for $n \geqslant 2$) as D increases, but ultimately decreases as the bubble approaches the drop surface. When the bubble is at the drop center, motion is predicted only for the P_1 mode [7].

Acknowledgement

This work was supported by the Materials Processing in Space Program Office of the National Aeronautics and Space Administration through a contract (NAS8-32944) from the Marshall Space Flight Center to Clarkson College of Technology. We are also grateful to the Dow-Corning Corporation for providing us with free samples of the DC-200 series silicone oils for the experiments.

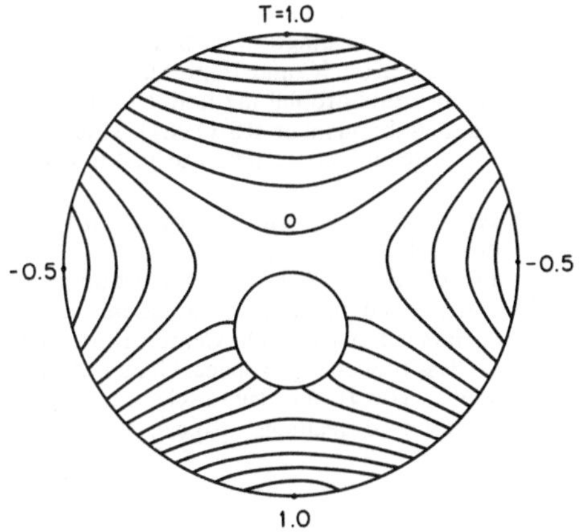

Figure 4. Isotherms in a liquid drop containing an eccentrically located bubble.

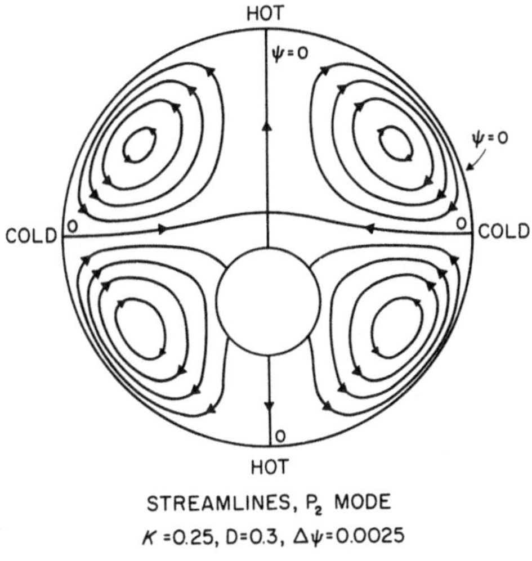

Figure 5. Flow streamlines in a liquid drop containing an eccentrically located bubble. Bubble motion toward the nearest warm pole (southern) is indicated.

References

1. Brabston DC and Keller HB (1975) Viscous flows past spherical gas bubbles. J Fluid Mech 69: 179–189.
2. Greenspan HP (1968) The theory of rotating fluids. London: Cambridge.
3. Jefferey GB (1912) On a form of the solution of Laplace's equation suitable for problems relating to two spheres. Proc Roy Soc (London) A87: 109–120.
4. Nielson GF and Weinberg MC (1977) Outer space formation of a laser host glass. J Noncrystalline Solids 23: 43–58.

5. Nolen RL, Downs RL, Miller NJ, Ebner MA, Doletzky NE and Solomon DE (1977) Inertial Confinement Fusion. Washington: Optical Society of America.
6. Rosencwaig A, Dressler JL, Koo JC and Hendricks CD (1978) Laser fusion hollow glass microspheres by the liquid-droplet method. UCRL-81421.
7. Shankar N, Cole R and Subramanian RS (1981) Thermocapillary migration of a fluid droplet inside a drop in a space laboratory. Int J Multiphase Flow 7: 581–594.
8. Stimson M and Jefferey GB (1926) The motion of two spheres in a viscous fluid. Proc Roy Soc (London) A111: 110–116.
9. Subramanian RS and Cole R (1979) Experiment requirements and implementation plan (ERIP) for "Physical phenomena in containerless glass processing," Experiment MPS 77F010, NASA Marshall Space Flight Center, Document 77FR010.
10. Weinberg MC (1978) Glass processing in space. Glass Industry, March: 22–27.
11. Woerner RL, Draper VC, Koo JC and Hendricks CD (1980) Conference on inertial confinement fusion, digest of technical papers. Washington: Optical Society of America, p 32.

Experimental and asymptotic study of nonspherical bubble collapse

G.L. CHAHINE

Hydronautics, Incorporated, 7210 Pindell School Road, Laurel, MD 20810, USA

Abstract. Observations of the behavior of spark-generated bubbles in the vicinity of solid and free boundaries are described. In all cases, the formation of a reentering region (microjet or constriction) occurs on the part of the bubble which has the most freedom of motion. Drag-reducing polymer additives are seen to significantly affect bubble departure from sphericity. Their presence weakens the influence of nearby solid boundaries, and seems to enhance that of a free surface. The relative importance of the acoustic pulses emitted during successive implosions and rebounds of the bubble is seen to be modified by the proximity of a solid wall. When the radius of the bubble is small compared to its distance from the closest boundary, a theoretical approach, using matched asymptotic expansion, is applied successfully to describe the nonspherical bubble behavior and the pressure field. This method is extended to the case of a multi-bubble system. It is very useful in determining the limiting distances of interaction. In the case of a free surface this distance is less than two bubble diameters. When applied to a solid wall covered with an elastic coating of finite thickness, or to a two-liquid interface this technique shows a selection process: bubbles closer than a limiting distance to the boundary are repelled during their collapse. The collapse is toward the boundary only for bubbles beyond this distance and is therefore less damaging.

Introduction

Since the early work of Rayleigh, the dynamics of cavities in general and bubbles in particular, have acquired a respectable place in cavitation research. Indeed, the modeling of cavitation erosion and noise, and the study of the related scaling effects require the knowledge of individual and collective bubble behavior. The collapse in high pressure regions of these cavities is held responsible for the damage despite controversial opinions about the mechanism by which the energy is transferred from the imploding cavity to the nearby boundaries. The early discussion on whether emitted shock waves or formed microjets are the direct sources of damage seems to resurface again and again. Strong evidence of both mechanisms exists, but in either case bubbles must be close enough to the wall for damage to occur. This stresses the need for more understanding of nonspherical bubble dynamics near boundaries to include the related effects in the erosion model. Presently, most approaches are of a statistical nature and are based on spherical bubble theories which have been extensively studied and have become increasingly sophisticated (see reviews in [15, 19], and in some of the papers of these proceedings). However, despite their great practical importance, nonspherical bubble dynamics theories are not very advanced, due to the complexity of the free boundary problem involved. Fortunately experimental observations and numerical computations have contributed to compensate for the lack of

Applied Scientific Research 38: 187–197 (1982) 0003–6994/82/0382–0187 $01.65.

analytical knowledge. Our intent, in this paper, is not to review the state of the art about these studies but to give an overall view of our contributions to the subject. The major part of the investigations reported here were conducted in Paris, while the author was a member of the 'Groupe Phénomènes d'Interfaces' at the Ecole Nationale Supérieure de Techniques Avancées. The other part was resumed since at Hydronautics.

In the first section of this paper high-speed photographic observations of the behavior of spark-generated bubbles in the presence of various types of boundaries are presented. These observations in distilled water are quantified and then, in the second section, compared to similar ones obtained with dilute drag-reducing polymer solutions. In the third section, some preliminary results on the noise emitted by the collapse of a bubble near a solid wall and its potential application to acoustical detection of erosion are discussed. In the last section, dimensional analysis and the method of matched asymptotic expansions are shown to be valuable tools to determine limiting distances of interaction and to describe relatively small deformations. The asymptotic method is used to successfully model the various situations described in the first section and is extended to a multiple bubble system. The analysis of the behavior of a bubble near a coated surface is also considered.

Nonspherical bubble collapse observations

In all the experiments reported here vapor bubbles (with some noncondensable gas) are generated in water by discharging a capacitor across a pair of platinum or tungsten electrodes for a very brief period of time. Two different generators are used, one capable of a maximum capacitor charge of 2000 V and the other of 10 kV. This system was first used by G.I. Taylor for underwater explosion studies in 1943. Its capabilities, and the validity of the analogy between the collapse of the bubbles it produces and cavitation bubbles, have been discussed by several authors [14]. The electrodes are mounted in a large vessel which is hermetically sealed and connected to a vacuum pump. Lowering the ambient pressure is used for degasing and, when desired, for increasing the bubble size, thus slowing down the phenomenon observed. This compensates for the relatively low framing rate of the high-speed camera utilized, a HYCAM model K20S4AW camera, whose maximum capability is 10 000 frames per second, or 20 000 half-fr/sec.

Behavior near a solid wall

It has been known for a long time now that a bubble near a solid surface collapses with the formation of a microjet. During its implosion the bubble first elongates perpendicular to the wall, then the side away from it flattens and a reentering region is formed initiating a microjet which can pierce the bubble and hit the wall. We repeated these observations in order to compare bubble behaviors in water and in dilute solymer solutions, and to attempt a correlation between erosion, noise and dynamics of isolated bubbles. In the first series of tests [5], an aluminium cylindrical specimen, used to record the damage due to the implosion, was fitted under the electrodes in a hole drilled in a plexiglass plate. It was observed later, while analyzing the motion pictures, that this specimen, not being tight enough in the hole, was being slightly sucked up towards the bubble during its growth and then refitted after the implosion. This secondary motion did not affect qualitatively the non-spherical collapse process. However, a closer look at the curves $R_A/R_{c,\,\mathrm{max}} = f(t/\tau_{\mathrm{sph}})$ (Figure 1) shows that the period of oscillation of the bubble decreases when $\eta = R_{c,\,\mathrm{max}}/l_0$ increases. This behavior, comparable to that

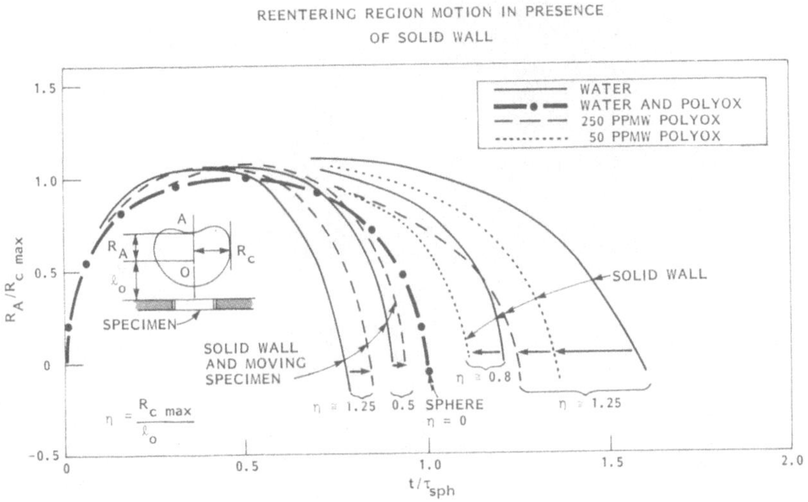

Figure 1. Motion of the reentering region in presence of a solid wall.

near a free surface, is the opposite of what happens near a fixed solid wall. See Figure 1 for definitions. τ_{sph} is the period of oscillation of the spherical bubble obtained with the same electrodes. The experiment was repeated later with a fixed wall, while studying the influence of a transverse velocity on the bubble collapse [17]. For very low transverse velocities, the shear effect is negligible and the bubble behaves as in the classical case near a solid wall. The lengthening effect on the bubble life is verified and increases with η (Figure 1). In both cases described above, the bubble is violently attracted towards the wall during its successive collapses and rebounds (Figure 2a).

Behavior between two solid walls

The collapse of bubbles between two solid walls is interesting for its practical applications in ultrasonic engineering, and in film lubrication cavitation. The great deformations involved are also of interest from the fundamental

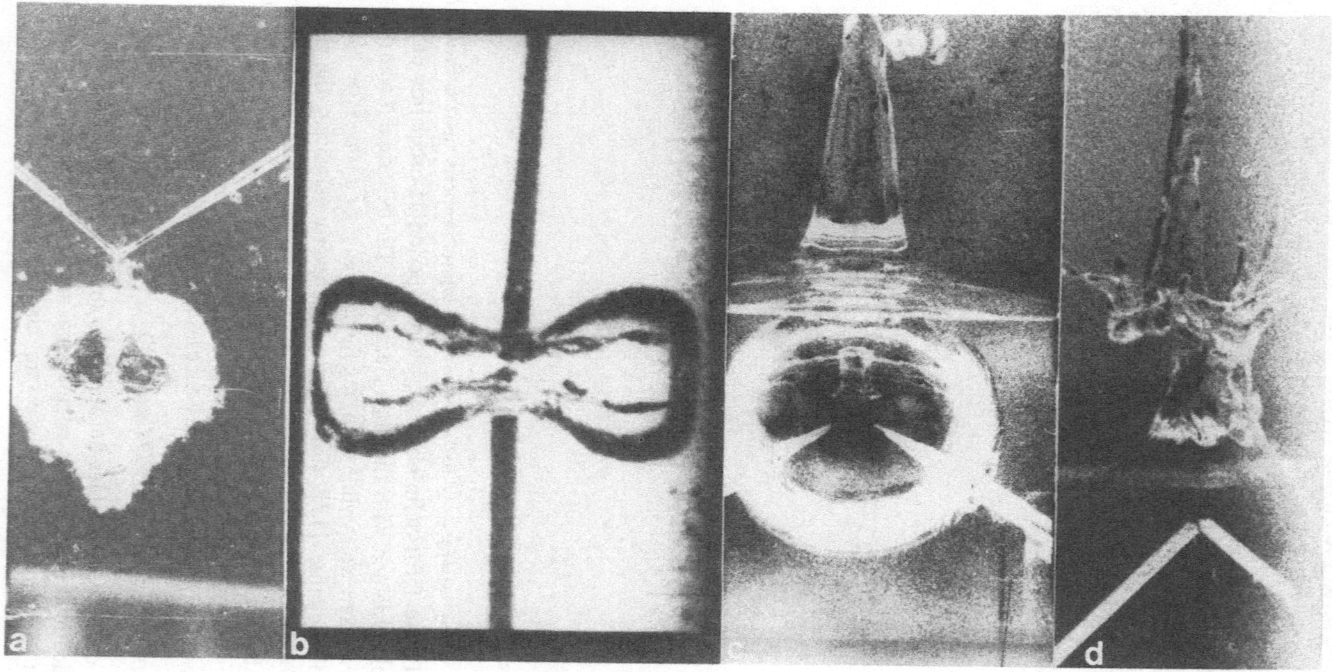

Figure 2. Bubble collapse: (a) near a solid wall, (b) between two walls, (c) near a free surface, (d) free surface with polymer additives.

dynamics point of view. When $\eta < 1$, a bubble at equal distance from the walls, first elongates parallel to the walls (direction of most freedom) during its growth, then perpendicularly when the implosion starts. Later the bubble constricts in the medium plane of symmetry and splits in two parts (Figure 2b). Each of the two bubbles which are formed later collapses with the formation of a microjet directed to the closer wall. When $\eta \gg 1$, the bubble behaves as a cylindrical cavity until the final stages of collapse where it constricts and splits in two parts [7, 9]. Quantitatively the presence of the two walls augments the bubble lifetime significantly. This lengthening effect increases dramatically with η (Figure 3). For $\eta \simeq 0.7$ the period increases by 50% (7% for one wall) and for $\eta \simeq 2$ it is doubled.

Behavior near a free surface and a two-liquid interface

In the case of a free surface the concept of a limiting distance of interaction is easy to define: for $\eta < 0.3$, the free surface remains undisturbed, and the implosion of the bubble is practically spherical. This observation is also confirmed by a theoretical approach [3]. For greater values of η, the upper side of the bubble, the 'most free to move', elongates toward the free surface then moves faster than the rest of the bubble during the collapse. A microjet moving away from the free surface is then formed, and pierces the bubble while a 'counter jet' arises from the free surface (Figure 2c). The curves $R_A = f(t)$ (Figure 3) show the same tendency as the moving solid wall: a shortening effect on the bubble period increasing with η.

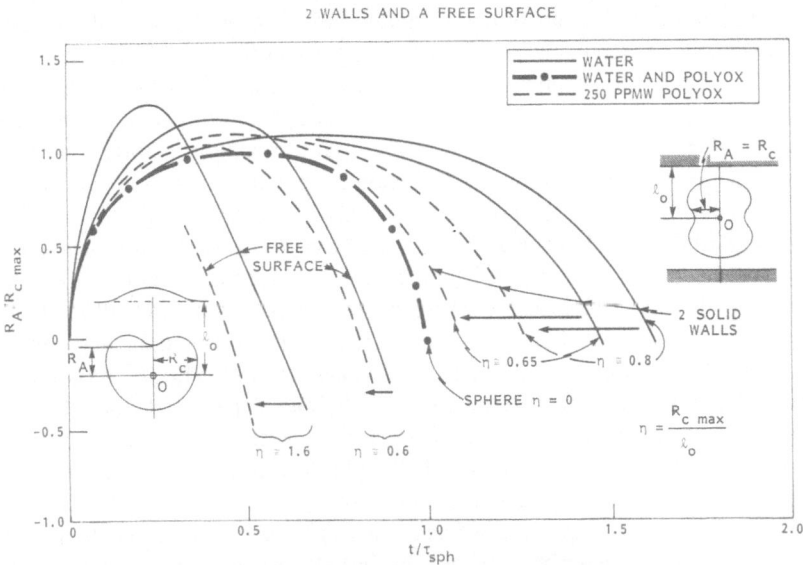

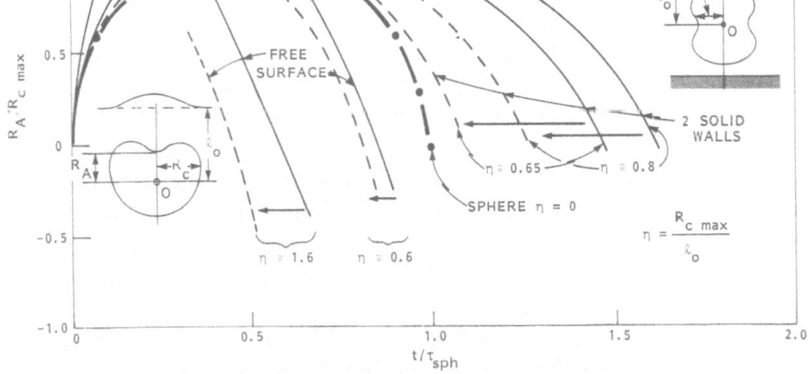

Figure 3. Motion of the reentering region in presence of two solid walls or a free surface.

A few liquid-liquid interfaces were investigated [2, 10]. Unlike the case of an air-liquid interface where the bubble is always repelled from the free surface, two types of behavior are observed depending on the relative magnitudes of η and two limiting numbers η_1, and η_2: $\eta < \eta_1$, no interaction; $\eta_1 < \eta < \eta_2$, the bubble is attracted to the interface and the microjet moves toward it; $\eta > \eta_2$, the bubble is repelled from the interface and the microjet moves away from it. This behavior is similar to that theoretically obtained near a coated solid wall [6]. Let us notice that in all the cases studied the bubble sizes and their period of oscillation are such that gravity has no time to intervene.

Influence of drag-reducing polymer additives

Drag-reducing polymers are known to greatly reduce the cavitation inception index for several types of flows. On axisymmetric bodies having laminar separation the inhibition is due to an early transition to a turbulent non-separating boundary layer [13]. The overall viscous flow is therefore modified by the presence of polymers. The onset of cavitation is also delayed in acoustic cavitation in a stagnant fluid [16]. In addition, cavitation erosion has been reported to be greatly modified (in both directions) with additives. This supports the idea that cavitation inhibition is also related to the inherent properties of polymer solutions. Experimental [12] and theoretical [2, 21] studies on a spherical bubble growth and collapse have shown no significant differences between a Newtonian and a viscoelastic fluid. This conclusion has been supported by our high-speed photographic observations of spark-generated bubbles in an unbounded fluid [4]. However, these observations showed that a Polyox WSR 301 solution has a noticeable influence on non-spherical bubble dynamics. Near solid walls, compared to a liquid having the same viscosity (water + glycerine), the effect of the presence of the additives is to bring the bubble behavior closer to that of a spherical cavity. Indeed, in the vicinity of a moving solid wall (Figure 1) for the same η, the addition of a 250 ppm of Polyox delays the creation of the microjet thus increasing the bubble lifetime and moving the curves $R_A = f(t)$ toward the spherical case curve. Near a fixed solid wall (Figure 1) or two solid walls (Figure 3) the apparently opposite effect (shortening of the period of oscillation) in the presence of polymers is also seen to reduce the differences between the considered case (given η) and the spherical case [9, 17]. This has prompted us to think that this stabilizing effect on bubble departure from sphericity, which is comparable to an increase in surface tension, may indicate a significant dynamical change in the bubble-liquid interface properties. However, the picture has been complicated by our recent analysis of the influence of additives on the collapse near a free surface. The first results indicate opposite effects to those described above. The bubble seems more unstable than in distilled water and the shortening effect is increased by an addition of 250 ppm of Polyox (Figure 3). The problem is complicated here by the behavior of the free surface itself. This interface is seen to be qualitatively very different than in the Newtonian case. The jet on the surface is smoother and thinner at the beginning. The surface looks stretched. Later the jet becomes distorted and loses its symmetry (Figure 2d), while in water the axial symmetry is conserved.

Collapse noise of a bubble near a solid wall

Cavitation is known as being an important source of noise. Usually undesirable, this effect can be useful as a means of cavitation detection, e.g., in fast neutron reactor cavitation. As for erosion, noise studies are of statistical nature and based on the dynamics of isolated cavities (for a review see [20]). The noise is principally emitted during the successive collapses and rebounds of the cavities. With spark-induced bubbles a first pressure pulse is emitted at the spark generation. This pulse is easily distinguishable from the bubble noise and can be eliminated for subsequent analysis. It is followed by a pressure decrease below the ambient during the bubble growth. Then, at the final stages of collapse, a very sharp positive pulse is emitted, and is followed by a negative one due either to a reflected wave in the micro-transducer or to an emitted shock wave. Similar pulses are emitted during successive bubble rebounds and collapses, decreasing each time in amplitude due to energy losses [4].

It is of practical importance to determine if any differences exist between the noise emitted by a bubble collapsing far from a solid wall (and thus spherical), and another one close enough to erode the wall. To investigate this, two hydrophones were placed at equal distance from the electrodes and were located one above the electrode-gap on the axis of symmetry perpendicular to the solid wall, and the other one on the same level as the electrode-gap parallel to the solid wall [8, 18].

The signals were recorded on a digital storage oscilloscope which can eventually be connected to a microprocessor. High-speed movies were simultaneously run for different values of η. For small values of η the bubble remains spherical and the two hydrophones detect identical signals. For $\eta > 0.5$ nonspherical effects become noticeable, especially for the relative importance of successive pulses and their number. As long as the bubble does not touch the wall during its first growth, the strength of the pulse due to the second collapse can exceed that of the first one. The reversal of this relative importance of the two spikes in comparison with the spherical case is, however, detected only by the probe located on the axis of symmetry. This is presumably due to a certain directivity of the emitted pulse and/or to a screening effect due to small bubbles formed near the wall during the rebound. When $\eta \geqslant 1$ only one very strong collapse pulse is noticeable. The first bubble collapse is totally on the wall and the successive rebounds are very weak. Frequency analysis of the signal seems to indicate a shift towards lower frequencies for higher η. However, the limitation of the frequency domain to 250 kHz restricts the validity of this conclusion. Amplitude analysis of pulses emitted by spherical collapse shows the relation $I \simeq 0.45\tau^{3/2}$ between the spike amplitudes and the period of oscillation of the bubbles. For non-spherical bubbles large deviations from this relation are observed [4].

Asymptotic approach

To study the various problems described above, a perturbation method is used. Basically, in all the considered cases, η is assumed to be small enough to admit in the first approximation that the bubble behaves spherically. The interaction can be shown to be weak if $\zeta = G\,\eta^3 \ll 1$, [3]. G, is related to the bubble period, τ, and its radius, a_m, when the volume rate is maximum, by the relation: $G = 4\pi \dot{a}_m \tau / a_m$ (with the Rayleigh model G is constant and for very weak interactions $\eta < 0.37$). Matched asymptotic expansions are used to describe the bubble behavior, 'inner region', and/or the boundary behavior (e.g. two-fluids interface, coated solid wall), 'outer region'. For the outer problem, up to the order η^2, the bubble can be replaced by a source of known strength, and one can concentrate on the interface (or the elastic solid) equations. In [4, 10] the equations of the two fluids interface are written and solved. The translation velocity of the spherical bubble of strength $q(t)$ is shown to be:

$$V(t) = -q^2/4 + \rho_1 q/2(\rho_2 - \rho_1).$$

This indicates a tendency of the bubble to move away from an air-liquid interface ($\rho_1 \ll \rho_2$). This tendency depends upon $q(t)$ for the two-liquid case. This corresponds to the experimental evidence described above. In [6] the equations for a solid wall coated with a material of Young's modulus E, compressibility K and thickness e_0, are considered. Kinematic and dynamical conditions are written at the coating-liquid interface and a condition of no-motion on the solid wall. A dimensional analysis of the equations of motion in the liquid and in the elastic medium shows that the solutions depend on ζ, K/E, and $\xi = E\tau^2/\rho_{\text{liq}} l_0^2$. For a weak interaction and a very compliant wall $\zeta \ll 1$, $\xi \ll 1$ and $K \gg E$. In this case the linearized equations show that the compliant wall acts as a free surface for bubbles below a distance, l^*, of the same order as the coating thickness. Conversely, it acts as a prolongation of the solid wall for bubbles beyond l^*. This result is interesting as it could explain the apparent resistance capability to cavitation damage of elastomeric coatings.

The study of the 'inner problem' is simplified by the fact that, in first approximation, the bubble behavior is described by the Rayleigh-Plesset equation. Simplified equations, soluble numerically, are derived for the higher orders of approximation. These equations take into account the presence of

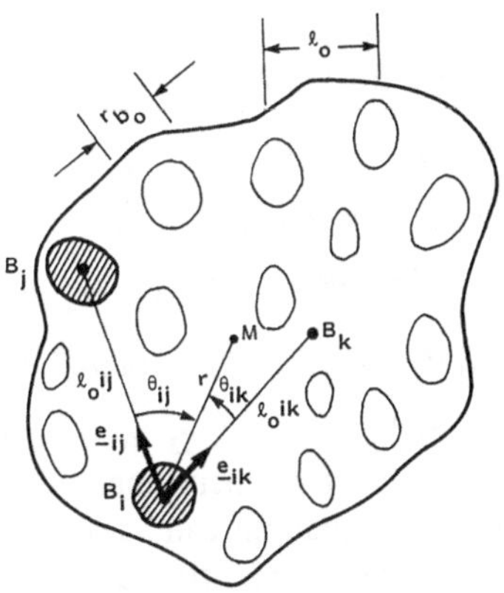

Figure 4. Sketch of geometrical characteristics for the multibubble system model.

various boundaries through matching conditions with the 'outer problem' in which all other bubbles or boundaries (through the method of images) appear as singularities. To present the problem in a general form, let us consider the collective behavior of N bubbles of characteristic interdistance l_0 and radius r_{b_0}. Now, we have $\eta = r_{b_0}/l_0$, and the configurations described above are only particular cases of this problem. If $P_\infty(t)$ is the imposed ambient pressure, the driving pressure for the collapse of the bubble B_i is:

$$P(B_i, t) = P_\infty(t) + \eta P_1(B_i, t) + \eta^2 P_2(B_i, t) + \ldots .$$

$P_1, P_2 \ldots$ are in the 'outer problem', the pressures due to the whole multi-bubble system. In the matched asymptotic expansion approach, the pressure in the vicinity of B_i, $p(B_i, r, t)$, is obtained by computing the motion of this bubble with the condition that p takes the limiting value $P(B_i, t)$ when $\tilde{r} = r/r_{b_0}$ tends to infinity. Adding the contributions of the $N-1$ remaining bubbles one can show that, at least up to the order η^2, these bubbles can be replaced by a unique equivalent bubble whose strength, q^{ik}, and position, B_k, (distance l_0^{ik} in the direction $\mathbf{e}_{ik}$, Figure 4) are defined by:

$$q_n^{ik}/l_0^{ik} = \sum_j q_n^j/l_0^{ij},$$

$$\mathbf{e}_{ik} \cdot q_n^{ik}/(l_0^{ik})^2 = \sum_j \mathbf{e}_{ij} \cdot q_n^j/(l_0^{ij})^2,$$

where n is the order of approximation, l_0^{ij} and $\mathbf{e}_{ij}$ the length and the unit vector of B_iB_j, and q_n^j the strength of the source representing the bubble B_j. The equation, of the nonspherical bubble can then be expressed as an expansion in powers of η, and as function of the polar angle, θ^{ik}, as follows:

$$r_b^i(t) = a_0^i(t) + \eta a_1^i(t) + \eta^2 [a_2^i(t) + f_2^i(t) \cos \theta^{ik}] + \ldots .$$

$a_0^i(t)$ is obtained by solving the Rayleigh-Plesset equation. The other orders are solutions of the following differential equations (superscript i omitted):

$$a_0\ddot{a}_1 + 3\dot{a}_0\dot{a}_1 + a_1 g_1(a_0, W_e, P_{g_0}) = -\dot{q}_0^{ik} l_0/l_0^{ik},$$

$$a_0\ddot{a}_2 + 3\dot{a}_0\dot{a}_2 + a_2 g_2(a_0, W_e, P_{g_0}) =$$

$$= -\dot{q}_1^{ik} l_0/l_0^{ik} + g_3(a_0, a_1, W_e, P_{g_0}),$$

$$a_0\ddot{d}_2 + 3\dot{a}_0\dot{d}_2 = -3(l_0/l_0^{ik})^2 (\dot{a}_0 q_0^{ik} + a_0\dot{q}_0^{ik}),$$

where $(\dot{d}_2 = \dot{f}_2 - \dot{l}^{ik})$, and g_1, g_2 and g_3 are known functions of a_0, a_1, the Weber number W_e and the nondimensional initial gas pressure inside the bubble, P_{g_0}. This system of equations can be solved numerically by a multiple Runge-Kutta procedure. The pressure field around the bubble, $p(B_i, r, t)$, can then be determined. $l_0^{ik} = l_0$ and $q_n^{ik} = sq_n^i$ correspond to the particular cases of a solid wall ($s = 1$) or a free surface ($s = -1$). These equations have been numerically solved [1, 11]. The computations, valid for values of $\eta < 0.5$, are able to describe the initial phases of the microjet formation, the pressure field generated and the shape of a nearby free surface. The pressures generated in the presence of a solid wall are orders of magnitude higher than in the spherical case. The influence of the initial gas pressure and of its law of compression, are seen to be as important as the proximity of the wall.

Acknowledgements

My thanks go to the members of the groupe GPI at the ENSTA, Paris who have contributed to the investigations presented above, to the Direction Recherches Etudes et Techniques who supported a great part of these studies, to Hydronautics for its current support and to M.P. Tulin who suggested the collective bubble collapse study.

References

1. Bovis AK and Chahine GL (1981) Etude asymptotique de l'interaction d'une bulle oscillante avec une surface libre voisine. J de mécanique 20 (3): 537–556.
2. Chahine GL (1974) Etude asymptotique et expérimentale des oscillations et du collapse des bulles. Docteur Ingénieur thesis. Paris VI: ENSTA rep. 042.
3. Chahine GL (1977) Interaction between an oscillating bubble and a free surface. J Fluids Engr 99: 709–716.
4. Chahine GL (1979) Etude locale de phénomène de cavitation. Analyse des facteurs régissant la dynamique des interfaces. Docteur d'Etat es-Sciences thesis. University Paris VI: ENSTA rep. 116.
5. Chahine GL and Fruman DH (1979) Dilute polymer solution effects on bubble growth and collapse. Phys fluids 22 (7): 1406–1408.
6. Chahine GL, Cohen D, Ducasse P and Ligneul L (1979) Influence d'un revêtement élastique sur le collapse d'une bulle de cavitation au voisinage d'une paroi solide. Proc 4th Int meeting on water-column separation. Cagliari, Italy: ENEL relazione n. 382, Nov. 1980: 204–223.
7. Chahine GL and Morine AK (1979) Collapse d'une bulle de caviation entre deux

parois solides. Proc 4th Int meeting on water-column separation. Cagliari, Italy: ENEL relazione n. 382, Nov. 1980: 79–101.
8. Chahine GL, Courbière C and Garnaud P (1979) Correlation between noise and dynamics of cavitation bubbles. Proc 6th conf on fluid machinery. Budapest, Hungary, pp 200–210.
9. Chahine GL and Morine AK (1980) The influence of polymer additives on the collapse of a bubble between two solid walls. ASME Cavitation and polyphase flow forum. New Orleans, Louisiana, pp 7–9.
10. Chahine GL and Bovis AG (1980) Oscillation and collapse of a cavitation bubble in the vicinity of a two-liquid interface. In: Cavitation and inhomogeneities in underwater acoustics, pp 23–30. New York: Springer Verlag.
11. Chahine GL and Bovis AG (1981) Pressure field generated by nonspherical bubble collapse. Proc ASME symp Cavitation erosion in fluid systems. Boulder, Colorado, pp 27–41.
12. Ellis AT and Ting LY (1974) Non Newtonian effects on flow cavitation and on cavitation in a pressure field. NASA SP-304, Vol. 1: 403–421.
13. Gates EM and Acosta AJ (1978) Some effects of several free stream factors on cavitation inception on axisymmetric bodies. Proc 12th Symp Naval Hydrodynamics N.A.S. Washington, pp 86–112.
14. Gibson DG (1972) The pulsation time of spark induced vapor bubbles. J Basic Engr, March: 248–249.
15. Hammitt FG (1980) Cavitation and multiphase flow phenomena. New York: McGraw Hill Int Book Company.
16. Hoyt JW (1977) Cavitation in polymer solution and fiber suspensions. ASME Cavitation and polyphase flow forum. Fort-Collins, 9–10.
17. Ligneul P (1980) Etude expérimentale de l'influence d'un champ de vitesse cisaillé sur la déformation d'une bulle de cavitation. ENSTA rep. 134.
18. Morine AK and Breuil A (1981) Etude de l'érosion due a l'implosion d'une bulle de cavitation. ENSTA rep. 141.
19. Plesset MS and Prosperetti A (1977) Bubble dynamics and cavitation. Annual Review J Fluid Mech 9: 145–185.
20. Ross D (1976) Mechanics of underwater noise. New York: Pergamon Press Inc.
21. Yang WJ and Lawson ML (1974) Bubble pulsation and cavitation in viscoelastic liquids. J Appl Phys 22 (7): 1406–1407.

Viscoelastic effect on the behaviour of an air bubble rising axially in a tube

MADELEINE COUTANCEAU and MOHAMED HAJJAM

Laboratoire de Mécanique des Fluides de l'Université de Poitiers, France

Introduction

In the present work, we consider the kinematic and dynamic behaviour of a single air bubble which rises along the axis of a vertical circular tube filled with a quiescent viscoelastic liquid. The volume V of the bubble is sufficiently large ($0.5\ \mathrm{cm}^3 < V < 40\ \mathrm{cm}^3$) for an internal motion to develop so that the bubble behaves effectively as a fluid body and not as a solid one. On the other hand, the mean apparent viscosity of the suspending liquid is very high ($\mu_a > 100\ \mathrm{Po}$) so that the effects of inertia and surface tension remain negligible; the Reynolds number Re remains less than $3 \cdot 10^{-2}$ and the Etvos number Eo greater than $2 \cdot 10^2$. The respective effects of elasticity and of the shear-thinning viscosity, on the shape and on the speed of the bubble as well as on the hydrodynamic field that it generates, are evaluated.

Our previous work

In the above conditions and in the case of a Newtonian liquid (silicone oil), we have previously shown [2, 6] how the wall effects affect the shape of the bubble, its rising speed and the hydrodynamic field that it generates. In particular, when the equivalent diameter d_e of the bubble is increasing compared with the tube diameter d_t, it has been pointed out that the bubble loses progressively its spherical shape; however it remains always symmetrical about its equatorial plane: at the beginning it takes the configuration of a prolate ellipsoid and then when $\lambda \geqslant 1$ ($\lambda = d_e/d_t$) it tends to be cylindrically shaped and bounded approximately by two spherical caps at both ends.

In this last case, we have found that the rising speed, as well as the equatorial radius of the bubble (i.e. the radius of its equatorial cross-section) tend towards limits.

On the other hand, the existence of superficial velocity on the bubble surface has been truly confirmed (they are increased by the wall proximity) and also the symmetry about the equatorial plane of the flow pattern. So 'viewed' in a fixed frame, the bubble appears to be 'accompanied' on either side by a pattern of closed streamlines forming two 'upstream-downstream' symmetrical eddies; the maximum length (expressed in terms of the axial bubble radius) of this disturbed domain is found to be greatly reduced by the wall effect.

Applied Scientific Research 38: 199–207 (1982) 0003–6994/82/0383–0199 $01.35.

Present purpose

Our present purpose was to determine how the phenomena are modified when the suspending liquid is viscoelastic and shear-thinning.

Then using the same experimental technique, previously described in [2], which is based upon precise visualization of the flow as well as of the bubble boundary, and upon rising speed measurements, we tried to evaluate separately the respective effects of the shear-dependent viscosity and of the elasticity. To this end, three categories of liquids have been used: a highly elastic liquid with constant viscosity that we call 'purely viscoelastic' liquid (Separan AP 30 in water-glucose solution) which was initially proposed by Boger and Hang Nguyenen [1]*, a weakly elastic liquid with a highly shear-thinning viscosity (C.M.C. in water) and a liquid with both a high shear-thinning viscosity and a relatively important elasticity (Polyox WSR 301 in water).

The rheological behaviours of the experimented liquids have been tested in a shear-viscometric flow; we have shown that the shear dependent viscosity is, for all these liquids, well characterized by a power law in the domain of shear-rate relative to our experiments. So for the 5% C.M.C. solution, $K = 205$ (dynes/cm^2) s^n with $n = 0.49$; the 6% C.M.C. solution, $K = 471$ (dynes/cm^2) s^n with $n = 0.35$; the 3.5% Polyox solution, $K = 506$ (dynes/cm^2) s^n with $n = 0.31$; the 0.1%, 0.075%, 0.05% Separan solutions, $K = 300$; 285; 219 (dynes/cm^2) s^n with $n = 1$.

Notations

In view of analyzing the results concerned with the shape of the bubble, we have characterized this shape by its main geometrical parameters namely: the equatorial radius R_{eq} which represents the radius of the greater cross-section of the bubble, the total axial length L_{Tax}, the axial length of the upper part of the bubble L_{ax}, i.e. the length of the part of the bubble situated above the equatorial plane.

Each of these parameters have been normalized by the tube radius R_t, so we have introduced

$$\lambda_{eq} = R_{eq}/R_t, \qquad \lambda_{ax} = L_{ax}/R_t, \qquad \lambda_{Tax} = L_{Tax}/R_t.$$

When the bubble is very elongated, the equatorial plane has been localized halfway of the cylindrical part of the bubble.

In addition it is to be reminded that λ represents the equivalent bubble diameter d_e to the tube diameter ratio.

The bubble speed U_0 has also been normalized by using the Poiseuille

*Certain particular properties of the Separan water-glucose solutions had been pointed out recently by D. Sigli and A. Maalouf and reported in J. Non Newtonian Fluid Mech. 1981, 9: 191–198.

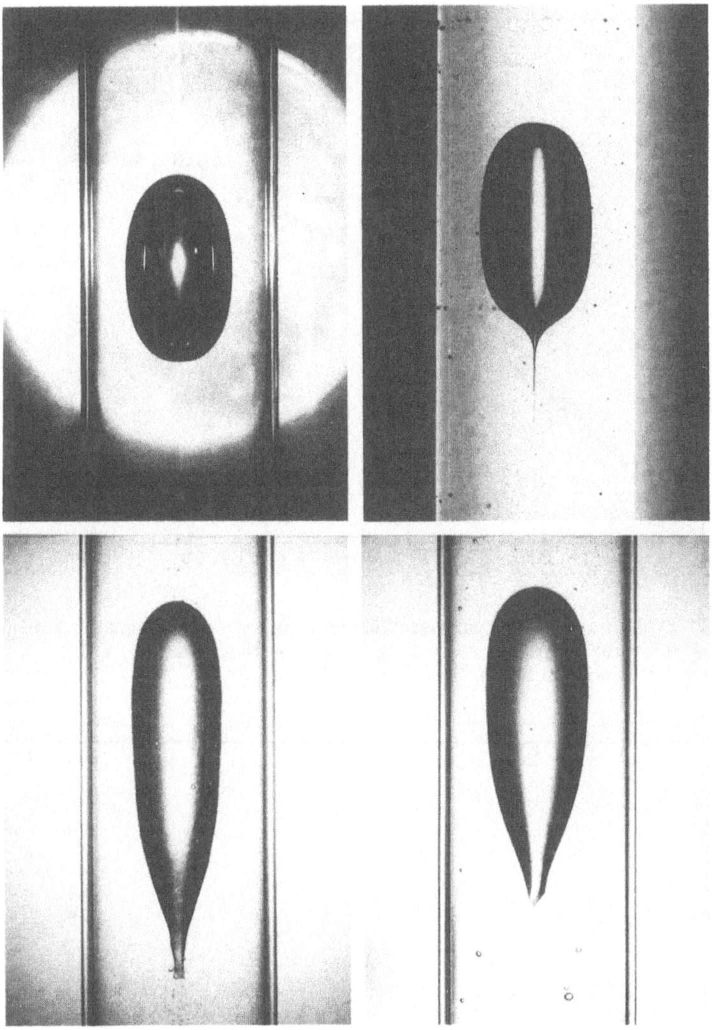

a b c d

Figure 1. The shape of a bubble rising axially in a tube filled with various liquids: (a) Newtonian oil and $\lambda = 0.79$; (b) 0.05% Separan solution and $\lambda = 0.80$; (c) 6% C.M.C. solution and $\lambda = 0.81$; (d) 3.5% Polyox solution and $\lambda = 0.82$.

number $Ps = U_0 \mu_r / \rho g d_e^2$, where μ_r is the reference viscosity defined for a shear rate which equals U_0 / d_e; the Poiseuille number characterizes the relative importance of the viscosity and gravity forces.

The main results

The main results deduced from this work can be summarized as follows.

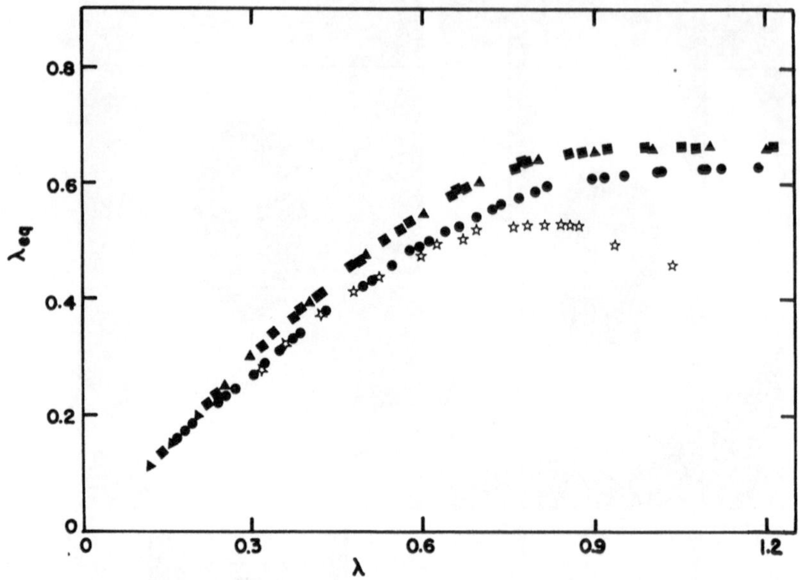

Figure 2. Evolution with the equivalent diameter ratio λ of the equatorial diameter ratio λ_{eq} for various sorts of suspending liquids: ▲ Newtonian oil, ▪ 0.1% Separan solution, ● 3.5% Polyox solution, ☆ 6% C.M.C. solution.

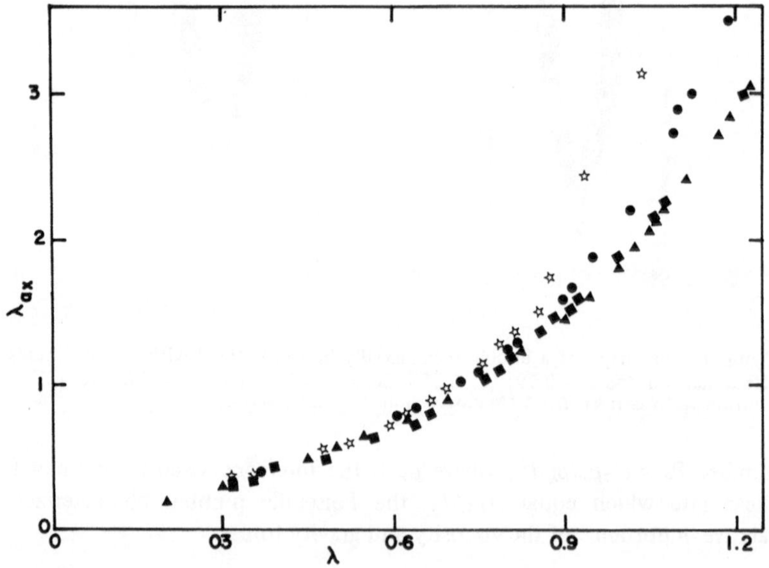

Figure 3. Evolution with the equivalent diameter ratio λ of the axial length of the upper part of the bubble to the tube radius λ_{ax} for various sorts of suspending liquids: ▪ 0.1% Separan solution, ▲ Newtonian oil, ● 3.5% Polyox solution, ☆ 6% C.M.C. solution.

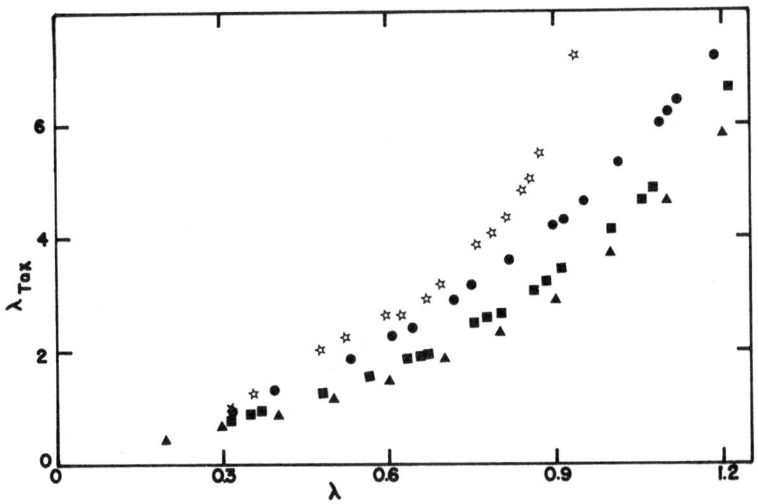

Figure 4. Evolution with the equivalent diameter ratio λ of the total axial length of the bubble to the tube radius ratio λ_{Tax} for various sorts of suspending liquids: ▲ Newtonian oil, ■ 0.1% Separan solution, ● 3.5% Polyox solution, ☆ 6% C.M.C. solution.

Concerning the shape of the bubble (Figures 1, 2, 3 and 4)

From numerous photographs of the shape of the bubble like those given as an example in Figure 1, it appears firstly that the elasticity does not modify sensibly the cross-section and the upper part of the bubble (see in Figures 2 and 3, the curves relative to 0.1% Separan solution and Newtonian oil).

Furthermore, in the purely viscoelastic liquid, the lower part of the bubble has been found to be symmetrical compared with the upper part, but is ended by an extremely thin tail (Figure 1b). Consequently, compared with the shape of the bubble rising in a Newtonian liquid, the effect of elasticity is practically limited to the addition of this needle-like tail.

Secondly, the effect of the shear-thinning viscosity is to thin down the bubble (Figures 1c and 1d), i.e. the cross-section is smaller (Figure 2), the upper part is somewhat elongated (Figure 3) and the lower part is clearly elongated (Figure 4) and progressively thinned down until it is reduced to a tail. Consequently the fore-aft dissymmetry of the bubble is accentuated compared with the case for which the suspending liquid is purely viscoelastic.

The results obtained in the Polyox solution show that the elasticity opposes the thinning of the bubble.

It has also been observed that in the C.M.C. solution the thinning of the bubble is abruptly accentuated when the wall effect becomes greater than a critical value (Figure 2).

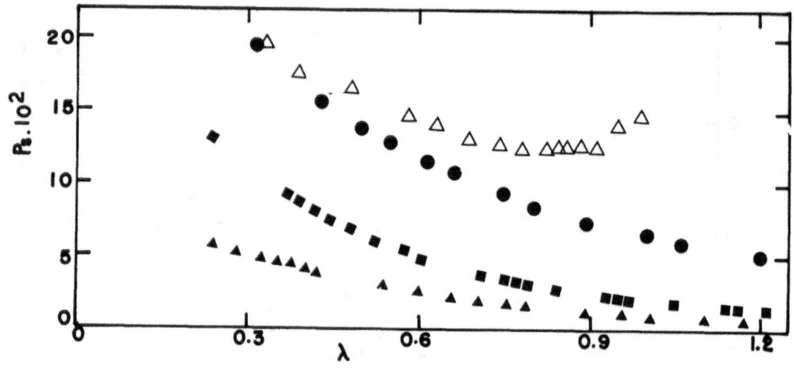

Figure 5. Evolution with the equivalent diameter ratio λ of the Poiseuille number Ps for various sorts of suspending liquids: ▲ Newtonian oil, ■ 0.1% Separan solution, ● 3.5% Polyox solution, ☆ 5% C.M.C. solution.

Concerning the speed of the bubble (Figure 5)

Elasticity effect. The elasticity increases the speed of the bubble. For example, compared with a bubble rising in a Newtonian liquid with the same viscosity and density and for the same value of λ, the speed of the bubble is increased by a factor of about 2 in the 0.1% Separan solution; of about 1.3 in the 0.075% Separan solution; of about 1.1 in the 0.05% Separan solution.

This result is remarkable because up to now, for the creeping motion, it appears that such an increase has never been pointed out.

Providing the wall effect is sufficient, this increase factor does not seem to be notably influenced by an increase of the wall effect. It depends only on the concentration of the solution.

As in the case of a Newtonian suspending liquid, the speed of the bubble rising in a purely viscoelastic liquid tends towards a limit when the equivalent radius ratio is increasing beyond a critical value ($\lambda \geqslant 1$).

Shear-thinning effect. For a given bubble in a given tube and compared with the speed of this bubble in a Newtonian liquid having the same density and the reference viscosity μ_r defined for $\gamma = U_0/d_e$, the effect of the shear-thinning viscosity is to increase the speed of the bubble and this effect is notably increased by the increase of the wall effect, the apparent viscosity being highly influenced by the corresponding increase of the shear-rate; in our experiments relative to the C.M.C. and Polyox solutions this factor was dominating.

In addition, it appears that in the shear-thinning liquids the speed of the bubble continues to increase when its equivalent radius is increasing compared with the tube radius: the limit observed in the Newtonian and purely viscoelastic liquids does not exist.

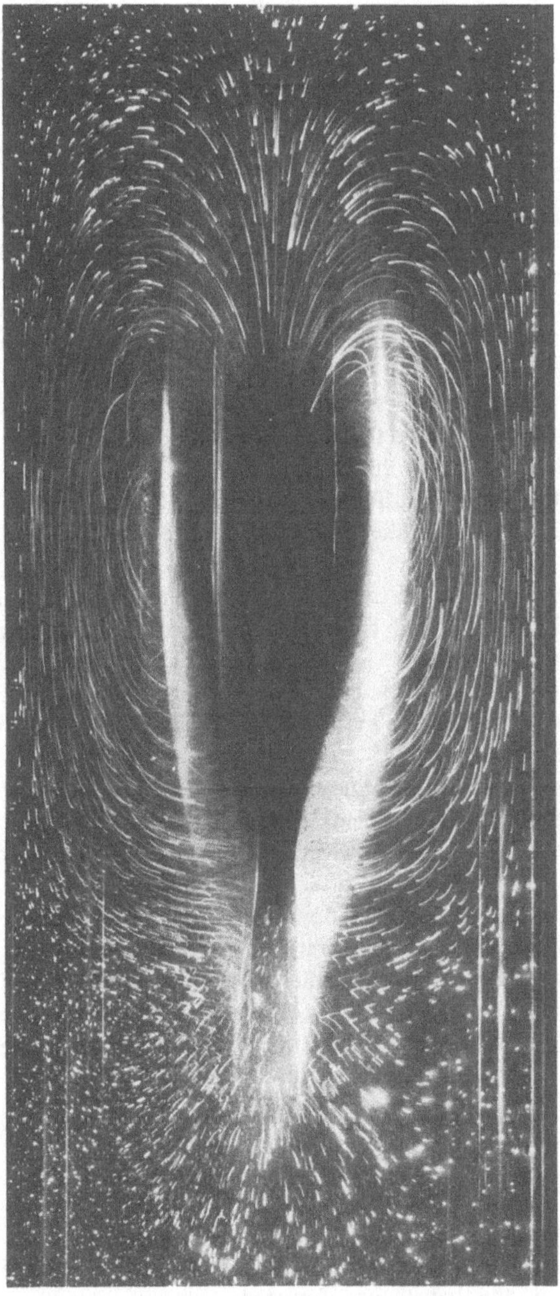

Figure 6. Visualization by a fixed camera of the flow pattern induced by a bubble rising axially in a tube filled with a 3.5% Polyox solution for $\lambda = 0.71$.

It has also been shown that the relative increase of the speed of the bubble with the wall effect due to the shear-thinning effect is slowed down by the elasticity effect.

Concerning the generated hydrodynamic field (Figure 6)

The visualization of the hydrodynamic field induced by the rising bubble 'viewed' by a fixed camera has shown that the upstream-downstream symmetry, observed in the Newtonian liquid, is destroyed and that, just behind the bubble and during the first part of the time of exposure, the visualization particles go up with the bubble as in the Newtonian case but, on the downstream axis and in the vicinity of this axis, during the second part of the time of exposure they more or less abruptly take an opposite direction and then they go down in a 'negative flow'. At a certain distance from the bubble, the particles participate only in this negative flow.

It is for the C.M.C. solution that the change in particle path direction is the least abrupt and for the purely viscoelastic liquid that it is the most abrupt.

The negative flow, which takes place from the rear axial part of the bubble itself, has also been found by means of velocity measurements (but not visualized) by Hassager et al. [3] and observed by Sigli and Coutanceau [5] and Hassager et al. [3] in the case of a rigid sphere moving in a viscoelastic liquid but then, the negative motion begins only at a certain distance downstream of the body.

It is to be remarked that the experiments of Hassager et al. [3] do not seem to correspond to a creeping flow and then the elasticity and inertia forces interact. We have shown that consequently the phenomenon is increased.

The existence of the negative flow explains the observed deformations of the bubble shape, this type of deformation depends upon the structure of this negative flow which is either parallel to the down axis (then it induces a needle-like tail as in the purely viscoelastic liquid), or fan-shaped (then the tail is conical as in the Polyox solution) or intermediate between the two above structures and the tail shape is also intermediate.

Conclusion

It may be thought that the significant results obtained from the present research can be useful from a rheological point of view because it provides information on the behaviour of various sorts of liquids under a non-viscometric flow for which there is a lack of information as was mentioned recently by Leal [4], as well as from a diphasic point of view because the knowledge of single bubble behaviour can serve as a basis for more complex and realistic situations such as bubbly flows and slug flows.

References

1. Boger DV and Hang Nguyen (1978) A model viscoelastic fluid. Polymer Eng and Science 18: 1037–1043.
2. Coutanceau M and Thizon P (1981) Wall effect on the bubble behaviour in highly viscous liquids. J Fluid Mech 107: 339–373.
3. Hassager O, Bisgaard C and Ostergaad K (1980) Measurements of velocity fields around objects moving in non-Newtonian liquids. Proceedings of VIIIth International Congress of Rheology, Naples.
4. Leal LG (1979) The motion of small particles in non-Newtonian fluids. J Non-Newtonian Fluid Mech 5: 33–78.
5. Sigli D and Coutanceau M (1977) Effect of finite boundaries on the slow laminar isothermal flow of a viscoelastic fluid around a spherical obstacle. J Non-Newtonian Fluid Mech 2: 1–21.
6. Thizon P and Coutanceau M (1977) Determination of wall effect on the shape and the kinematic and dynamic behaviour of a gas bubble rising in a viscous liquid. Euromech Colloquium 98.

Acoustic cavitation noise spectra

E. CRAMER* and W. LAUTERBORN

Drittes Physikalisches Institut, Universität Göttingen, Bürgerstrasse 42–44,
D-3400 Göttingen, Fed. Rep. of Germany

Abstract. Experiments are described which allow to measure and visualize the dynamics of the total acoustic cavitation noise spectrum. A distinct threshold for broad-band noise much higher than the threshold for the first subharmonic has been found. At this noise threshold a broadening of all spectral lines is observed. A simple theoretical model which consists in adding up the main parts of the sound emitted by differently sized bubbles is reported. Similar noise spectra as observed experimentally are obtained.

1. Introduction

Many attempts have been made up to now to explain the spectrum of acoustic cavitation noise, but a complete insight has not yet been found. The various lines in the noise spectrum are closely related to the frequency f_0 of the driving sound field. They consist of its harmonics nf_0 $(n = 2, 3, \ldots)$ and of its subharmonics f_0/m $(m = 2, 3, 4, \ldots)$ and their harmonics. The $f_0/2$ subharmonic was first observed by Esche [5], who also found the $f_0/3$ subharmonic and its harmonics, and Bohn [2] showed an example with the $f_0/4$ subharmonic together with its harmonics. The harmonics nf_0 $(n = 2, 3, \ldots)$ are explained in terms of forced nonlinear bubble oscillations, and the noise background is explained in terms of shock waves emitted by collapsing bubbles. First explanations of the subharmonics have been given by Güth [6], which were confirmed by Neppiras [9], Eller [4] and others. A bubble in a liquid is a nonlinear resonant system which is capable of oscillating at its own frequency f_B in a driving sound field of frequency f_0 with $f_B = f_0/m$ $(m = 2, 3, \ldots)$. But no bubbles of suitable size could be found even with high speed photographic and holographic methods. A numerical analysis by Lauterborn [8] showed that also bubbles driven at frequencies far below their resonant frequency may produce subharmonic lines in the spectrum of their oscillation. Other calculations with a bubble model containing sound radiation [3] lead to the same result.

The aim of the work presented here is to measure the time development of the total cavitation noise spectrum when the sound field intensity is raised linearly and to make a first step towards a numerical analysis of the sound radiated by a cavitation bubble field.

2. The experiment

The experimental set-up is shown in Figure 1. A piezoceramic cylinder with a resonance frequency of 23.56 kHz produces the cavitation bubble field. The

**Present address:* Physikalisch-Technische Bundesanstalt – IB, Berlin, Germany.

Applied Scientific Research 38: 209–214 (1982) 0003–6994/82/0383–0209 $00.90.

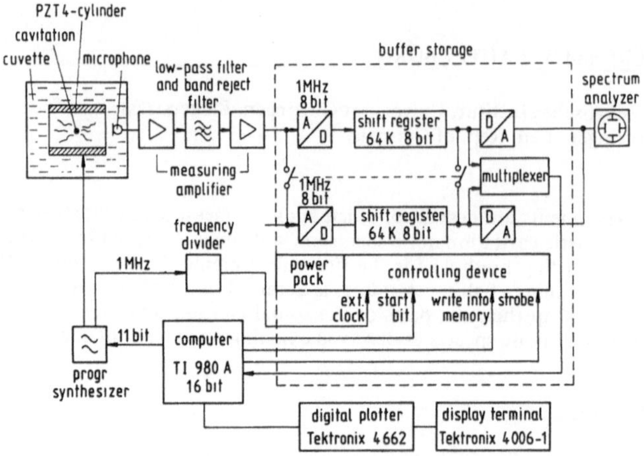

Figure 1. Experimental set-up for recording and processing acoustic cavitation noise. The generation of the cavitation bubble field and the recording of the cavitation noise are controlled by a minicomputer. The noise is digitized with high sampling rates and buffered in a 128 k words storage. From there the computer can read the data for processing.

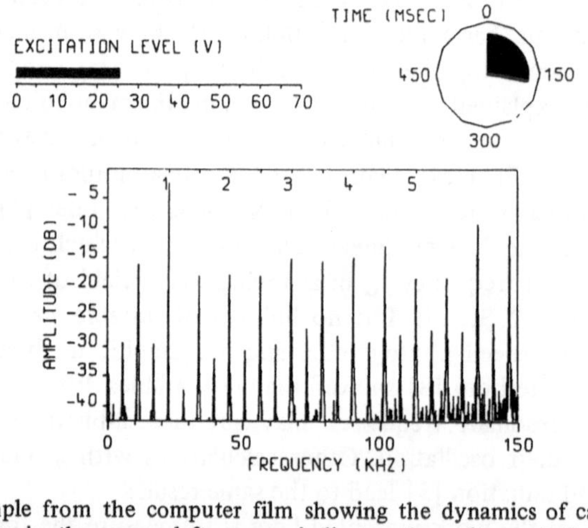

Figure 2. Example from the computer film showing the dynamics of cavitation noise spectra. The bar in the upper left corner indicates the voltage at the piezoceramic cylinder producing the cavitation bubble field. The clock in the upper right corner tells the time after the beginning of the experiment.

experiment is driven by a minicomputer, which modulates the amplitude of a programmable synthesizer and also starts a buffer storage of twice 64 k words length with two A/D converters. The sampling rate for the experiments reported here is 500 kHz but may be as high as 2 MHz. The sound emitted by the cavitation bubble field is picked up by a broad-band microphone. After suitable filtering the signal is digitized and stored in the buffer storage.

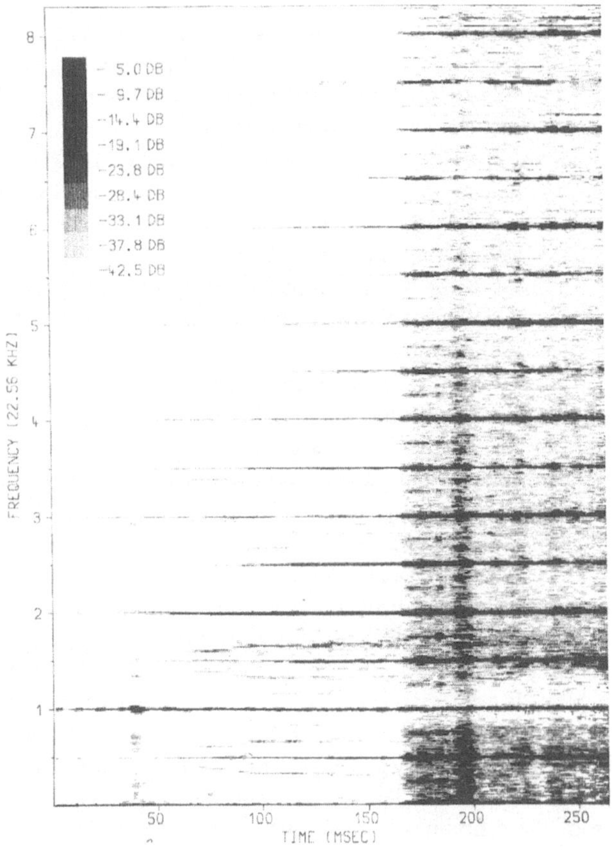

Figure 3. Example of 'visible cavitation noise' from an experiment with linearly increasing amplitude (0 V–60 V) at the piezoceramic cylinder producing the cavitation bubble field. The noise spectrum is plotted versus the time of 370 short time spectra. The amplitudes of the spectral lines are figured in a grey scale using a 3 × 3 dot matrix.

From these data many overlapping short time spectra of 4 k samples each are calculated using the FFT algorithm. To show the full dynamics of the noise spectra two methods are employed. The first is to produce a computer film of 1024 overlapping short time spectra. An example of such a spectrum is given in Figure 2. The film shows sudden bursts of the various spectral lines and of broad-band noise. Also fluctuating phenomena and oscillations of lines can be observed. The film also shows the change between stable oscillating bubble fields (sharp line spectra, little-noise amplitude) and unstable oscillating bubble fields (diffusion of single lines, large-noise amplitude).

The other method is to plot all short time spectra in a two dimensional picture, called 'visible cavitation noise', as shown in Figure 3. The frequency of the spectra is plotted versus time. The amplitudes of the spectral components are figured in a grey scale using a 3 x 3 dot matrix. The example in Figure 3 shows the outcome of an experiment with linearly increasing voltage at the piezoelectric cylinder (0 V–60 V). From one data set of 128 k samples 370 overlapping short time spectra are calculated.

With this method it is possible to present the dynamics of the whole noise spectrum in one picture. In the beginning of the measurement in Figure 3 we only have a line at the driving frequency f_0. After 40 ms some low-frequency noise appears for a short time with spurious $2f_0$, $3f_0$, $1/2f_0$ and $3/2f_0$ signals. Then the harmonics of the driving frequency become more pronounced and lines at $f_0/2$ and their harmonics and somewhat later lines at $f_0/3$ and some of its harmonics set constantly in.

After 170 ms the onset of broad-band noise is observed together with the onset of the $f_0/4$ subharmonic and its harmonics, whereas the $f_0/3$ subharmonic disappears together with its harmonics. At the same time a sudden broadening of all spectral lines occurs. The $nf_0/4$ lines are only stable until a short intense burst of noise appears at 195 ms and do not show up again later.

The change in the appearance of the noise spectrum at 170 ms may be interpreted as the change between stable oscillating bubble fields and unstable oscillating bubble fields. The broadening of the spectral lines may then be due to the erratic motion of bubbles thereby introducing Doppler shifts. Once a strong noise amplitude is set up also various frequency mixing effects may come into play. In any case there seems to be a sharp threshold between deterministic (or nearly so) and chaotic behaviour. As the chaotic behaviour is driven by a single frequency a strong periodic substructure remains, and the noise spectrum might be called as one resulting from 'periodic chaos'. The implication of these findings in relation to the new developments on chaotic behaviour of nonlinear dynamical systems has not yet been worked out but is pursued and recommended for further work.

3. Numerical analysis of the sound radiation from a bubble field

To compare the experimental results with theory is very difficult, because the various bubble models introduced up to now cannot handle bubble fields with thousands of bubbles with strong interaction between them. Therefore first investigations have been made of sound radiation from many single spherical bubbles with different sizes and no interaction. Bubbles in a certain range of radii at rest are taken uniformly distributed. The example in Figure 4 shows the noise spectrum of a bubble distribution with radii at rest from 10 μm to 50 μm. The bubbles are driven by a sound field with a frequency of 23.56 kHz and a sound pressure amplitude of 0.8 bar. The emitted sound of these bubbles is simply summed up in a certain distance, the same for all bubbles. The radiated sound amplitude is taken from calculations which also resulted in the resonant curves as given in [3], Figure 4b. To handle the great many data only the peaks of the pressure distribution around each bubble are

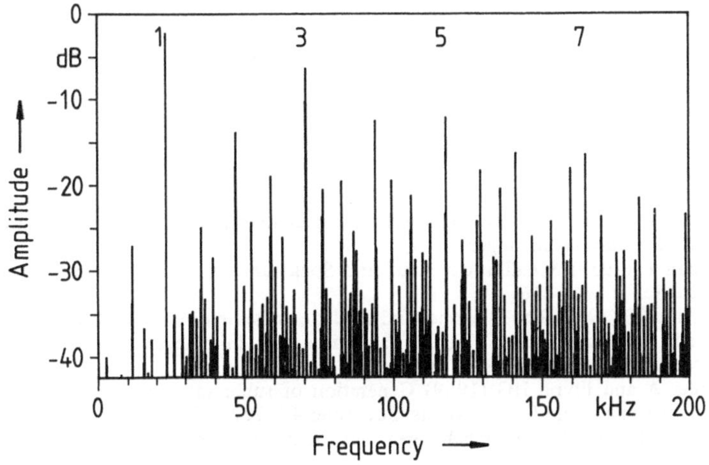

Figure 4. Numerically obtained spectrum of the cavitation noise from a bubble field. The radii at rest of the bubbles are assumed to be uniformly distributed between 10 μm and 50 μm. The frequency of the driving sound field is 23.56 kHz, and the pressure amplitude is 0.8 bar.

taken into account. Even with this very simple model similar spectra as those measured are obtained. Strong subharmonics are produced from a bubble distribution with radii at rest far below the resonance size.

4. Conclusion

With the experimental method described above it is possible to look at all frequencies of the cavitation noise spectrum simultaneously while the driving sound pressure amplitude is altered. Therefore this method allows to observe the total dynamics of the spectrum. This should open up new possibilities to answer the question of the cavitation noise generation mechanism in a bubble field. Another important but unanswered question is the bubble size distribution during cavitation. To answer this question holographic methods are developed since some time [1, 7].

Existing theory can be said to describe the behaviour of a single spherical bubble satisfactorily for many purposes. Based on this theory a very simple model describing the sound radiation from a bubble field is presented. But this may be considered only as a first step. The next step should take into account bubble interaction. Investigations along these lines are under way. A complete theory of acoustic cavitation noise spectra demands a self-consistent set of equations where the noise output also acts as input to drive the bubbles. We are far from achieving this goal.

Acknowledgment

This work has been sponsored by the Fraunhofer Gesellschaft, München, Germany.

References

1. Bader F (1973) Kurzzeitholografische Untersuchungen von Kavitationsblasenfeldern, PhD dissertation Göttingen.
2. Bohn L (1957) Schalldruckverlauf und Spektrum bei der Schwingungskavitation. Acustica 7: 201–216.
3. Cramer E (1980) The dynamics and acoustic emission of bubbles driven by a sound field. In: Lauterborn W (ed) Cavitation and inhomogeneities in underwater acoustics, pp 54–63. Berlin: Springer.
4. Eller A and Flynn HG (1969) Generation of subharmonics of order one-half by bubbles in a sound field. J Acoust Soc Amer 46: 722–727.
5. Esche R (1952) Untersuchungen der Schwingungskavitation in Flüssigkeiten. Acustica 2: AB 208–218.
6. Güth W (1956) Nichtlineare Schwingungen von Luftblasen in Wasser. Acustica 6: 532–536.
7. Haussmann G and Lauterborn W (1980) Determination of size and position of fast moving gas bubbles in liquids by digital 3-D image processing of hologram reconstructions. Appl Opt 19: 3529–3535.
8. Lauterborn W (1976) Numerical investigations of nonlinear oscillations of gas bubbles in liquids. J Acoust Soc Amer 59: 283–293.
9. Neppiras EA (1969) Subharmonic and other low frequency emission from bubbles in sound-irradiated liquids. J Acoust Soc Amer 46: 587–601.

The growth and collapse of bubbles near deformable surfaces

D.C. GIBSON* and J.R. BLAKE**

*CSIRO Division of Energy Technology, Highett, Victoria 3190, Australia;
**Department of Mathematics, University of Wollongong, Wollongong, NSW 2500, Australia

Abstract. Results from recent theoretical and experimental studies of the interaction of pulsating bubbles with nearby deformable surfaces are presented. The bubble impulse is defined and shown to be an important indicator of the nature of collapse. Experiments have revealed an entirely new form of collapse in the vicinity of finite impedance surfaces and useful parametric descriptions of surface inertia and stiffness have been found.

1. Introduction

Pulsating cavitation and underwater explosion bubbles migrate towards a rigid surface and away from a free surface provided buoyancy forces are small [6]. The forces which promote significant bubble migration during the collapse are also responsible for bubble distortion, which is invariably of the same form. The bubble becomes involuted from the rear, and a high speed liquid jet threads the bubble in its direction of migration. This jet is now considered to be a prime mechanism for cavitation damage to solid bodies immersed in a cavitating liquid [2, 7, 10].

Some years ago Gibson [7] speculated that there may be a useful deformable coating, with characteristics between those of a free surface and a rigid surface in its response to a pulsating bubble. Such a coating, which might be a sponge-like material, with gas bubbles trapped in its pores, could be used to cover rigid structures and thereby repel pulsating bubbles and prevent damage to the structure. Recently we have collaborated in a theoretical and experimental study aimed at defining the types of surface that repel cavitation bubbles [4, 8]. In this paper we shall report some of our most recent calculations and observations.

2. Theoretical developments

Theoretical developments include a study of conservative properties (i.e. mass, momentum, angular momentum and energy) and simulations of bubble growth and collapse in the neighbourhood of different types of boundary. One of the most valuable quantities amenable to calculation is the bubble impulse $\mathbf{I}$ defined by

$$\mathbf{I} = \int_S \phi \, \mathbf{n} \, \mathrm{d}S \tag{1}$$

Applied Scientific Research 38: 215–224 (1982) 0003–6994/82/0383–0215 $01.50.

where S is the bubble surface and $\mathbf{n}$ is the outward normal to the liquid (i.e. into the bubble). This was shown [5] to be equal to

$$\mathbf{I} = \int_0^t \mathbf{F}_e(t)\, \mathrm{d}t \tag{2}$$

with $\mathbf{F}_e(t)$ defined by

$$\mathbf{F}_e(t) = \rho \int_{\Sigma_b} \left\{ \frac{1}{2} |\nabla\phi|^2 \mathbf{n} - \frac{\partial \phi}{\partial n} \nabla\phi \right\} \cdot \mathrm{d}S, \tag{3}$$

where ρ is the liquid density, ϕ the velocity potential and Σ_b the deformable boundary surface. A picture of the theoretical problem is shown in Figure 1.

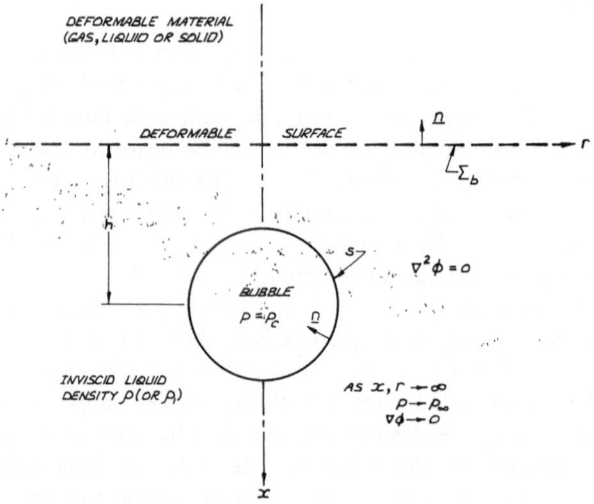

Figure 1. Coordinates and parameters used in numerical models and impulse calculations.

The surface integral expression in (3) has been used [5] to calculate the impulse due to a point source of strength $q(t)$ in the presence of different types of (linear) boundaries. With the source initially a distance h away from a plane boundary, the expression for the impulse in the x-direction (see Figure 1) is

$$I_x(t) = \frac{\rho H}{4\pi h^2} \int_0^t q^2(t')\, \mathrm{d}t', \tag{4}$$

where the parameter H has the following values for the boundaries studied:

(i) Rigid boundary $H = -\frac{1}{4}$
(ii) Free surface $H = \frac{1}{4}$

(iii) Interface between liquids of different density (with the source in liquid of density ρ_1) $\quad H = \frac{1}{4}\frac{(\rho_1 - \rho_2)}{(\rho_1 + \rho_2)}$

(iv) Inertial boundary $\quad H = H(\alpha)$

The inertial boundary has no stiffness, but it has inertia due to its mass, described by the surface density σ. The parameter α is defined as ($\alpha = \rho h/\sigma$).

The function $H(\alpha)$ may be evaluated from the expression

$$H(\alpha) = \alpha - \tfrac{1}{4} - 2\alpha^2 \, e^{2\alpha} E_1(2\alpha), \tag{5}$$

where E_1 is an exponential integral (see e.g. [1]). $H(\alpha)$ has as its two bounds the rigid boundary limit as α tends to zero and the free surface limit as α tends to infinity. It has one zero at $\alpha = \alpha_0 = 0.7798057$.

These results for a simple source may have profound implications in our bubble studies because it appears from our experiments that the liquid jet goes in the direction of the impulse, at least for the rigid and free surface limits [4, 8]. This behaviour was inferred by Benjamin and Ellis [2], although they did not calculate the impulse.

Several numerical models have been developed to simulate bubble-boundary interaction, with the accent on fast approximate techniques that can be used to analyse a wide boundary parameter space at reasonable cost. The approaches used so far include (i) a line distribution of sources and dipoles, with appropriate images along the axis of symmetry [3, 8], (ii) several ring distributions of sources inside the bubble and above the boundary surface [4], and (iii) a multipole located midway between the two axial extremities of the bubble together with image singularities. Other approaches being examined involve the use of spherical bipolar and tangent plane co-ordinates [12, 13] and the use of a surface integral technique similar to that of Hess and Smith [9].

Our simulations vary from previous studies [e.g. 3, 10] in several important respects. The growth as well as the collapse phase of the bubble motion has been considered in the model. This is essential when energy may be transmitted to and regained from an adjacent surface. The time step is obtained by specifying a maximum increment in the dimensionless potential $\Delta\phi$ from one time step to the next, instead of using the usual displacement restrictions. Using this and the dimensionless velocities at the present step, Δt is then calculated from the relationship

$$\Delta t = \frac{\Delta\phi}{\mathrm{Max}_S \, [1 + \frac{1}{2}|\mathbf{u}|^2]} \, . \tag{6}$$

Thus Δt depends inversely on the square of the maximum velocity.

3. An empirical approach

The numerical models predict all the salient features of free surface and rigid boundary interactions [4, 8]. However, they have done little to reduce the parameter space we must search to define our protective coating. The evaluation of $\alpha = \alpha_0$ for an inertial boundary [5] is the one theoretical guide that suggests a change from rigid boundary response to free surface response.

Further insight has been gained from the realization that, if a bounding surface responds in exactly the same way as the infinite liquid it replaces, a nearby bubble will grow and collapse spherically and will neither move towards nor away from the surface. Clearly, this is another guide to the change-over from a rigid boundary to a free surface response. Starting from this point, a plane of particles was marked in an infinite liquid, a short distance from the point of creation of a vapour bubble. The bubble was then allowed to grow to its maximum size and collapse again, Figure 2(a). Using Rayleigh's original analysis [11], the normal force on and the normal velocity of the surface of marked particles were calculated throughout the bubble pulsation. Concentrating on an initially plane surface $0.5 R_m$ away from the point of inception of the bubble, and averaging over a circular area of radius R_m, the total force $F(t)$ acting on the marked particles from the bubble side of the liquid, and the average normal displacement $\bar{\eta}$ of the particles subtended by the circular area of radius R_m were calculated. These quantities are shown as a function of time in Figure 3 for the whole life of a Rayleigh bubble.

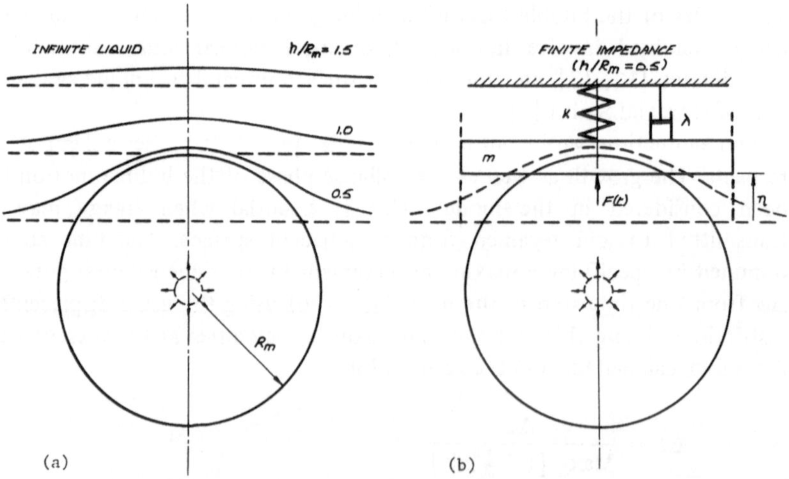

Figure 2. (a) Displacement of planes of particles in an infinite liquid during growth of a nearby spherical bubble; (b) Lumped parameter description of a deformable surface that replaces the infinite liquid beyond $h/R_m = 0.5$.

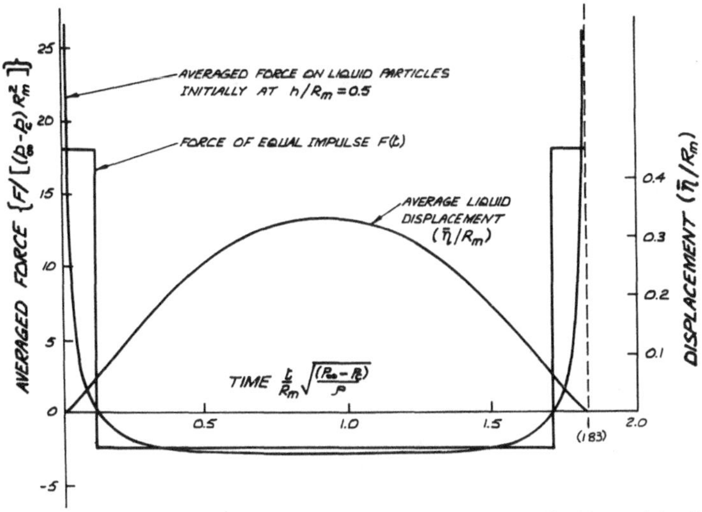

Figure 3. Average force on and displacement of the infinite liquid particles initially laying in a plane a distance $h/R_m = 0.5$ from a pulsating Rayleigh bubble.

If it is now presumed that the averaging circle is the face of a piston of mass m, held in position by a spring with stiffness k and a dashpot with damping λ, as shown in Figure 2(b), this simple system can be subjected to the force $F(t)$ to see whether its displacement $\eta(t)$ is similar to the average displacement of the infinite liquid it has replaced. The governing equation is

$$m\ddot{\eta} + \lambda\dot{\eta} + k\eta = F(t) \tag{7}$$

and it remains to find the region in the (m, λ, k) space where $\eta(t)$ is similar in shape to $\bar{\eta}$ in Figure 3.

The continuous force $F(t)$, which is infinite at the moment of inception and annihilation of a Rayleigh bubble, was replaced with a piecewise constant force of equal impulse (see Figure 3) and then equation (6) was evaluated analytically.

When equation (6) is cast in dimensionless form, the relevant boundary parameters are

$$\begin{aligned} m^* &= m/(\rho R_m^3) = \pi\sigma/(\rho R_m), \\ \lambda^* &= \lambda/(R_m^2\sqrt{\rho(p_\infty - p_c)}), \\ k^* &= k/((p_\infty - p_c)R_m), \end{aligned} \tag{8}$$

where p_∞ is the pressure in the liquid before the bubble is grown, p_c is the saturation vapour pressure of the liquid and $2R_m$ is the maximum horizontal width of the bubble.

In this work we have only explored the plane $\lambda^* = 0$, i.e. boundaries with no damping, and have defined a domain (m^*, k^*) worthy of closer investigation, which is shown in Figure 4.

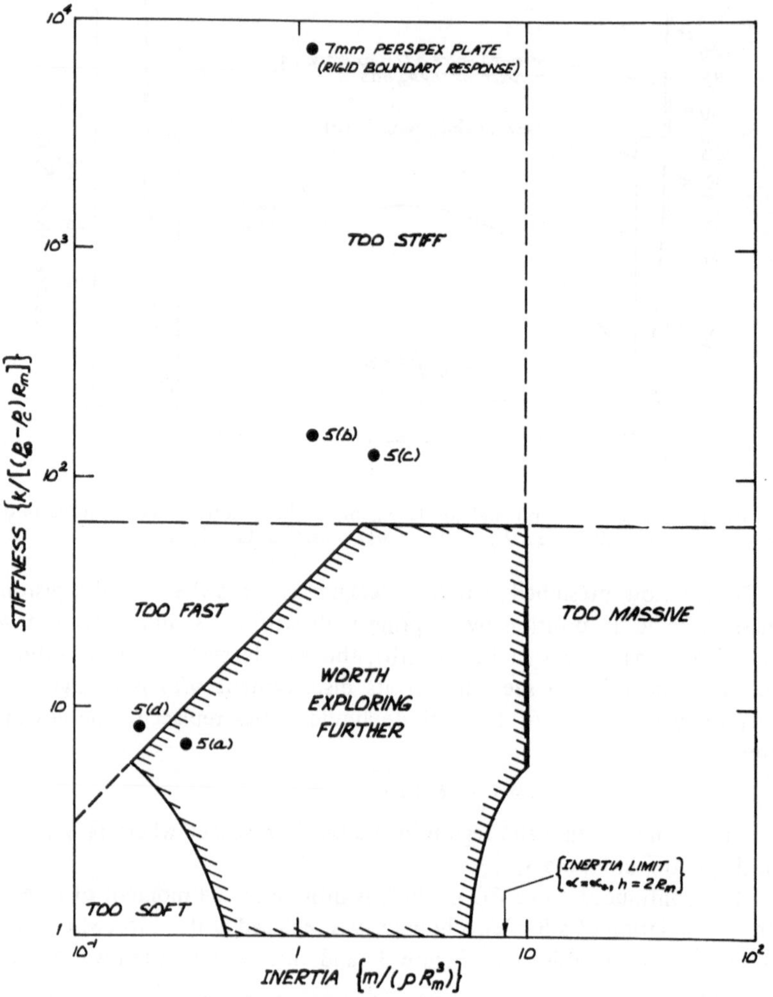

Figure 4. The necessary inertia and stiffness of a deformable surface for prevention of jet impact during collapse of nearby cavitation bubbles.

4. Experiments

The experimental apparatus is described in [4]. The central part is a 370 mm deep, 260 mm internal diameter, sealed perspex tank, filled with distilled water to a height of approximately 240 mm. A high voltage spark probe, that can be traversed along the centre line, extends into the tank from the base. The surface under test is mounted in a frame that can be lowered down onto the water from above.

The tank is mounted on a horizontal platform suspended between verticle guide rails by an electromagnet. A HYCAM cine-camera, fitted with a half-frame

16 mm prism, and a PAL 2.4 kW continuous light source are also mounted on the platform. The vapour bubble is generated by an electric spark discharged from a 0.25 microfarad capacitor at approximately 9 kV. Camera speeds of 11–19 frames per millisecond have been used.

In order to study relatively large, slowly pulsating bubbles, without them being influenced by buoyancy, the experiment is performed with the tank in free fall. Each experiment follows a similar pattern. The spark probe is set at a prescribed distance h beneath the surface under test, the tank pressure p_∞ is fixed and the camera set in motion. When the camera reaches the desired framing rate, it triggers the spark and releases the electromagnet. The platform then falls from a state of rest, while the bubble grows and collapses, and the camera records its motion. After development the film negative is analysed on a standard microfiche reader with a 24:1 magnification.

Experiments have involved a systematic examination of the growth and collapse of bubbles close to stretched natural and vulcanised rubber sheets, with thicknesses from 1.3 to 7.0 mm, and surface densities σ from 1.9 to 10.8 kg/m^2. Each sheet was stretched over a 150 mm diameter frame, with radial strains ϵ_r varying from 0 to 0.14. The effective spring constant k was obtained by driving pistons of different diameters into the stretched sheets, and measuring the reactive load versus displacement. Highlights of the experimental results are shown in Figure 5.

Two phenomena deserve special mention. For low values of (m^*, k^*), typically (0.3, 6.6) for the surface in Figure 5(a), the bubble is completely repelled during the collapse and the jet that forms is then directed along the axis of symmetry away from the loosely tensioned, lightweight 1.3 mm thick natural rubber sheet. Increasing (m^*, k^*) to (1.2, 158), by inserting a taut 4 mm thick vulcanised rubber sheet into the frame, reveals the entirely new phenomenon shown in Figure 5(b). The expansion is much the same as before. However, during the collapse, no jet forms in either direction. Instead, the bubble collapses from the sides forming an hour glass shape, gradually becoming more and more elongated along its axis of symmetry, and eventually collapsing into separate parts, one which sticks to the boundary and the other which moves away. This necking phenomenon is still evident for a much heavier and tighter, 7 mm thick vulcanised rubber sheet, with (m^*, k^*) now (2.2, 125) Fgiure 5 (c).

Rigid boundary response has only been achieved with a 7 mm thick Perspex (Lucite) plate mounted on the frame (see Figure 5 of [8]). This sheet has (m^*, k^*) of typically (1.1, 7400). Clearly there is a large range of boundary stiffness k^* to work with, where no liquid jet strikes the boundary in the collapse phase. We note in passing that the empirical model gives a similar upper limit on σ, or inertia, to that corresponding to the zero impulse value of $\alpha = \alpha_0$, when the source $q(t)$ of Blake and Cerone [5] is placed a distance $h = 2R_m$ from the boundary (see Figure 4).

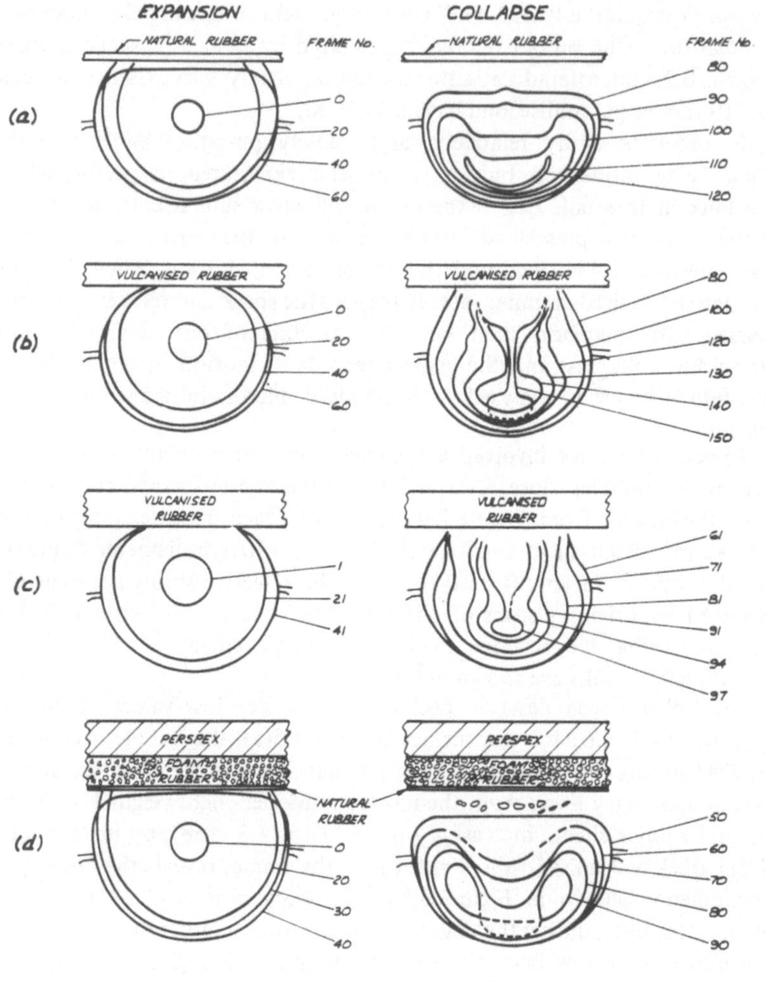

Figure 5. Examples of the interaction of pulsating bubbles with various deformable surfaces described in Table 1. All bubbles generated at frame zero. Camera speeds (frames/ms) were: (a) 11.7, (b) 13.5, (c) 12.9, (d) 12.3.

Very recent experiments have involved tests with 7 and 14 mm thick, open cell, foam rubber coatings glued between the 7 mm thick Perspex backing plate and a 0.8 mm thick, untensioned, natural rubber covering sheet. These composite coatings have low (m^*, k^*) for small surface displacements. Tests have been conducted with collapse pressure $(p_\infty - p_c)$ varying from 5000 to 50,000 Nm^{-2} and maximum bubble diameters varying from 50 to 20 mm. In every case the bubble is repelled during the collapse, and the jet that forms is directed away from the coated Perspex plate. A particular example for the 8 mm thick composite foam coating is given in Figure 5 (d). This composite has (m^*, k^*) values of (0.2, 8.0), which are similar to the values for the stretched 1.3 mm thick natural rubber sheet (Figure 5a).

Table 1 shows the critical stiffness and inertia parameters for the range of deformable surface studied to date. It also identifies the location of published bubble/surface interaction pictures for each surface.

Table 1. Dimensionless surface stiffness and inertia properties and bubble/surface interaction records presented here or elsewhere.

Deformable surface type	Boundary distance (h/R_m)	Radial strain ϵ_r	Boundary inertia m^*	Boundary stiffness k^*	Record of bubble/boundary interaction
7 mm thick Perspex plate	0.94	0	1.13	7386	Figure 5 (b) of [8]
4 mm thick vulcanised rubber	0.64	0.133	1.20	158	Figure 5 (b)
7 mm thick vulcanised rubber	0.64	0.143	2.18	125	Figure 5(c)
1.3 mm thick natural rubber	0.58	0.05	0.31	6.62	Figure 5 (a)
8 mm thick composite (natural/foam) rubber coating	0.56	0	~ 0.19	8.0	Figure 5 (d)
Free surface	0.56	0	0	0	Plate 1 of [4]

5. Discussion and conclusions

Approximate numerical methods have been found to accurately model the growth and collapse of bubbles close to a rigid boundary and a free surface. Simple models, imposing a linearised boundary condition have been used to examine the liquid impulse caused by bubble motions close to various types of interface. The impulse is directed towards a rigid boundary, away from a free surface, away from a lighter liquid and away from an inertial boundary provided the parameter $\alpha = \rho h/\sigma$ is greater than $\alpha_0 = 0.78$. This last result may have profound implications, for we know that rigid boundaries attract and free surfaces repel pulsating bubbles, and the jet formed during the collapse is in the same direction as the impulse.

Since, in practice, we may model a finite bubble by a distribution of sources over the bubble surface, circumstances will arise where some of the sources will be in regions close to an inertial boundary where $\alpha < \alpha_0$, and they will be attracted to it; whereas, others will be in regions further away, where $\alpha > \alpha_0$, and they will be repelled by it. These ideas may help to explain the sticking, elongation and bubble splitting phenomena illustrated in Figure 5 for increasing values of m^*.

One of the most satisfying results of the experimental studies to date, is the demonstration that the boundary 'inertia' and 'stiffness' (m^*, k^*) may be achieved in a variety of ways, each leading to the same form of interaction with nearby pulsating bubbles. It is clear that there is quite a large parameter space for our surface coatings, where jets can be prevented from striking the coating during the collapse of nearby pulsating bubbles. We therefore have no hesitation in repeating the words that close [7]. This is: '. . . an interesting, and perhaps rewarding, topic for further investigation'.

Acknowledgements

We wish to acknowledge the contributions of Dr P. Cerone and Dr T.S. Horner to our theoretical work; and Lieutenant M. Ghani to our experimental work. We are particularly grateful to our photographer Mr N.B. Hamilton for recording the fine details of the bubble motions described here.

References

1. Abramowitz M and Stegun IA (eds) (1965) Handbook of mathematical functions. New York: Dover.
2. Benjamin TB and Ellis AT (1966) The Collapse of cavitation bubbles and the pressures thereby produced against solid boundaries. Trans R Soc Ser A 260: 221–240.
3. Bevir MK and Fielding PJ (1975) Numerical solution of incompressible bubble collapse with jetting. In: Ockendon JR and Hodgkins WR (eds) Moving boundary problems in heat flow and diffusion. Oxford: Clarendon.
4. Blake JR and Gibson DC (1981) Growth and collapse of a vapour cavity near a free surface. J Fluid Mech 111: 123–140.
5. Blake JR and Cerone P (1981) A note on the impulse due to a vapour bubble near a boundary. J Aust Math Soc (in press).
6. Cole RH (1965) Underwater explosions. New York: Dover.
7. Gibson DC (1968) Cavitation adjacent to plane boundaries. Proc 3rd Aust Hydraulics and Fluid Mech Conference, Sydney, pp 210–214.
8. Gibson DC and Blake JR (1980) Growth and collapse of vapour bubbles near flexible boundaries. Proc 7th Aust Hydraulics and Fluid Mech Conference, Brisbane pp 283–286.
9. Hess JL and Smith AMO (1962) Calculation of potential flow about arbitrary bodies. In: Kuchemann D (ed) Progress in aeronautical sciences Vol 8. London: Pergamon.
10. Plesset MS and Chapman RB (1971) Collapse of an intially spherical vapour cavity in the neighbourhood of a solid boundary. J Fluid Mech 47 Pt 2: 283.
11. Rayleigh Lord (1917) On the pressure developed in a liquid during the collapse of a spherical void. Philos Mag 34: 94.
12. Small RD and Weihs D (1975) Axisymmetric potential flow over two spheres in contact. J Appl Mech 42: 763–765.
13. Weihs D and Small RD (1975) An exact solution of the motion of two adjacent spheres in axisymmetric potential flow. Israel J Technol 13: 1–6.

Acoustic emission of single laser-produced cavitation bubbles and their dynamics

W. HENTSCHEL and W. LAUTERBORN

Drittes Physikalisches Institut, Universität Göttingen, Bürgerstrasse 42–44,
D-3400 Göttingen, Fed. Rep. of Germany

Abstract. Cavitation bubble dynamics is investigated by the method of 'optic cavitation', i.e. the formation of single cavities in liquids by light. From the sound waves radiated upon collapse the pressure-time curve is obtained. Maximum bubble size and shock wave amplitudes are evaluated and an energy balance is considered. Numerical calculations with a modified Gilmore model taking into account the mass loss of the cavity can explain the rapid damping of the bubble oscillation observed in the experiments.

1. Introduction

One of the main problems in cavitation physics is bubble dynamics and the acoustic emission closely connected with it. Liquid jets together with the shock waves radiated upon collapse appear to play the most essential roles in the damage to solids in cavitating liquids. The clarification of this mechanism is one of the main tasks of our work.

The single bubble experiments are typical model experiments to reduce the complexity encountered in bubble fields. Q-switched ruby laser pulses are used to achieve breakdown and cavity formation in the liquid [8, 10]. The main advantage of optic cavitation is that bubbles can be made to appear at a given instant of time at a given location in the bulk of the liquid. The bubble motion has been investigated by high-speed photography with framing rates up to one million frames per second [8, 12] and by high-speed holo-cinematography with framing rates up to 40 000 holograms per second [5, 9, 11].

In this paper are described our attempts to establish the relations between the dynamics of single bubbles and the sound and shock waves radiated by them using the pressure-time curves picked up with a microphone [6, 7].

2. Experiments

The cavities necessary for the investigations are produced with the aid of intense laser light. The experimental set-up is given in Figure 1. A passively Q-switched ruby laser emits giant pulses of up to a few joules total energy at a pulse duration of about 30 to 50 ns. The light is focused by means of a single lens of short focal length into a cuvette (50 mm × 50 mm × 50 mm) filled with distilled water. The bubbles then appear in the vicinity of the focal point of the lens. A broadband microphone placed at a distance of

Applied Scientific Research 38: 225–230 (1982) 0003–6994/82/0383–0225 $00.90.

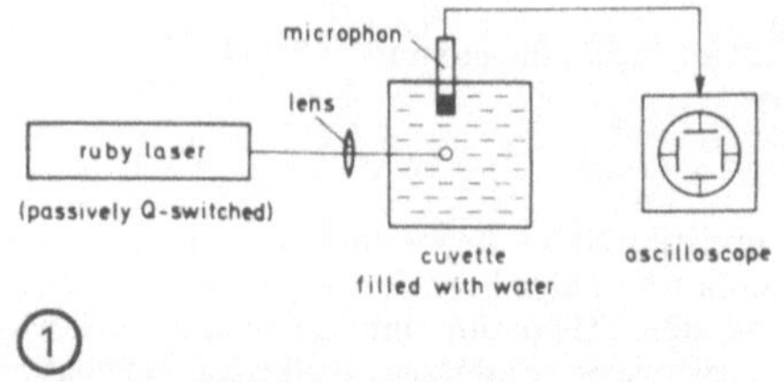

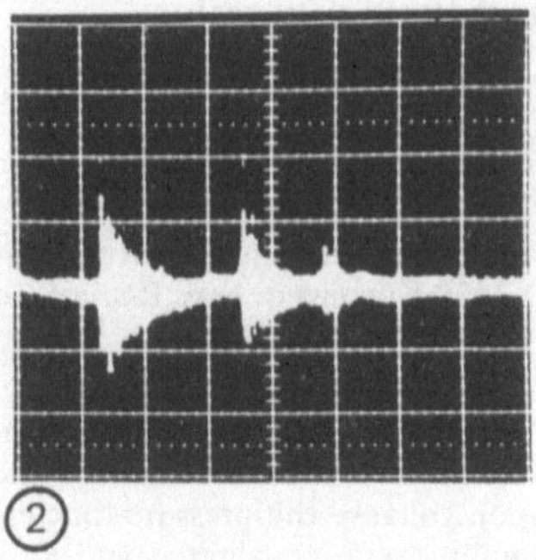

Figure 1. Set-up for cavitation noise recording of single laser-produced cavitation bubbles in water.
Figure 2. Pressure-time curve of a single laser-produced cavity in water.

11 to 21 mm from the point of breakdown picks up the sound emitted by the cavity. The pressure-time history is displayed on an oscilloscope and stored by photographing the display.

A typical example of pressure-time curves obtained with this arrangement is shown in Figure 2. The main features of the pressure-time curves are that a strong shock wave is radiated upon breakdown and, with decreasing strength, at each of the successive collapses of the cavity initially created. These shock waves are reflected at the cuvette walls and at the free liquid surface and are therefore followed by additional pulses of decreasing strength. A great number of frames are evaluated to get the collapse times (one half of the time between two pressure pulses).

3. Experimental results

The collapse time T_c is connected with the maximum bubble radius R_{max} before a collapse by the Rayleigh relation

$$T_c = 0.915\, R_{max} \sqrt{\frac{\rho}{P_\infty - P_v}}$$

with ρ the density of the liquid, P_∞ the external pressure and P_v the vapor pressure of the liquid. These data give a linear relationship between the maximum bubble size before and after collapse for cavities of 1 to 6 mm in radius. Figure 3 shows that the maximum radius after the collapse has about half the value of the maximum cavitation bubble radius before the collapse.

With a shock wave width of 30 ns according to Brinkmeyer [1] and using the technical data of the microphone the strength of the shock wave is calculated and plotted versus the maximum radius R_{max} before the collapse (Figure 4). To get comparable conditions all values for the pressure amplitude are normalized to a microphone distance of 10 mm from the centre of breakdown. A quadratic relationship is obtained for cavities of 1 to 4 mm in radius.

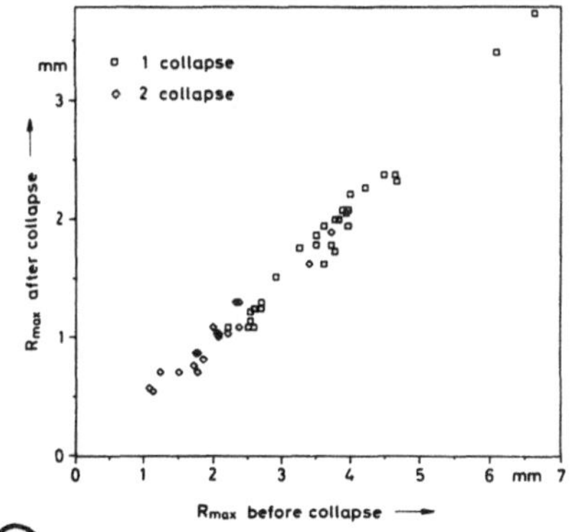

3

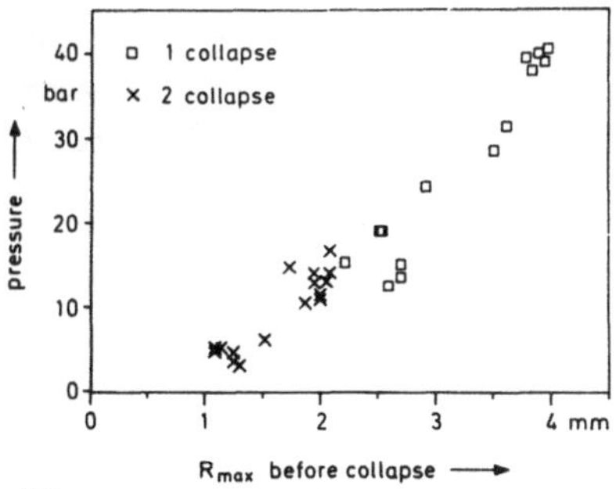

4

Figure 3. Maximum radii of single laser-produced cavitation bubbles before and after collapse in water. The radius values are calculated from the collapse times via the Rayleigh relation.

Figure 4. Experimental values of pressure amplitudes of collapsing laser-produced cavitation bubbles depending on the size of the bubble before collapse, normalized to a microphone distance of 10 mm from the breakdown centre.

These experimental values for the maximum radii before and after collapse together with the pressure amplitudes radiated upon collapse form the basis for calculating some aspects of energy balance in cavitation bubble dynamics. The total energy E of a cavity is given as the potential energy at the time of maximum radius $R_{\max}$ by

$$E = \tfrac{4}{3}\pi P_{\infty} R_{\max}^3 .$$

Therefore the energy dissipated during one cycle of the oscillation is given by the potential energy difference between two successive maxima. The energy of the cavities before and after collapse is plotted versus the maximum radius before the collapse in Figure 5. Only about 12% of the initial energy is left in the bubble after collapse.

The shock wave energy E_s radiated upon collapse is calculated according to Cole [2] by

$$E_s = \frac{4\pi R_m^2}{\rho c} \int P^2 \, dt$$

with c the velocity of sound and R_m the distance of the collapsing bubble centre to the point where the pressure amplitude P is measured. A rectangular shock wave profile with a width of 30 ns is assumed. In Figure 5 the shock wave energy versus the maximum bubble radius before the collapse is also given. As can be seen the amount of energy radiated by sound waves is on the average only 1.2% in our experiments. Additionally, the sum of shock wave energy and bubble energy after collapse is marked by a horizontal bar on top of the open rectangles in Figure 5. By comparing the loss of total energy with the acoustically radiated energy it is seen that sound radiation only accounts

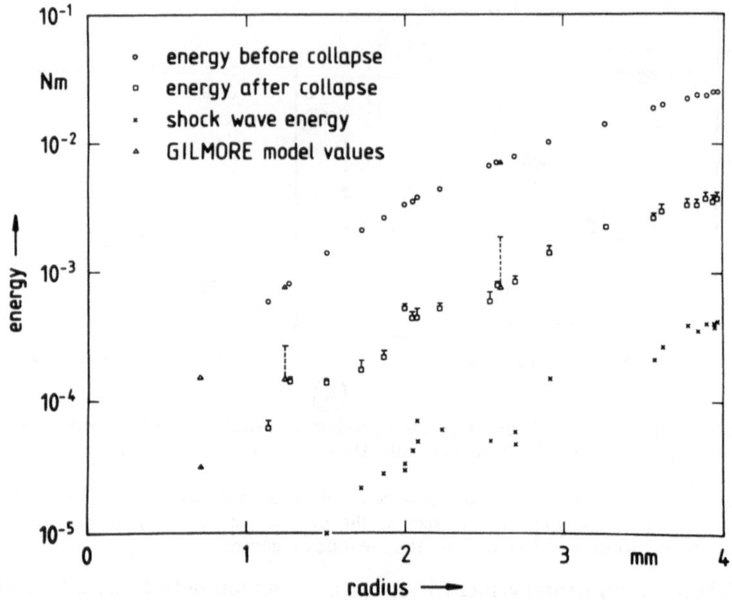

Figure 5. Energy balance of single cavitation bubbles. The energy before and after collapse and the shock wave energy radiated upon collapse is plotted versus the maximum radius before collapse. For comparison some values calculated with the Gilmore model are added.

for a small part of the energy loss. Other much more significant damping mechanisms must be present.

Theoretical values taken from the Gilmore model (see next section) for the shock wave energy are much greater (about 16% of the total energy) than the experimental ones. This is attributed to the non-spherical collapse of the bubbles in the experiments. It is conjectured that in addition to the known damping mechanisms of heat conduction and diffusion, damping may also be provided by the nonsphericity of the collapse.

4. Numerical calculations

Bubble models that take into account only the compressibility of the liquid and therefore the damping of the oscillation by only the radiation of sound waves like the Gilmore model cannot explain the rapid damping observed in our experiments. The Gilmore model neglects the mass loss by condensation and diffusion of gas and vapour contained in the bubble. The gas content of the cavity decreases during the oscillation by diffusion (in our case the production process of the cavity may lead to a gas content not in equilibrium with the surrounding liquid) and separation of tiny bubbles, the vapour content changes by condensation. Therefore the Gilmore model was modified

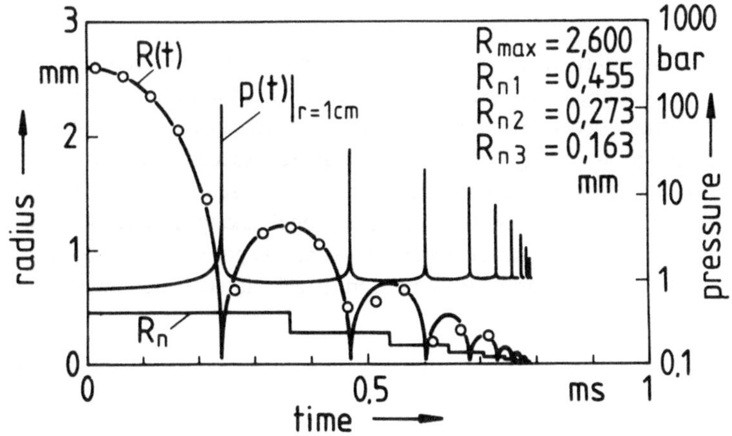

Figure 6. radius-time curve of the damped bubble oscillation for a spherical cavity in water. The damping by mass loss is taken into account in a modified Gilmore model by a simple percentage decrease of the equilibrium radius of the bubble.

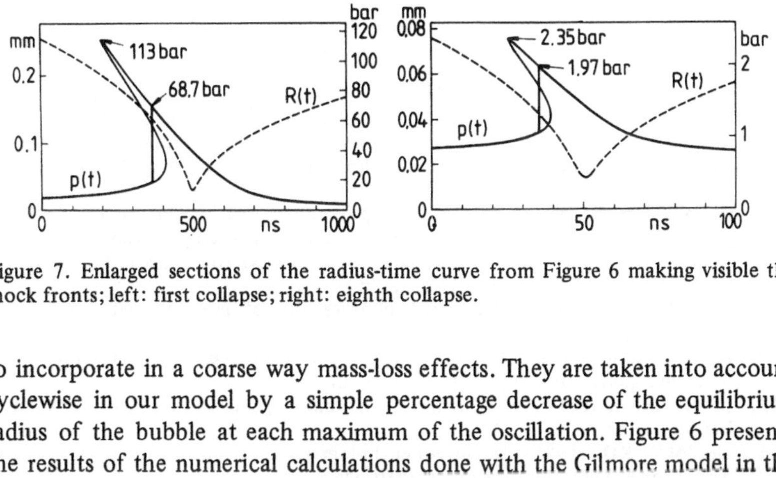

Figure 7. Enlarged sections of the radius-time curve from Figure 6 making visible the shock fronts; left: first collapse; right: eighth collapse.

to incorporate in a coarse way mass-loss effects. They are taken into account cyclewise in our model by a simple percentage decrease of the equilibrium radius of the bubble at each maximum of the oscillation. Figure 6 presents the results of the numerical calculations done with the Gilmore model in the version of Cramer [3, 4]. The radius of the bubble $R(t)$, the equilibrium radius R_n and the pressure amplitude $P(t)$ at a distance of 10 mm are plotted versus time. A good agreement can be obtained with radius-time values (open circles) as measured by rotating-drum camera experiments when a percentage decrease of R_n to 60% of the preceeding value is used at each maximum.

Two sections of the drawing near the first and eighth collapse are enlarged in Figure 7. The shock front is calculated for equal areas at the left and at the right side of the triple values pressure curve. One gets a pressure amplitude of about 69 bars for the first collapse whereas a value of about 20 bars is measured in the experiments for a bubble of equal size.

5. Conclusion

The starting point for all considerations has been the experimental results of the acoustic emission of laser-produced cavitation bubbles in water. From the pressure-time curves the maximum radii and the shock wave amplitudes are evaluated and an energy balance is considered. According to our findings the radiation of shock waves seems not to play an essential role in damping the bubble oscillation. A modified Gilmore model taking into account the mass loss by condensation and diffusion of gas and vapour contained in the cavity can be made to give a good agreement in its calculated radius-time curves with experimental results.

Acknowledgment

The work was sponsored by the Fraunhofer Gesellschaft, Munich.

References

1. Brinkmeyer E (1976) Kohärent-optische Verfahren zur Untersuchung kurzer Stoßwellen bei akustisch erzeugter Kavitation. PhD dissertation, Göttingen.
2. Cole RH (1948) Underwater explosions. Princeton.
3. Cramer E (1980) The dynamics and acoustics of bubbles driven by a sound field. In: [10], pp 54–63.
4. Cramer E, Lauterborn W (1981) Zur Dynamik und Schallabstrahlung kugelförmiger Kavitationsblasen in einem Schallfeld. Acustica 50 (in press).
5. Ebeling KJ, Lauterborn W (1978) Acoustooptic beam deflection for spatial frequency multiplexing in high speed holocinematography. Appl Opt 17: 2071–2076.
6. Hentschel W (1979) Akustische und optische Untersuchungen zur Dynamik holografisch erzeugter Kavitationsblasen-Systeme. Masters thesis, Göttingen.
7. Hentschel W (1980) Zur Dynamik lasererzeugter Kavitationsblasen. Fortschritte der Akustik, DAGA '80, pp 415–418. Berlin.
8. Lauterborn W (1972) High-speed photography of laser-induced breakdown in liquids. Appl Phys Lett 21: 27–28.
9. Lauterborn W, Ebeling KJ (1977) High speed holography of laser-induced breakdown in liquids. Appl Phys Lett 31: 663–664.
10. Lauterborn W (ed) (1980) Cavitation and inhomogeneities in underwater acoustics. Berlin: Springer.
11. Merboldt KD (1981) Hochfrequenzholografie mit dem auskoppelmodulierten Argon-Ionen Laser. Masters thesis, Göttingen.
12. Timm R (1980) Bubble collapse studied at a million frames per second. In: [10], pp. 42–46.

Surface tension driven oscillations of a bubble in a viscoelastic liquid

CLAES INGE and FRITZ H. BARK

Department of Mechanics, Royal Institute of Technology, 100 44 Stockholm, Sweden

Abstract. Small amplitude surface tension driven oscillations of a spherical bubble in a dilute polymer solution are considered. The rheological properties of the liquid are modelled by using a 3-constant constitutive equation of the Oldroyd type. The Laplace transform of the solution of the initial value problem is inverted numerically. As in the Newtonian fluid case, both a discrete and a continuous spectrum occurs. In addition to the non-dimensional parameters in the corresponding problem for a Newtonian fluid, the results depend on two other parameters: the ratio of the relaxation time of the polymer solution and the time scale of the flow (the Deborah number) and the product of the polymer concentration and the intrinsic viscosity. For small bubbles in an aqueous solution having a small relaxation time, significant additional damping is found even for dilute solutions.

Introduction

Erosion of solid surfaces by cavitation is a very important and so far unexplained phenomenon in fluid mechanics. Naudé and Ellis [13] demonstrated that a microjet is formed inside a bubble collapsing in the vicinity of a solid wall. The impingement of this microjet against the solid wall was suggested as the eroding mechanism. A theoretical explanation of the mechanism leading to the formation of a microjet was given by Plesset [16], who showed that a symmetrically collapsing bubble in an incompressible, inviscid and unbounded fluid is linearly unstable with respect to asymmetric disturbances. The qualitative validity of his results for finite deviations from the spherical shape of a bubble collapsing in the neighbourhood of a wall was verified by the numerical computations of Plesset and Chapman [17].

During the last decade, some interesting properties of cavitation in dilute solutions of long chain polymers have been discovered. Ellis, Waugh and Ting [6] and Hoyt [9] found that such polymer additives delay flow-generated cavitation in the sense that the inception cavitation number decreases. Furthermore, experimental investigations have shown that cavitation noise can be considerably reduced by dissolved polymers [10, 14]. Several studies have therefore been undertaken to investigate the effects of polymer additives on the dynamics of bubbles. The experiments by Ellis and Ting [5, 22] showed no significant effects on the symmetric growth and collapse of a single bubble. The experiments by Chahine and Fruman [4], however, showed a weak but distinct retarding effect on the initial development of the microjet. There are also theoretical investigations of the spherically symmetric collapse

Applied Scientific Research 38: 231–238 (1982) 0003–6994/82/0383–0231 $01.20.

of a bubble in a dilute polymer solution, modelled as a viscoelastic liquid in the literature [8, 12, 23]. Although different rheological models were used the results are rather similar. The effect of viscoelasticity was shown to be weak except when the strain rate is very large, which is the case for small radii and large velocities. Large normal stresses are then developed and these cause the bubble to rebound. However, for typical parameter values, this rebound does not take place until the radius is several orders of magnitude smaller than the initial radius. In fact, it is quite possible that rebound caused by other phenomena, such as the pressure due to gas or uncondensed vapour in the bubble, will take place before the viscoelastic stresses become strong enough to reverse the motion.

The modifications caused by small Newtonian viscous effects on the linear instability mechanism for a collapsing bubble have been computed by Prosperetti and Seminara [19]. A small viscosity was shown to give a slightly reduced growth rate of the unstable motion. The corresponding problem for a bubble in a viscoelastic liquid would be a relevant extension of the work of Prosperetti and Seminara. Unfortunately, that problem turns out to be very complicated. The present work on the simpler problem of surface tension driven oscillations is therefore a preliminary study of asymmetric bubble motions in a viscoelastic liquid. The same problem for Newtonian liquids has been treated by Prosperetti [18, 20, 22].

Statement of the problem

Consider a gas-filled bubble of constant radius R in an unbounded liquid. The pressure in the gas balances the pressure in the liquid and the surface tension, whereby a state of static equilibrium prevails. The dynamic effects of the gas will be neglected in what follows. At some initial instant, the equilibrium shape of the bubble is slightly disturbed by external means and the bubble is left to move freely. The problem to be considered is to calculate the free, damped oscillatory motion of the bubble. In the nondimensional formulation of the problem the bubble radius is used as the length scale. The quantity $T = (\rho R^3/\gamma)$, where ρ is the density of the liquid and γ is the surface tension, is used as the time scale. This choice is motivated by the fact that the surface tension is the restoring force causing the oscillations of the bubble.

The motion of the liquid can be calculated from the dimensionless momentum equations and the equation of continuity:

$$\frac{\partial \mathbf{v}}{\partial t} + \mathbf{v} \cdot \nabla \mathbf{v} = -\nabla p + \nabla \cdot \boldsymbol{\sigma}, \tag{1}$$

$$\nabla \cdot \mathbf{v} = 0, \tag{2}$$

where t is the dimensionless time. $\mathbf{v}$, p and $\boldsymbol{\sigma}$ are the dimensionless velocity vector, pressure and deviatoric stress tensor, respectively, in the liquid. The viscoelastic properties of the liquid are modelled by using the following 3-constant constitutive equation of the Oldroyd type [3, 11, 15]:

$$\boldsymbol{\sigma} + \mathrm{D}\left(\frac{\partial \boldsymbol{\sigma}}{\partial t} + \mathbf{v}\cdot\nabla\boldsymbol{\sigma} - \boldsymbol{\sigma}\cdot\nabla\mathbf{v} - (\nabla\mathbf{v})^{*}\cdot\boldsymbol{\sigma}\right) =$$

$$= \frac{2}{\alpha\,\mathrm{Re}}\left[d + \alpha\mathrm{D}\left(\frac{\partial d}{\partial t} + \mathbf{v}\cdot\nabla d - d\cdot\nabla\mathbf{v} - (\nabla\mathbf{v})^{*}\cdot d\right)\right], \quad (3)$$

where d is the dimensionless rate of strain tensor. The * superscript in (3) stands for matrix transposition. The dimensionless parameters Re, D and α are defined by

$$\mathrm{Re} = \frac{\rho R^2}{\eta T} \qquad \text{Reynolds number,}$$

$$\mathrm{D} = \frac{\lambda}{T} \qquad \text{Deborah number, and}$$

$$\alpha = (1 + c\,[\eta])^{-1},$$

where η is the dynamic viscosity of the solvent, $[\eta]$ is the intrinsic viscosity, λ is the terminal stress relaxation time of the dissolved polymer molecules and c is the concentration. The constitutive equation (3) can be derived under the assumption that the polymer molecules can be modelled as pairs of small beads connected with linear springs [3, 11].

The boundary conditions to be fulfilled by the solution to (1)–(3) are

$$\mathbf{n}\times(\boldsymbol{\sigma}\cdot\mathbf{n}) = 0, \quad (4)$$

$$\mathbf{n}\cdot\boldsymbol{\sigma}\cdot\mathbf{n} = \nabla\cdot\mathbf{n} + p - p_g, \quad (5)$$

$$\lim_{|\mathbf{x}|\to\infty} \mathbf{v}, p = 0, \quad (6)$$

where $\mathbf{n}$ is the unit normal vector to the bubble surface, $\mathbf{x}$ is the radius vector and p_g is the gas pressure. A spherical coordinate system, (r, θ, φ), whose origin is located at the centre of the bubble in its equilibrium state will be used. Following Plesset [16], the perturbed surface of the bubble is written as

$$r = 1 + a(t)\,Y_n^m(\theta, \varphi), \quad (7)$$

where Y_n^m is a spherical harmonic of degree $n \geqslant 2$. Only linear motions will be considered and it is consequently assumed that $|a| \ll 1$.

It is convenient to decompose the velocity field into an irrotational part and a rotational part [18]. The vorticity field is decomposed into a toroidal part and a poloidal part. The poloidal part of the vorticity is neglected since it does not influence the radial motion and is therefore dynamically insignificant in the present problem. After some rather tedious manipulations, analogous to those in [18], one finds the following expression for the Laplace transform $\tilde{a}(s)$ of $a(t)$.

$$\tilde{a} = \frac{1}{s}\left[a(0) + \frac{(1+\mathrm{D}s)(\dot{a}(0)s - n_1 a(0))}{(1+\mathrm{D}s)(s^2+n_1) + \frac{2s}{\alpha\,\mathrm{Re}}(1+\alpha\mathrm{D}s)(n_2+n_3 Q_n(z))}\right], \tag{8}$$

$$n_1 = (n-1)(n+1)(n+2), \qquad n_2 = (n+2)(2n+1),$$

$$n_3 = n(n+2)^2, \qquad z = \left[\frac{\alpha\,\mathrm{Re}\,s(1+\mathrm{D}s)}{(1+\alpha\mathrm{D}s)}\right]^{1/2},$$

$$Q_n(z) = -\frac{2K_{n-1/2}(z)}{zK_{n+1/2}(z) + 2K_{n-1/2}(z)}.$$

$K_\nu(z)$ is the modified Bessel function of the second kind of order ν. The dot denotes differentiation with respect to time. In the limit c approaching zero, i.e. vanishing concentration of polymer additive, (8) reduces, as it should, to the expression derived by Prosperetti [18] for the corresponding problem for a Newtonian liquid.

It can be shown that the corresponding problem for a drop of a visco-elastic liquid, whose rheological properties are given by (3), leads to the same expression (8) for the Laplace transform of the function $a(t)$ if the quantities n_1, n_2, n_3 and Q_n are redefined as

$$n_1 = n(n-1)(n+2), \qquad n_2 = (n-1)(2n+1),$$

$$n_3 = (n+1)(n-1)^2, \qquad Q_n = \frac{2I_{n+3/2}(z)}{2I_{n+3/2}(z) - zI_{n+1/2}(z)},$$

where $I_\nu(z)$ is the modified Bessel function of the first kind of order ν.

In the problem for a bubble, $\tilde{a}(s)$ is not a unique function of the complex transform variable s. The non-uniqueness can be removed by e.g. two branch cuts along the negative part of the real s-axis. One from $s = 0$ to $s = -1/\mathrm{D}$ and the other from $s = -1/(\alpha\mathrm{D})$ to $s = -\infty$. As in the Newtonian fluid case [20, 21], the spectrum for the initial value problem for a bubble thus consists of a discrete part (the poles of $\tilde{a}(s)$) and a continuous part. In the initial value problem for a drop only a discrete spectrum occurs. Despite the complicated form of (8), approximative analytical solutions for $a(t)$ can be obtained by using asymptotic methods in certain regions of Re-D-space. Such results will be presented elsewhere. The results presented in this paper were obtained by inverting the Laplace transform numerically. The numerical inversion method

[2] was tested by comparison with the results obtained by Prosperetti for Newtonian liquids and with exactly invertible functions. It was found that the method is reliable for inversion of the type of functions dealt with in this work.

Results and discussion

Before presenting some numerical results, a brief qualitative discussion of the behaviour of the polymer molecules, modelled as bead-springs, in an arbitrary viscous potential flow may be illuminating. In flows with vorticity, the dumbbells are affected by both straining and rotation of the fluid elements and their effect on the flow may be somewhat complicated. The discussion is therefore restricted to flows with negligible vorticity. Also, the case of very large Reynolds numbers will be excluded for the moment.

The relaxation time of the dumbbells is proportional to the size of the beads and to the inverse of the spring constant [11]. If e.g. the bead size is fixed, a small (large) relaxation time corresponds to stiff (soft) springs. Small Deborah numbers indicate that the relaxation time is small compared to the time scale of the flow, which means that the viscous forces on the beads are not strong enough to essentially extend the stiff springs. The viscous resistance on the flow around the beads will, from the macroscopic point of view, manifest itself as an increased viscosity. This effect can, of course, also be deduced from the constitutive equation (3). The limit $D \to 0$ gives the constitutive equation for a Newtonian liquid having a viscosity η/α where $\alpha < 1$.

If the Deborah number is large, the viscous force on the beads is large compared with the spring force provided that the extension of the spring is not large. On the flow-time scale, initially undistorted dumbbells will thus deform in the same way as the fluid elements. For an oscillatory motion having a period which is significantly smaller than the relaxation time, large spring forces will never develop and the flow will be essentially unaffected by the dumbbells. If the contribution from the dumbbells to the stress at $t = 0$ is assumed to be zero, this behaviour of the fluid can also be deduced from (3) by taking the limit as $D \to \infty$, which results in the equation for the Newtonian solvent.

In the intermediate case $D \sim 1$, the effect of the dumbbells will, roughly speaking, be a compromise between the two previous cases. Both viscous dissipation and extension of the dumbbells large enough to create significant elastic stresses may be expected. This would increase both the damping and the frequency of an oscillatory motion.

For very large Reynolds numbers, the constitutive equation (3) indicates that the viscous stresses in the solvent as well as the effects of the dumbbells are weak. However, largely extended springs may have strong effects. But for the kind of small amplitude oscillatory motion considered here, large extension of the dumbbells will never take place and a large Reynolds number will

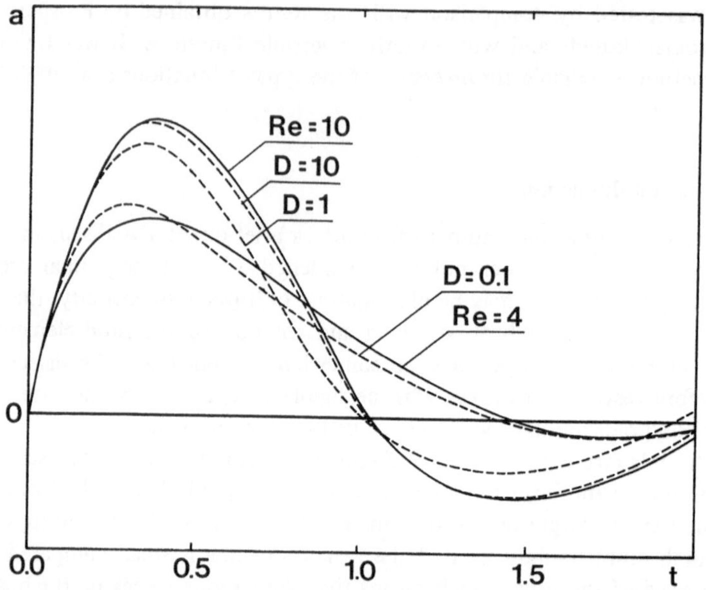

Figure 1. The oscillation amplitude of a bubble for $n = 2, a(0) = 0$ and $\dot{a}(0) \neq 0$. The dashed lines represent viscoelastic liquids with Re = 10, $\alpha = 0.4$ and different Deborah numbers. The full lines represent Newtonian liquids ($\alpha = 1$) with Re = 10 and Re = 4. The former represents the Newtonian solvent, and the latter is equivalent to a viscoelastic liquid with Re = 10, $\alpha = 0.4$ and D = 0.

indicate small effects of both the solvent viscosity and suspended dumbbells. For given material parameters, large bubble radii correspond to large Reynolds numbers and therefore small effects of the dumbbells. Small bubble radii leads to large Deborah numbers, which also implies small effects of the dumbbells. This means that viscoelastic effects will be significant only in a limited range of bubble radii. According to Ferry [7] the terminal relaxation time in dilute polymer solutions may be as small as 10^{-5} s. Assuming a relaxation time of the order 10^{-4}–10^{-5} s, viscoelastic effects in dilute solutions will occur for bubble radii of the order 10^{-4}–10^{-5} m.

Figure 1 shows the amplitude function $a(t)$ for the distortion of the free surface of the bubble for $n = 2$. The initial conditions are $a(0) = 0$ and $\dot{a}(0) \neq 0$. The curves for three different Deborah numbers, the curve for the Newtonian solvent and the curve for a Newtonian liquid with the viscosity η/α are shown in the graph. It can be seen that the damping of the oscillations increases monotonically with decreasing values of the Deborah number. A slightly increased frequency can also be detected for the D = 1 case. As the Deborah number becomes small, the behaviour of the liquid tends to that of a Newtonian liquid with the increased viscosity η/α, and the motion becomes almost aperiodic. The same behaviour, although less pronounced, was obtained for larger values of the Reynolds number. This behaviour is in qualitative agreement with the heuristic arguments given above for arbitrary potential flows. This is hardly surprising since the only source of vorticity in the present case is the shear stress constraint (4) at the free surface, which, for oscillatory motions at Reynolds numbers of the order of magnitude considered here, is known [1] to give only weak vorticity in a boundary layer at the free surface.

The present calculation shows that the effect of small amounts of dissolved polymers on the linear surface tension driven oscillations of a bubble at moderately large Reynolds numbers is mainly of viscous character. Only small elastic effects were found. For a collapsing bubble, however, the effect of viscous dissipation is usually rather weak [19], and the extra dissipation due to the polymers is not likely to affect the fluid motion very much. On the other hand, large and rapid extension of the liquid elements do often occur near collapsing bubbles. This may create large elastic normal stresses in the liquid. As shown by Folger and Goddard [8] and Moore [12], such stresses may be large enough to cause rebound of a symmetrically collapsing bubble. In the case of an asymmetrically collapsing bubble such stresses may also be expected in the microjet, which could then be significantly retarded. Future work will verify or disprove this conjecture.

References

1. Batchelor GK (1970) An introduction to fluid dynamics. Cambridge University Press.
2. Bellman RE, Kalaba RE and Lockett J (1966) Numerical inversion of the Laplace transform. Rand Corporation.
3. Bird RB, Hassager O, Armstrong RC and Curtiss CF (1977) Dynamics of polymeric liquids. John Wiley & Sons, Inc.
4. Chahine GL and Fruman DH (1979) Dilute polymer solution effects on bubble growth and collapse. Phys Fluids 22 (7): 1406–1407.
5. Ellis AT and Ting RY (1974) Non-Newtonian effects on flow-generated cavitation and on cavitation in pressure fields. NASA-SP-304 1: 403–417.
6. Ellis AT, Waugh JG and Ting RY (1970) Cavitation suppression and stress effects in high-speed flows of water with dilute macromolecule additive. Trans ASMED, J Basic Engng 92: 459–466.
7. Ferry JD (1970) Viscoelastic properties of polymers. John Wiley & Sons, Inc.
8. Folger HS and Goddard JD (1970) Collapse of spherical cavities in viscoelastic fluids. Phys Fluids 13 (5): 1135–1141.
9. Hoyt JW (1976) Effect of polymer additives on jet cavitation. Trans ASMED, J Fluids Engng 98 (1): 106–112.
10. Lagerstedt T and Bark G (1978) Influence of polymer on propellor cavitation noise. The Swedish Shipbuilding Experimental Tank, report 2251-1.
11. Lumley JL (1971) Applicability of the Oldroyd constitutive equation to flow of dilute polymer solutions. Phys Fluids 14 (11): 2282–2284.
12. Moore K (1978) PhD thesis DAMTP, Cambridge University.
13. Naudé CF and Ellis AT (1961) On the mechanism of damage by nonhemispherical cavities collapsing in contact with a solid boundary. Trans ASMED, J Basic Engng 83: 648–656.
14. Oba R, Ito Y and Uranishi K (1978) Effect of polymer additives on cavitation and noise in water flow through an orifice. Trans ASMED. J Fluids Engng 100: 493–499.

15. Oldroyd JG (1950) On the formulation of rheological equations of state. Proc Roy Soc (A) 200: 523–541.
16. Plesset MS (1954) On the stability of flows with spherical symmetry. J Appl Phys 25: 96–98.
17. Plesset MS and Chapman RB (1971) Collapse of an initially spherical vapour cavity in the neighbourhood of a solid boundary. J Fluid Mech 47: 283–290.
18. Prosperetti A (1977) Viscous effects on perturbed spherical flows. Quart Appl Math 35: 339–352.
19. Prosperetti A and Seminara G (1978) Linear stability of a growing or collapsing cavity in a slightly viscous liquid. Phys Fluids 21 (9): 1465–1470.
20. Prosperetti A (1980) Normal-mode analysis for oscillations of a viscous drop in an immiscible liquid. J Mec 19 (1): 149–182.
21. Prosperetti A (1980) Free oscillations of drops and bubbles. J Fluid Mech 100: 333–347.
22. Ting RY and Ellis AT (1974) Bubble growth in dilute polymer solutions. Phys Fluids 17 (9): 1461–1462.
23. Ting RY (1975) Viscoelastic effect of polymers on single bubble dynamics. AIChE J 21 (4): 810–815.

The local measurement of the size and velocity of bubbles rising in liquids

W.W. MARTIN and G.M. CHANDLER

Department of Mechanical Engineering, University of Toronto, Toronto, Ontario, Canada

Abstract. An experimental study of the rise of small air bubbles (0.1 to 1.0 mm in diameter) in a quiescent pool is described. Local measurements of rise velocity were obtained as a function of height above the source nozzle using a laser-Doppler method. In addition, the bubble diameter was determined simultaneously from the same optical signals. Data are presented for various bubble diameters and spacings in bubble columns for both distilled water and a dilute polymer solution.

It was found that for distilled water the rise velocity near the nozzle reaches the maximum observed in other studies before decelerating to its terminal velocity due to surfactant accumulation at its interface. The maximum rise velocity in dilute polymer was much lower for the same bubble diameter and reached its terminal velocity much faster. The results are shown to be in closer agreement with predictions for a solid sphere in this case.

1. Introduction

The laser-Doppler anemometer is an effective opto-electronic system for local nonintrusive measurement of flow velocity. For single phase fluids LDA has been employed productively in flows where previous techniques could only provide inaccurate data, e.g., in separated flow. Although the application of LDA to two-phase flows is inherently more difficult, it was shown in several explorative studies that it is possible to get good quality signals on which to base velocity measurements of both phases. For example a recent review [1] presents results for a turbulent flow with solid spheres up to 0.8 mm in diameter. LDA schemes were also suggested for determining the diameter of spherical particles, bubbles and droplets by several authors [1, 2, 4, 6].

Recently a systematic study of LDA signals from single bubbles rising in a quiescent liquid [3] was undertaken by directly digitizing the photomultiplier output to obtain the signal characteristics of light scattered in the forward and 90° directions. It was shown that the visibility function of the forward scatter bursts is uniquely related to the bubble diameter over a range of size which depends on the diameter of an aperture placed in the collecting optics. Ambiguity due to the bubble trajectory through the beam intersection region was removed by only considering bursts with a Gaussian pedestal shape as described in [3].

In this paper we apply the same LDA optical and electronic signal processing system but employ a modified data analysis scheme for determining bubble size based on the forward – scatter bursts. Measurements of the local size and

Applied Scientific Research 38: 239–246 (1982) 0003–6994/82/0383–0239 $01.20.

velocity are presented for bubbles rising in a quiescent pool of distilled water and dilute polymer (50 ppm, Polyox dissolved in distilled water). Columns of bubbles ranging in diameter between 0.1 and 1.0 mm were generated by injecting air through a nozzle. In addition to measuring velocity as a function of height above the nozzle, the effect of bubble spacing was also studied.

2. Experimental arrangement

Single vertical bubble columns were generated by injecting air through glass nozzles in a quiescent pool 76.2 mm square and 500 mm high. It was found that stable bubbles between 0.1 mm and 1.0 mm in diameter could be formed by drawing micropipette tubing to a very long taper. A given nozzle produced constant diameter bubbles whose spacing in the column was controlled by varying the delivery pressure of the air.

The bubble velocity and diameter were measured as a function of height above the nozzle in distilled water and dilute polymer (50 ppm Polyox). The optical arrangement and electronic signal processing equipment employed in the measurements is fully described in [3] and is shown in Figure 1. A standard single channel, helium-neon forward scatter optics was modifed to include an adjustable circular aperture on the light-collecting lens in order to optimize the photomultiplier signals for data processing and bubble-size range. The test section was mounted on a three-dimensional traverse to allow the bubble trajectory to be accurately located in the region of the focal point of the forward-focusing lens and to permit the adjustment of the position of measurement above the nozzle. With this arrangement the optics could be held fixed throughout the experiments.

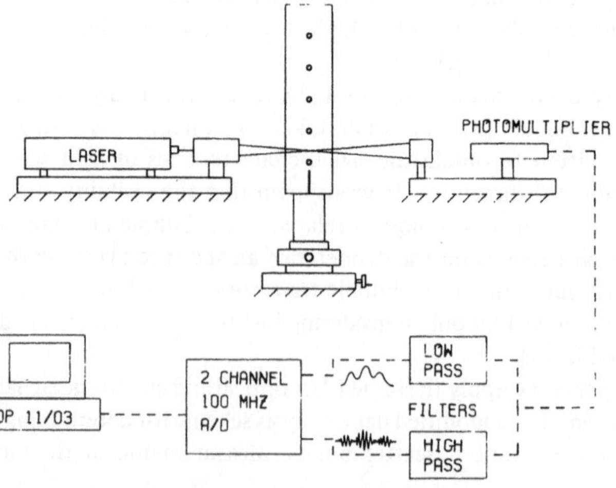

Figure 1. Sketch of experimental arrangement and instrumentation.

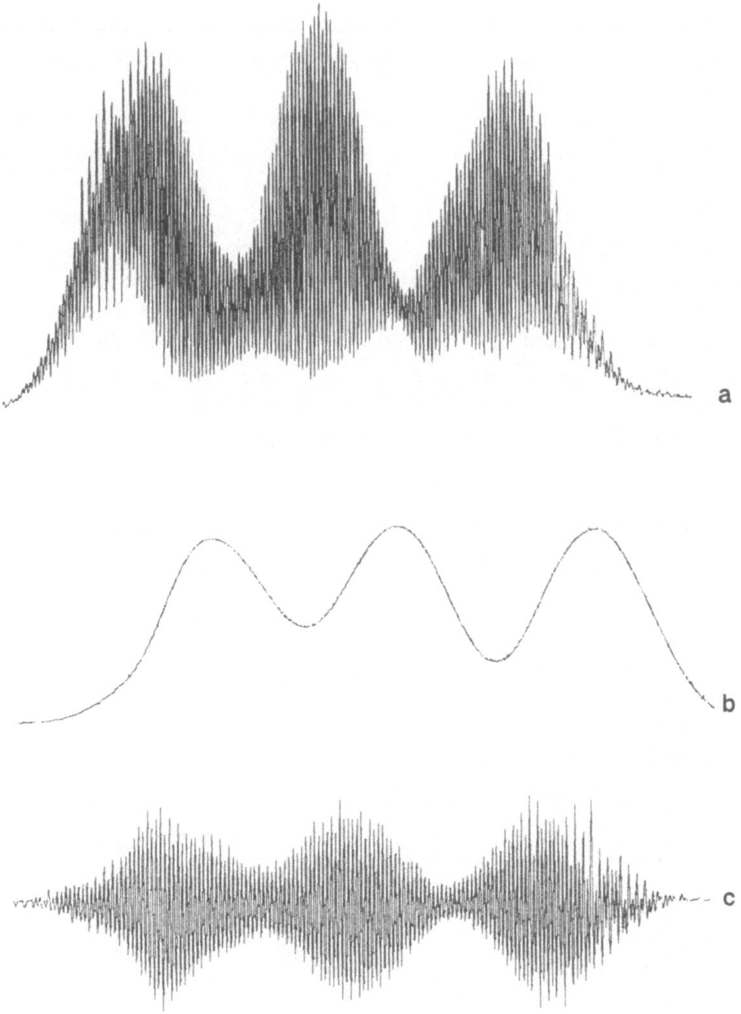

Figure 2. Doppler signal characteristics; (a) photomultiplier output; (b) low-pass filtered signal; (c) high-pass filtered signal.

3. Data collection and analysis

The velocity of a bubble is obtained from the frequency of the scattered light. Information on the size (or more precisely the radius of curvature) is contained in the amplitude distribution of the scattered light. It was shown in [3] that the visibility of the forward scattered signal (i.e. the ratio of the Doppler amplitude to the pedestal amplitude can be used to uniquely determine the bubble size. However, the visibility was computed from a direct digital recording of the photomultiplier signal. This technique has the disadvantage that calibration is required and that the computation for each burst is lengthy due to the discrimination process employed.

An alternative procedure was first proposed in [6] and is easily implemented in the digital processing scheme. As shown in Figure 1, the photomultiplier signal (Figure 2a) is first divided and the two channels are separately high-pass and low-pass filtered (Figures 2b and 2c, respectively). The resulting signals are then input to the two channel Biomation 8100 transient recorder for digital recording and transfer to the PDP 11/03 minicomputer. The high-pass filtered signal is analyzed to determine its frequency which is used to compute the bubble velocity. The low-pass filtered signal is analyzed to determine the time between the two side peaks generated by the bubble just entering and just leaving the beam focus. The velocity divided by this residence time yields an estimate of the

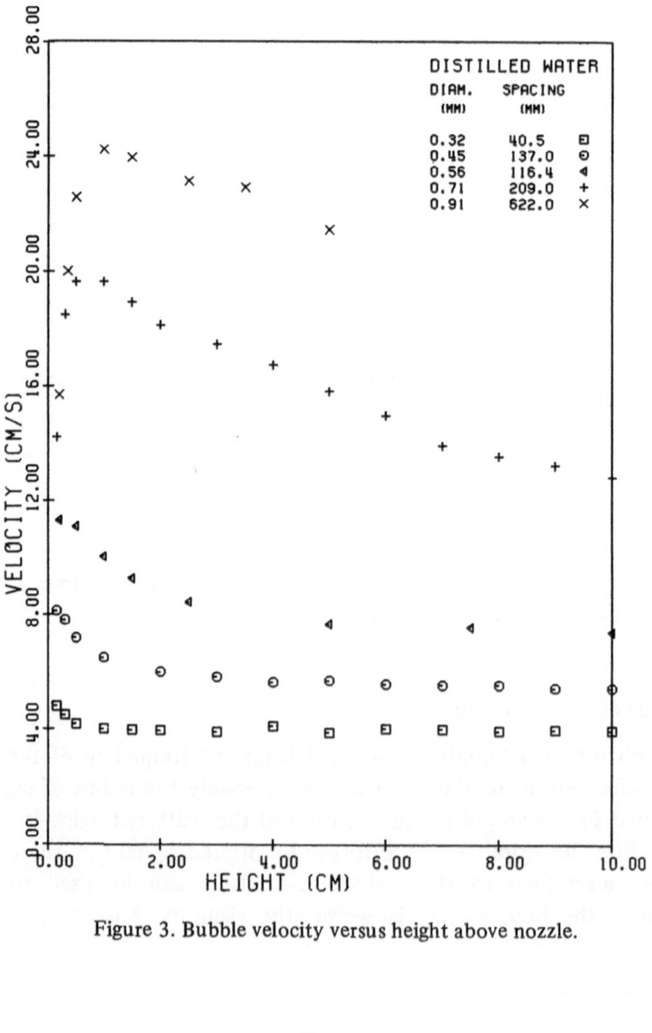

Figure 3. Bubble velocity versus height above nozzle.

bubble diameter with an accuracy of better than 10%. This technique requires no calibration and much less computational time and was implemented on the PDP 11/03 minicomputer.

4. Results

The primary data obtained in the present study are bubble velocities as a function of height above the nozzle. Typical results are presented in Figures 3 and 4 for distilled water and dilute polymer, respectively. A comparison of these two figures reveals that the unsteady phase during which the bubbles accelerate to a peak velocity and then decelerate towards a terminal velocity is markedly different for the two liquids. Since the two liquids have nearly equal density and viscosity, the difference is

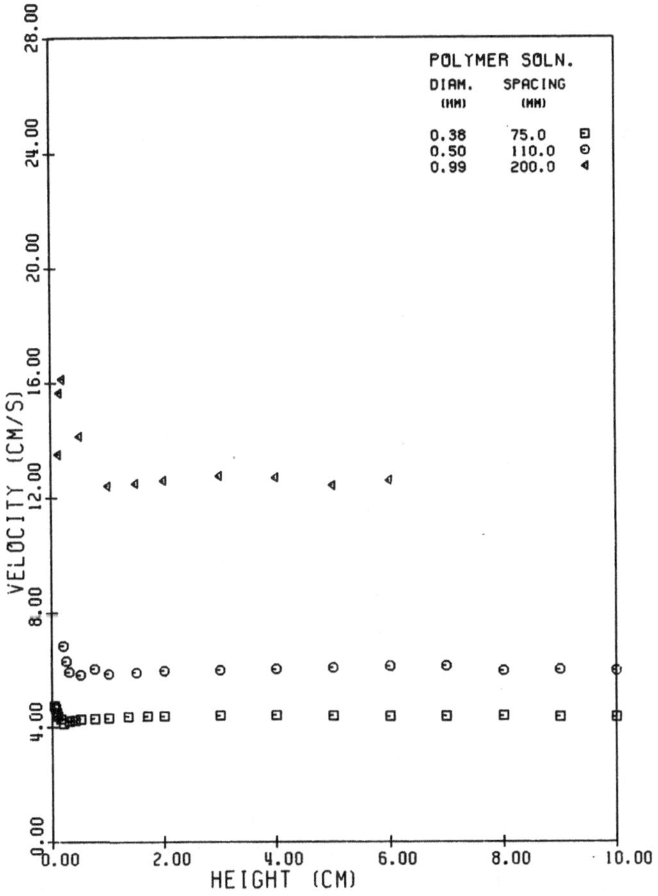

Figure 4. Bubble velocity versus height above nozzle.

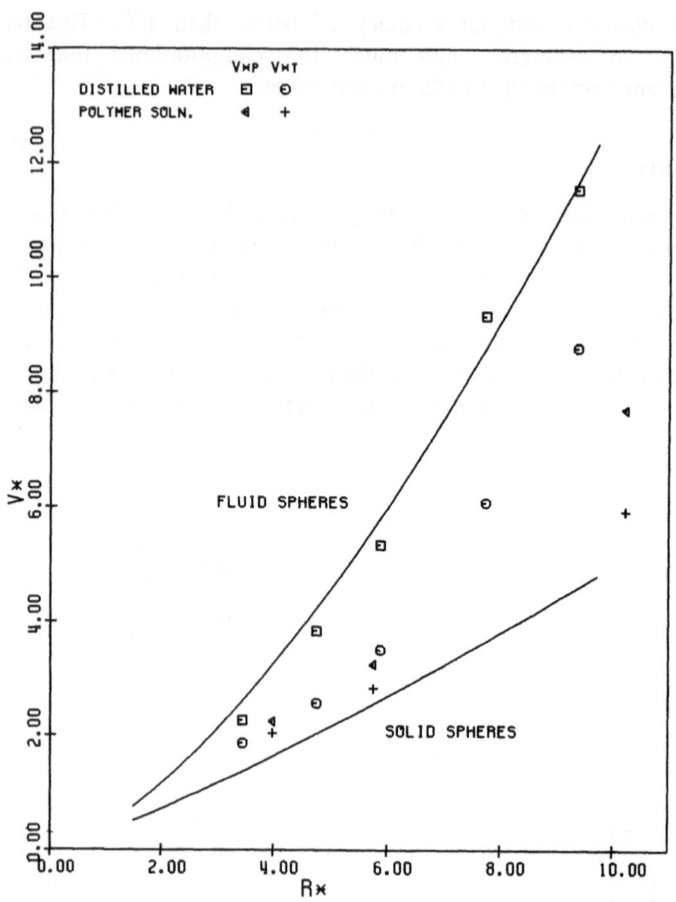

Figure 5. Dimensionless peak and terminal velocities as a function of dimensionle bubble radius.

apparently due to a difference in fluid properties associated with th sheared interface region if it assumed that the bubbles are spherical.

The review by Wallis [5] of previous data show that the bubble ris velocity lies between a correlation presumed to correspond to a flui sphere with internal circulation and the predictions for a solid sphere. Thi was also found in the present experiments as indicated in Figure 5 where bot the dimensionless peak velocities and terminal velocities are shown as function of dimensionless radius. As for the data in Figures 3 and 4, th bubble spacing was large enough not to influence the rise velocity. Th dimensionless variables are $V^* \equiv V[\rho^2/\mu g \Delta\rho]^{1/3}$ and $R^* \equiv R\,[g\rho\Delta\rho/\mu^2]^{1/3}$ where ρ is the liquid density, $\Delta\rho$ is the density difference between the tw phases and μ is the liquid viscosity. Figure 5 shows clearly that bubbles i dilute polymer behave rather more as solid spheres. In distilled water th bubbles reach much higher velocities near the maximum values previously reported and have terminal velocities well above the solid sphere prediction.

The effect of bubble spacing on the terminal rise velocity is shown in Figure 6. For the lowest spacings (~ 10 diameters) some induced circulation in the liquid is possible in which case the relative velocity between the bubble and liquid would be lower than the observed absolute velocity. However, the liquid velocity was not measured in this study. The optical arrangements proposed by Durst [1] to measure the velocity of both phases can be used in future work.

Acknowledgement

This research was carried out under Grant No. A9196 of the National Sciences and Engineering Research Council of Canada.

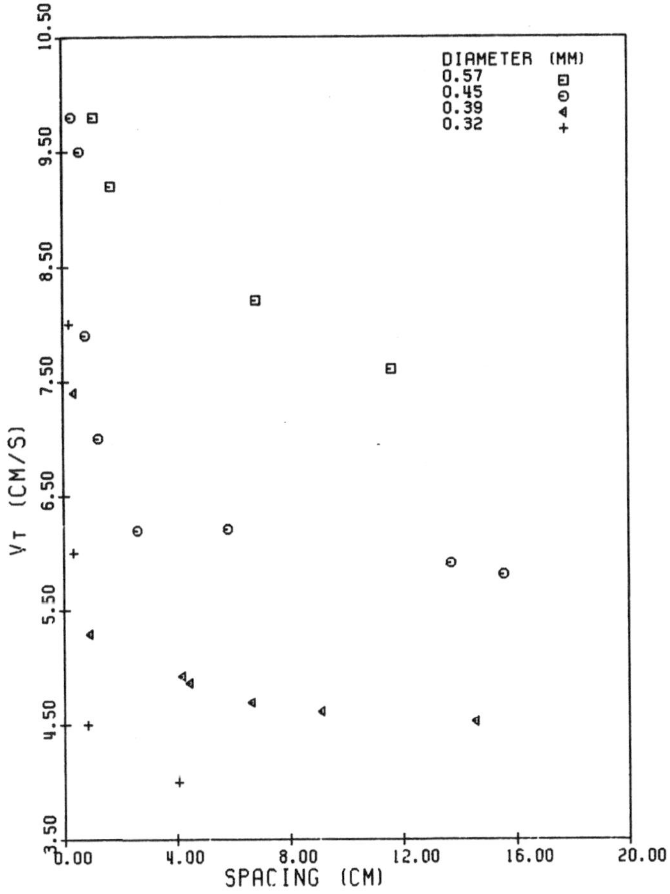

Figure 6. Terminal velocity in distilled water versus bubble spacing.

References

1. Durst F (1978) Studies of particle motion by laser-Doppler techniques. Proc Dynamic Flow Conf 1978, Marseille and Baltimore: 345–372.
2. Lee SL and Srinivasan J (1978) Measurement of local size and velocity probability density distributions in two-phase suspension flows by laser-Doppler technique. Int J Multiphase Flow 4: 141–155.
3. Martin WW, Abdelmessih AH, Liska JJ and Durst F (1981) Characteristics of laser-Doppler signals from bubbles. Int J Multiphase Flow 7: 439–460.
4. Ungut A, Yule AJ, Taylor DS and Chiqier NA (1978) Particle size measurement by laser anemometry. J Energy 2: 330–336.
5. Wallis GB (1974) The terminal speed of single drops or bubbles in an infinite medium. Int J Multiphase Flow 1: 491–508.
6. Wigley GJ (1978) The sizing of large particles by laser anemometry. J Physics E – Scient Instr 11: 639–642.

The unsteady drag on a spherical bubble at large Reynolds numbers

PHAM DAN TAM

Département des Réacteurs à Eau, Service des Transferts Thermiques,
Centre d'Etudes Nucléaires de Grenoble, Grenoble, France

Abstract. The flow past a spherical bubble undergoing a rectilinear motion in the unsteady flow of an unbounded liquid medium is investigated. The liquid velocity field at infinity is assumed to be uniform and the Reynolds number to be large. The Strouhal number is taken to be of order unity. The velocity distribution is sought by superposition of a perturbation field on the potential flow past the bubble so that the flow field is divided into four regions, i.e. the external flow field where the potential flow holds, the boundary layer, the rear stagnation point region and the wake. The flow in the rear stagnation point region and the wake is assumed to be essentially inertial. The unsteady drag experienced by the bubble is calculated from the mechanical energy balance of the liquid.

1. Introduction

The viscous drag force on a spherical gas bubble undergoing a steady motion has been investigated by several authors using a boundary layer argument. Levich [3, 4] first calculated the drag experienced by the bubble from the viscous dissipation in the irrotational flow past the bubble. Chao [1] has obtained the drag on a fluid sphere with negligible separation by solving both internal and external flow equations. Moore [5] has improved Levich's result by extending the dissipation method to include the contribution from the boundary layer and the wake.

The case of unsteady rectilinear motion has been studied by Petrov [6], Chen [2] and Slavchev and Simeonov [7]. Petrov has shown that the viscous drag on the bubble is given at the first order by Levich's formula, the dependence on the flow history appearing in small terms of order $0(\mathrm{Re}^{-1/2})$ where Re is the Reynolds number. Using Moore's matching method, Chen has solved the case of a bubble starting impulsively from rest and evaluated the drag from the mechanical energy balance. More recently, Slavchev and Simeonov have considered the initial motion of a bubble the velocity of which is a power function of time and they have solved the stream function equation up to the fourth order by the method of matched asymptotic expansions.

For all the papers mentioned above, the bubble velocity is prescribed. The purpose of the present paper is to derive a general formula for the drag valid for a wider class of unsteady motions.

2. Statement of the problem and strategy of solution

Let a gas bubble move through the unsteady flow of a viscous liquid. Let ρ and ν denote respectively the density and the kinematic viscosity of the liquid,

Applied Scientific Research 38: 247–254 (1982) 0003–6994/82/0383–0247 $01.20.

σ the interfacial tension, and $V_{\infty 0}$ a certain velocity scale representing the order of magnitude of the velocity slip between the bubble and the liquid.

The general assumptions are the following:

– the bubble retains a spherical shape (i.e. the Weber number We = $RV_{\infty 0}^2 \rho/\sigma$ is small), the bubble radius R is constant and there is no mass transfer at the interface which is assumed to be free of surfactants

– the bubble motion is rectilinear with a velocity $U(t)$

– the liquid medium is unbounded and the liquid velocity field at infinity $V_\infty(t)$ is uniform and has the same direction as the bubble motion

– the liquid is incompressible and the variations of the physical properties of the liquid and the gas are negligible, i.e. the temperature variations are very small

– the flow past the bubble is axisymmetric and the body forces field is irrotational and consistent with the hypothesis of axisymmetry

– the Reynolds number Re = $V_{\infty 0}R/\nu$ is large.

The velocity distribution is sought by superposition of a perturbation field on a basic flow which is the potential flow past the bubble. Moore's procedure will be adopted.

The different steps of the solution will be the following:

– when the Reynolds number is large, dimensional analysis of the liquid equations shows that the tangential boundary condition at the interface requires the introduction of a boundary layer (BL)

– the BL equations will be solved and the solution will be shown to exhibit a singularity in the neighbourhood of the rear stagnation point (RSP) so that the assumptions which have been used for the derivation of the BL equations will break down

– the RSP region will be modelled by assuming that the flow in the neighbourhood of the RSP is essentially inertial and that the streamlines of the actual flow are only slightly displaced from those of the potential flow

– the same assumptions will be used to model the near wake and the velocity perturbation (VP) will be obtained

– the VP in the distant wake where the viscous effects cannot be neglected any more will be evaluated

– the force exerted by the liquid on the bubble during the unsteady motion will be calculated from the mechanical energy balance of the liquid.

3. The boundary layer

In order to avoid the problem of BL reversal, the velocity slip $U^0(t) = U(t) - V_\infty(t)$ is assumed to retain the same sign at all times t. The equations are derived in a reference frame linked to the centre of the bubble and a system of spherical coordinates (r, θ, φ) is used, where θ is measured from the front stagnation point. For the sake of convenience, the coordinate $x = r - R$ will be used instead of r. The perturbation equations are obtained by substracting the basic flow equations from the Navier-Stokes equations.

The equations are made dimensionless by introducing a scale for each quantity. Let t_0 denote a certain time scale characterizing the transient regime, and let δR and p_0 denote respectively the scales for the length in the r-direction and the pressure perturbation (PP). The length scale in the θ-direction is assumed to be a characteristic geometrical dimension of the bubble, e.g. the radius R. The continuity equation yields a first relationship between the scales. If the dynamic viscosity of the gas is much smaller than that of the liquid, the tangential boundary condition gives rise to the so-called tangential stress-free condition which yields another relationship between the scales.

Far from the bubble, the potential flow is consistent with the boundary conditions. Near the interface, viscous terms have to remain in the equations in order to satisfy the tangential boundary condition, even if the Reynolds number is very large. Substituting the different scales in the perturbation equations and taking into account the previous requirement, the condition $\delta = 0(\mathrm{Re}^{-1/2})$ is obtained. It can be shown that the Euler number $\mathrm{Eu} = p_0/\rho V_{\infty 0}^2$ is $0(\mathrm{Re}^{-1})$ and the time-derivative term will remain if the Strouhal number $\mathrm{St} = t_0 V_{\infty 0}/R$ is $0(1)$.

Retaining the first-order terms with respect to $\mathrm{Re}^{-1/2}$, the following BL equations are obtained:

$$\frac{\partial v_r}{\partial x}\sin\theta + \frac{\partial}{\partial\theta}(v_\theta \sin\theta) = 0; \qquad 3U^0(t)v_\theta \sin\theta = \frac{\partial p}{\partial x};$$

$$\frac{\partial v_\theta}{\partial t} - 3U^0(t)x\frac{\partial v_\theta}{\partial x}\cos\theta + \tfrac{3}{2}U^0(t)\frac{\partial}{\partial\theta}(v_\theta \sin\theta) = \frac{\partial^2 v_\theta}{\partial x^2}; \tag{1}$$

$$v_r(x,\theta,0) = v_{r0}(x,\theta); \qquad v_\theta(x,\theta,0) = v_{\theta 0}(x,\theta); \tag{2}$$

$$v_\theta(\infty,\theta,t) = 0; \tag{3}$$

$$v_r(0,\theta,t) = 0;$$

$$\frac{\partial v_\theta}{\partial x}(0,\theta,t) = 3U^0(t)\sin\theta; \tag{4}$$

$$p(\infty,\theta,t) = 0;$$

where v_θ, v_r and p are respectively the components of the VP and the PP and where $\delta = \mathrm{Re}^{-1/2}$, $\mathrm{St} = 1$ and $\mathrm{Eu} = \mathrm{Re}^{-1}$ have been adopted for the sake of simplicity. The same symbols as for the dimensional quantities have been used to denote their dimensionless counterparts. The quantities v_{r0} and $v_{\theta 0}$ are the prescribed initial conditions.

The solution of the problem (1) to (4) is:

$$v_\theta(x,\theta,t) = v_{\theta i}(x,\theta,t) + v_{\theta l}(x,\theta,t)$$

with

$$v_{\theta i}(x,\theta,t) = \frac{\gamma^3}{\sqrt{4\pi\alpha}} \int_0^\infty v_{\theta 0}(h,\boldsymbol{\theta}) \left\{ \exp\left[-\frac{(x-h\gamma^2)^2}{4\alpha}\right] + \exp\left[-\frac{(x+h\gamma^2)^2}{4\alpha}\right] \right\} dh$$

$$v_{\theta l}(x,\theta,t) = -\int_0^t 3U^0(\tau)\sin\theta \frac{r^4}{\sqrt{\pi\beta}} \exp\left(-\frac{x^2}{4\beta}\right) d\tau \tag{5}$$

$$\boldsymbol{\theta}(\theta,t) = 2\ \mathrm{Arctg}\left\{\mathrm{tg}\frac{\theta}{2}\exp\left[-\tfrac{3}{2}H(t)\right]\right\};$$

$$H(t) = \int_0^t U^0(\tau)\,d\tau$$

$$r(\theta,t,\lambda) = \frac{1}{\sin\theta}\,\frac{1}{\mathrm{ch}\{\tfrac{3}{2}[H(t)-H(\lambda)] + \mathrm{Argthcos}\,\theta\}};$$

$$\alpha(\theta,t) = \int_0^t r^4(\theta,t,\lambda)\,d\lambda$$

$$\beta(\theta,t,\tau) = \int_\tau^t r^4(\theta,t,\lambda)\,d\lambda;$$

$$\gamma(\theta,t) = r(\theta,t,0).$$

The solutions for v_r and p are then easily obtained. The component $v_{\theta i}$ can be shown to vanish with time. Henceforth, only the component $v_{\theta l}$ will be taken into consideration.

4. The rear stagnation point region

From (5), it can be shown that v_θ has a singularity for $\theta = \pi$. Because of the continuity requirement, v_r also increases sharply. Thus the BL equations become invalid near the RSP.

Now it is assumed that the viscosity is unimportant in the vicinity of the RSP. Taking into account the asymptotic behaviour of the BL solutions near the RSP, the analysis of the orders of magnitude of the different terms in the perturbation equations shows the existence of an overlap region between the BL and the RSP region. Then the matching procedure consists in considering that the vorticity convected from the BL to the RSP region is given by its value in the last stages of the BL. Moreover, it is assumed that the actual streamlines are only slightly displaced from those of the basic flow, so that at the same order of approximation than previously, the vorticity is essentially convected by the basic flow.

In the vicinity of the RSP, the BL thickness is $0(\delta/\epsilon^2)$ where $\epsilon = \pi - \theta$, and the distance of a point to the axis of symmetry is $0(\epsilon)$. Equating these two orders of magnitude yields the size of the RSP region, i.e. $0(\mathrm{Re}^{-1/6})$. Therefore the vorticity is $0(\mathrm{Re}^{-1/6})$ and the VP $0(\mathrm{Re}^{-1/3})$. The size of the RSP region gives an estimate for the location of the separation point corresponding to an angle $\theta_s(t)$.

The curvature of the interface is unimportant and it is possible to adopt a system of cylindrical coordinates (s, φ, z) linked to the RSP with z directed downstream. Let ω, v_s and v_z denote respectively the vorticity, the transverse and axial components of the VP.

The component v_z is solution of Poisson's equation:

$$\Delta v_z = \frac{1}{s}\frac{\partial}{\partial s}(s\omega)$$

with

$$v_z(s,0,t) = v_z(\infty,z,t) = v_z(s,\infty,t) = 0,$$

and the solution is:

$$v_z(s,z,t) = \tfrac{1}{2}\int_0^\infty\int_0^\infty \omega(s',z',t)\,\{[(s-s')^2 + (z-z')^2]^{-3/2}$$

$$- [(s-s')^2 + (z+z')^2]^{-3/2}\}\,(s-s')s'\,ds'\,dz' \qquad (6)$$

with

$$\omega(s,z,t) = -\int_0^t 3U^0(\tau)s\,\frac{\partial}{\partial\tau}\operatorname{erfc}\left(\frac{s^2 z}{2b_s}\right)d\tau;$$

$$b_s^2(t,\tau) = \int_\tau^t r^4\,[\theta_s(t),t,\lambda]\,d\lambda.$$

The component v_s is easily obtained from the continuity equation and yields:

$$v_s(s,z,t) = -\frac{1}{2s}\int_0^\infty\int_0^\infty \omega(s',z',t)$$

$$[f(s,s',z,z') - f(s,s',z,-z')]\,s'\,ds'\,dz' \qquad (7)$$

with

$$f(s,s',z,z') = \frac{[(s-s')^2 + (z-z')^2]^{3/2} - (s-s')^3 + s'(z-z')^2}{(z-z')[(s-s')^2 + (z-z')^2]^{3/2}}.$$

5. The wake

The vorticity created in the BL is convected downstream to form a wake which is fed by the RSP region. Due to the ability of the interface to move, the BL separation is retarded so that the wake will be very thin. The modelling of the wake requires the introduction of two regions, i.e. the near wake where the same assumptions as for the RSP region are used, and the distant wake where the viscous effects become important.

The near wake is characterized by distances from the interface which are of order unity. Equating this order of magnitude with the typical distance from the interface in the RSP region yields the order $0(\mathrm{Re}^{-1/4})$ for the wake width. Therefore the vorticity is $0(\mathrm{Re}^{-1/4})$ and the VP $0(\mathrm{Re}^{-1/2})$. The analysis of the orders of magnitude of the different terms involved in the z-momentum equation yields the domain of validity of the inertial flow assumption, i.e. $0 < z' \ll \mathrm{Re}^{1/2}$ where z' is the coordinate z referred to the bubble radius.

Assuming that the actual streamlines are practically those of the basic flow, one obtains from the vorticity:

$$v_z(s,z,t) = -\int_0^t \frac{9}{2\sqrt{\pi}} U^0(\tau) \frac{r_s^4}{b_s} \exp\left(-\frac{s^4}{36 b_s^2}\right) \mathrm{d}\tau \qquad (8)$$

with

$$r_s(t,\tau) = \sin\theta_s(t) r[\theta_s(t), t, \tau].$$

The distant wake is characterized by the fact that the viscous terms balance the inertia terms. Therefore, the relevant length scale along the axis is $0(\mathrm{Re}^{1/2})$. Assuming that the axial component of the VP and the PP have the same scale as in the near wake, one obtains:

$$\frac{\partial v_z}{\partial t} + U^0(t)\frac{\partial v_z}{\partial z} = \frac{\partial^2 v_z}{\partial s^2} + \frac{1}{s}\frac{\partial v_z}{\partial s},$$

$$v_z(s,z,0^-) = \frac{KU_0}{4\pi z} \exp\left(-\frac{U_0 s^2}{4z}\right),$$

$$v_z(\infty,z,t) = 0, \qquad U_0 = U^0(0^-),$$

where K is a constant to be determined by the z-boundary condition. The solution is:

$$v_z(s,z,t) = \frac{KU_0}{4\pi[z + U_0 t - H(t)]} \exp\left\{-\frac{U_0 s^2}{4[z + U_0 t - H(t)]}\right\}.$$

6. The drag on the bubble

Taking into account the interfacial stress-free condition and the behaviour of the basic flow field at infinity, the mechanical energy balance in a reference frame linked to the liquid velocity at infinity can be written, after having removed the contribution of the basic flow:

$$D(t)U^0(t) = -\frac{\partial e}{\partial t} - \int_{\mathscr{D}} \Phi \, d\tau - \int_{\mathscr{D}} \varphi \, d\tau - 4\mu \int_{\mathscr{S}} \mathbf{v} \cdot (\mathbb{D} \cdot \mathbf{n}) \, d\sigma \tag{9}$$

where $D(t)$ is the viscous drag, e the kinetic energy of the perturbation flow, Φ and φ respectively the dissipation functions of the basic and perturbation flows, μ the dynamic viscosity of the liquid, $\mathbf{v}$ the VP, $\mathbb{D}$ the rate-of-strain tensor corresponding to the basic flow, $\mathscr{D}$ the liquid domain, $\mathscr{S}$ the bubble surface and $\mathbf{n}$ the unit normal vector directed outwardly with respect to the liquid.

Now the orders of magnitude of the different terms involved in (9) and issued from the different flow regions have to be determined. For the external flow region, the dissipation term is $0(\mathrm{Re}^{-1})$ and yields Levich's result $\hat{\varphi}_E = 12\pi\mu R[U^0(t)]^2$ where $\hat{\varphi}$ denotes the dissipation term. For the BL and the wake, both $\partial e/\partial t$ and $\hat{\varphi}$ are of the same order $0(\mathrm{Re}^{-3/2})$. In the PSA region, because of the VP of order $0(\mathrm{Re}^{-1/3})$ and the size of order $0(\mathrm{Re}^{-1/6})$, the $\partial e/\partial t$ term is $0(\mathrm{Re}^{-7/6})$. Lastly, the surface term $P_{\mathscr{S}}$ is $0(\mathrm{Re}^{-3/2})$ for the BL.

Finally, the viscous drag is given [8] by:

$$-\frac{D(t)U^0(t)}{\rho R^2 V_{\infty 0}^3}\mathrm{Re} = 12\pi U^{*2}(t) + \mathrm{Re}^{-1/6}\left(\frac{\partial e}{\partial t}\right)_{\mathrm{RSP}}$$

$$+ \mathrm{Re}^{-1/2}\left[\left(\frac{\partial e}{\partial t}\right)_{\mathrm{BL}} + \hat{\varphi}_{\mathrm{BL}} + \left(\frac{\partial e}{\partial t} + \hat{\varphi}\right)_{\mathrm{W}} + P_{\mathscr{S}}\right]$$

with

$$U^*(t) = \frac{U^0(t)}{V_{\infty 0}},$$

$$\left(\frac{\partial e}{\partial t}\right)_{\mathrm{RSP}} = \pi \frac{\partial}{\partial t}\int_0^\infty\int_0^\infty (v_s^2 + v_z^2)\, s \, ds \, dz,$$

where v_s is given by (7) and v_z by (6)

$$\left(\frac{\partial e}{\partial t}\right)_{\mathrm{BL}} = \pi \int_0^\pi\int_0^\infty \frac{\partial}{\partial t}\left[\int_0^t 3U^0(\tau)\frac{r^4}{\sqrt{\pi\beta}}\right.$$

$$\left.\exp\left(-\frac{x^2}{4\beta}\right) d\tau\right]^2 dx \sin^3\theta \, d\theta,$$

$$\hat{\varphi}_{\mathrm{BL}} = \tfrac{9}{2} \int_0^\infty \int_0^\pi \left[\int_0^t U^0(\tau) \frac{r^4}{\beta^{3/2}} \exp\left(-\frac{x^2}{4\beta}\right) \mathrm{d}\tau \right] x^2 \sin^3\theta \, \mathrm{d}x \, \mathrm{d}\theta ,$$

$$\left(\frac{\partial e}{\partial t} + \hat{\varphi}\right)_{\mathrm{W}} = \tfrac{81}{4} U^0(t) \int_0^\infty \left[\int_0^t U^0(\tau) \frac{r_s^4}{b_s^4} \exp\left(-\frac{s^4}{36 b_s^2}\right) \mathrm{d}\tau \right]^2 s \, \mathrm{d}s ,$$

$$P_{\mathscr{S}} = -36\sqrt{\pi}\, U^0(t) \int_0^t U^0(\tau) \int_0^\pi \frac{r^4}{\sqrt{\beta}} \sin^3\theta \, \mathrm{d}\theta \, \mathrm{d}\tau .$$

7. Conclusion

An expression for the viscous drag on a spherical bubble with an arbitrary velocity slip of constant sign has been derived up to the $\mathrm{Re}^{-3/2}$ order. The most surprising thing is the $\mathrm{Re}^{-7/6}$ term arising from the rate of change of kinetic energy in the RSP region. The formula takes into account a dependence on the history of the velocity slip but unfortunately, it is not exploitable in its present form. However, it could be used to improve the results concerning the models with prescribed slip. An interesting case could be for the velocity field at infinity the superposition of a sinusoidal oscillation on a basic flow and for which it seems reasonable to assume that the velocity slip will be of the same type.

References

1. Chao BT (1962) Motion of spherical gas bubbles in a viscous liquid at large Reynolds numbers. Phys Fluids 5: 69–79.
2. Chen JLS (1974) Growth of the boundary layer on a spherical gas bubble. Trans ASME J Appl Mech 41: 873–878.
3. Levich VG (1949) Motion of a bubble at large Reynolds numbers (in Russian). Zhur Eksp Teoret Fiz 19: 18–24.
4. Levich VG (1962) Physicochemical hydrodynamics. Prentice-Hall.
5. Moore DW (1963) The boundary layer on a spherical gas bubble. J Fluid Mech 16: 161–176.
6. Petrov AG (1971) Nonstationary boundary layer at spherical gas bubble (in Russian). Vestnik Moskovskogo Universiteta, Mekhanika 26: 69–76.
7. Slavchev SG and Simeonov S (1979) The unsteady boundary layer on a spherical gas bubble. ZAMM 59: 43–50.
8. Pham Dan Tam (1981) De la traînée instationnaire sur une petite bulbe. Thèse de doctorat d'Etat ès sciences, Université Scientifique Médicale de Grenoble et Institut National Polytechnique de Grenoble, Grenoble.

On the dynamics of bubbles in polymer aqueous solutions

A. SHIMA and T. TSUJINO

Institute of High Speed Mechanics, Tohoku University, Sendai, Japan

Abstract. By analysing experimentally the rheological behaviors of polyethylene oxide solutions, polyacrylamide solutions, carboxymethylcellulose solutions, and hydroxyethylcellulose solution, a rheological model is proposed. The effect of various polymer additives on an air-liquid surface tension is made clear. The equation of motion for a bubble and the pressure equation are derived by using the new rheological expression, and then the effects of a kind of polymer and a polymer concentration on the bubble radius-time history and impulse pressure are numerically examined.

1. Introduction

Investigations on cavitation phenomena in polymer solutions have been carried on, since it was found that a cavitation occurrence was inhibited by addition of polymers to water [4]. That is, the cavitation suppression due to polymer additives were made clear for various flow conditions, for example, jet flow [5], flows around hemispherically nosed bodies [1, 3], flow on a rotating disk [12], in a venturi tube [6], and in an orifice [7]. In relation to these studies, investigations on the behavior of cavitation bubbles in non-Newtonian fluids attract much attention. Theoretical work has hitherto been conducted for a power-law model [8, 13], a Powell-Eyring model [9], an Oldroyd model [11, 14], etc., and the effects of the rheological parameters in these models on the bubble motion have been examined.

Although various drag reducing polymers are used as additives in cavitation experiments of polymer solutions, not only in the experimental studies, but also in the theoretical studies of bubbles in non-Newtonian fluids, the dependence on which kind of polymer, on the polymer concentration, etc. have never been fully clarified. One of the reasons is that rheological analyses for polymer solutions have not been sufficiently made. In this paper, therefore, first of all the rheological measurements for various polymer aqueous solutions are reported. Since the surface tension of liquid is often an important parameter in cavitation problems, a measurement of the surface tension of polymer solutions was conducted.

Second, the equation of motion for a bubble and the pressure equations were theoretically analysed by the obtained rheological equation. Further, according to numerical calculations, the bubble radius-time history and the impulse pressure occurring during the bubble collapse in polymer solutions were clarified, and the effects of the kind of polymer and the polymer concentration were compared and discussed.

Applied Scientific Research 38: 255–263 (1982) 0003–6994/82/0383–0255 $01.35.

2. Results of rheological measurement for polymer solutions

By using a rheometer of cone-plate type the relation between an apparent viscosity η and a shear rate $\dot{\gamma}$ is measured in the present experiment.

Measured values of apparent viscosities for a PEO solution, a Polyox solution, a PAM (A-20P) solution, a CMC-4H solution, a CMC-BS solution, and a HEC solution are plotted and compared in Figure 1. As shown in this figure, the difference of the viscosities of these polymer solutions appears

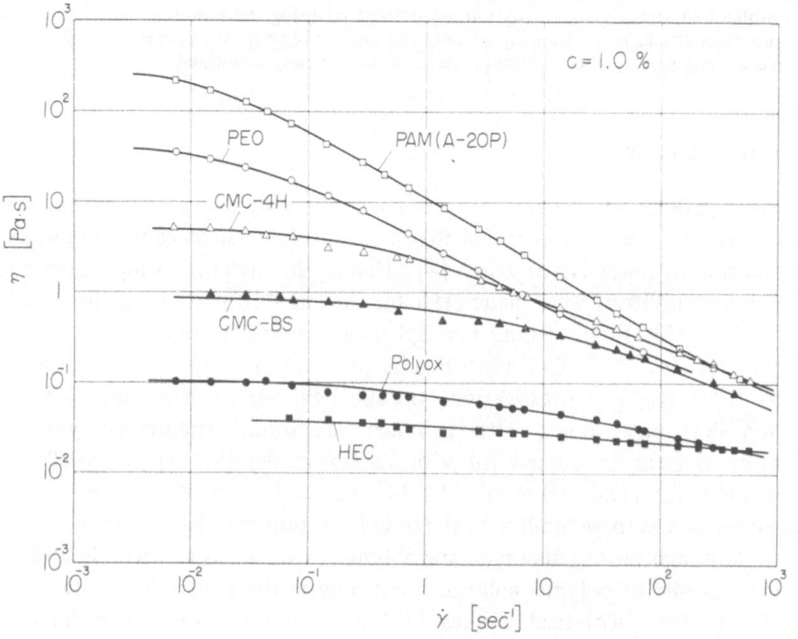

Figure 1. Relations of $\eta - \dot{\gamma}$ in various polymer solutions.

remarkably in a comparatively small range ($\dot{\gamma} \lesssim 10$). The value of η for the PAM (A-20P) solution is largest, second comes η for the PEO solution and then follow the carboxymethylcellulose solutions. The values of η for the Polyox and HEC are relatively small. In the larger range of $\dot{\gamma}$, the value of η for each polymer solution often crosses because of the difference of the inclination of $\eta - \dot{\gamma}$. We shall propose the following four-parameter model as a model which represents non-Newtonian viscosity such as the above type.

$$\eta - \eta_\infty = \frac{\eta_0 - \eta_\infty}{(1 + \lambda|\dot{\gamma}|)^n}, \tag{1}$$

where η_0 viscosity at zero shear rate, η_∞ viscosity at infinite shear rate, λ time constant, n material contant. The solid lines in Figure 1 denote the relation of $\eta - \dot{\gamma}$ of equation (1). Good agreements are found between the solid lines and the measured values, so it may be noted that the apparent viscosities of those polymer solutions are well represented by equation (1). For example the values of the parameters η_0, η_∞, λ and n in equation (1) are listed in Table 1.

Table 1. Values of η_0, η_∞, λ and n.

Polymer solution	Concentration [%]	η_0 [Pa·s]	η_∞ [Pa·s]	λ [s]	n	t_e [K]
PEO solution	0.5	2.983	1.49×10^{-2}	1.63×10	0.600	291.95
	1.0	4.119×10	6.15×10^{-2}	3.19×10	0.719	292.05
Polyox solution	0.5	2.814×10^{-2}	4.52×10^{-3}	3.90	0.302	292.65
	1.0	1.092×10^{-1}	1.08×10^{-2}	3.02	0.306	292.45
PAM(N-110) solution	0.2	2.405	1.57×10^{-2}	2.66×10	0.644	291.55
	0.5	1.354×10	2.19×10^{-2}	6.22×10	0.713	293.75

3. Surface tension

In this section let us clarify the relation between the surface tension σ and the polymer concentration c by the measurement of the surface tension of each polymer solution. The values of the surface tension of PEO solutions, Polyox solutions, HEC solutions, and water are plotted in Figure 2. It is found from the figure that the surface tension of water is reduced by the addition of

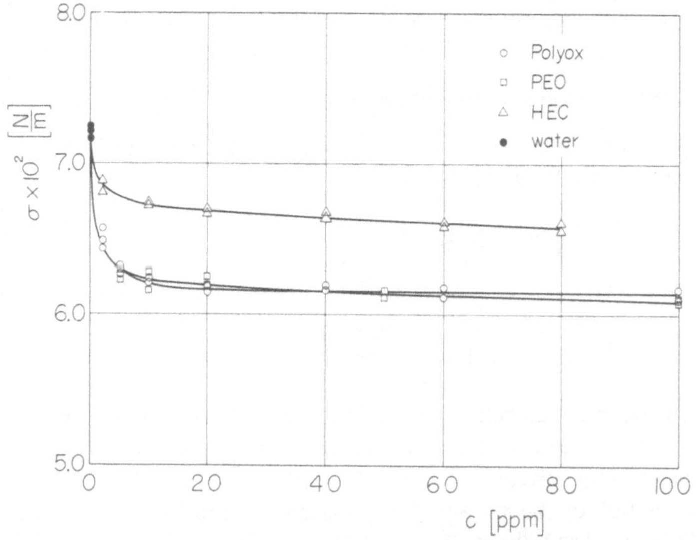

Figure 2. Relations between surface tension and concentration of polymer solutions.

small quantities of these polymers. Figure 3 shows the results for the cases of CMC-4H, CMC-BS, PAM (A-20P), and water. As shown in the figure, the effects of the carboxymethylcellulose and polyacrylamide on the surface tension are smaller, and the values of the surface tension decrease a little, while the polymers are added in fair quantity in water.

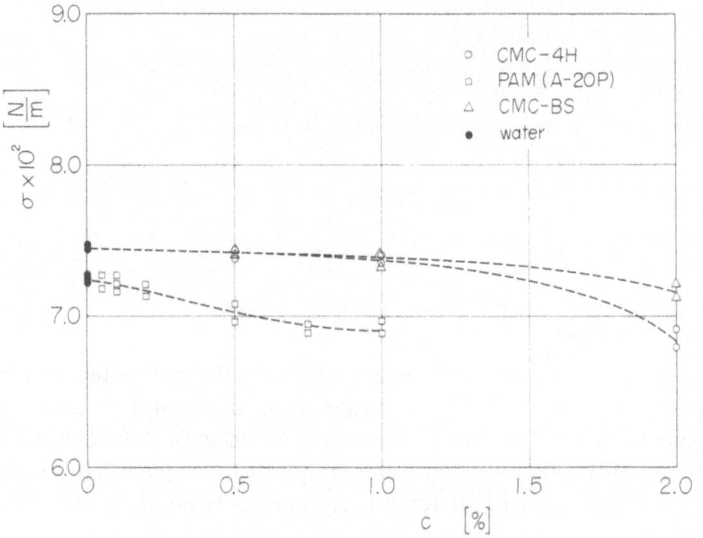

Figure 3. Relations between surface tension and concentration of polymer solutions.

4. Theoretical analysis of bubble behavior

Using spherical co-ordinates, the equations of continuity and of motion for an incompressible fluid are given by

$$\frac{1}{r^2}\frac{\partial}{\partial r}(r^2 v_r) = 0, \tag{2}$$

$$\rho_l\left(\frac{\partial v_r}{\partial t} + v_r\frac{\partial v_r}{\partial r}\right) = \frac{\partial \sigma_{rr}}{\partial r} + \frac{2\sigma_{rr} - \sigma_{\theta\theta} - \sigma_{\varphi\varphi}}{r} \tag{3}$$

where, v_r denotes the velocity component in the r direction, t time, ρ_l liquid density, and σ_{rr}, $\sigma_{\theta\theta}$, and $\sigma_{\varphi\varphi}$, the normal stress components in the r, θ, and φ directions, respectively.

The relation of the viscous stress tensor $\boldsymbol{\tau}$ and the deformation rate tensor $\boldsymbol{\Delta}$ is represented as follows [2]:

$$\boldsymbol{\tau} = -\eta\boldsymbol{\Delta}\,. \tag{4}$$

Substituting equation (1) into equation (4), a general expression of the rheological equation for polymer solutions may be written as follows:

$$\boldsymbol{\tau} = -\left\{\eta_\infty + \frac{\eta_0 - \eta_\infty}{[1 + \lambda|\sqrt{\frac{1}{2}(\boldsymbol{\Delta}:\boldsymbol{\Delta})}|]^n}\right\}\boldsymbol{\Delta}. \tag{5}$$

Hence the motion of a spherical gas bubble in polymer solutions may be governed by the above equations (2), (3), and (5).

As in the former analytical method [10], the following equation of motion for the bubble is derived:

$$\begin{aligned} \beta\frac{d^2\beta}{d\tau^2} + \frac{3}{2}\left(\frac{d\beta}{d\tau}\right)^2 &= \frac{q}{\beta^{3\gamma}} - 1 - \frac{c_3}{\beta} - \frac{c_0}{\beta}\frac{d\beta}{d\tau} \\ &\quad - \frac{\dfrac{c_1}{\beta}\dfrac{d\beta}{d\tau}}{\left[1 + c_2\left|\dfrac{1}{\beta}\dfrac{d\beta}{d\tau}\right|\right]^n} + n\frac{c_1}{c_2}\left|\frac{1}{\beta}\frac{d\beta}{d\tau}\right|^{-1}\cdot\left(\frac{1}{\beta}\frac{d\beta}{d\tau}\right) \\ &\quad \times\left\{\frac{1}{1-n}\left[1 - \left(1 + c_2\left|\frac{1}{\beta}\frac{d\beta}{d\tau}\right|\right)^{1-n}\right]\right. \\ &\quad \left. + \frac{1}{n}\left[1 - \frac{1}{\left(1 + c_2\left|\dfrac{1}{\beta}\dfrac{d\beta}{d\tau}\right|\right)^n}\right]\right\}, \end{aligned} \tag{6}$$

where

$$\beta = \frac{R}{R_0}, \quad \tau = \frac{t}{R_0}\sqrt{\frac{p_\infty}{\rho_l}}, \quad q = \frac{p_0}{p_\infty}, \quad c_0 = \frac{M_0}{p_\infty}\left(\frac{1}{R_0}\sqrt{\frac{p_\infty}{\rho_l}}\right),$$

$$c_1 = \frac{M_1}{p_\infty}\left(\frac{1}{R_0}\sqrt{\frac{p_\infty}{\rho_l}}\right), \quad c_2 = M_2\left(\frac{1}{R_0}\sqrt{\frac{p_\infty}{\rho_l}}\right), \quad c_3 = \frac{2\sigma}{R_0 p_\infty},$$

$$M_0 = 4\eta_\infty, \quad M_1 = 4(\eta_0 - \eta_\infty), \quad M_2 = 2\sqrt{3}\,\lambda,$$

and R is the bubble radius, R_0 the initial bubble radius, p_∞ the pressure far from the bubble, p_0 the initial gas pressure in the bubble, σ surface tension γ the ratio of specific heats for the gas in the bubble.

For the pressure on the bubble wall is obtained the following equation

$$\frac{p_w}{p_\infty} = \frac{q}{\beta^{3\gamma}} - \frac{c_3}{\beta} - \frac{c_0}{\beta}\frac{d\beta}{d\tau} - \frac{c_1}{\beta}\frac{d\beta}{d\tau}\Big/\left[1 + c_2\left|\frac{1}{\beta}\frac{d\beta}{d\tau}\right|\right]^n, \tag{8}$$

where p_w is the pressure on the bubble wall.

The maximum impulse pressure p_{max} may be obtained as $R = R_{min}$ (minimum radius of the bubble) and $d\beta/d\tau = 0$ in equation (8). Also, the initial conditions in numerical analyses are $\beta = 1$ and $d\beta/d\tau = 0$ at $\tau = 0$.

5. Calculated results and discussion

According to the numerical analyses of the derived equations, comparisons of the bubble radius-time histories and the maximum impulse pressures occurring during the bubble collapse in PEO solutions, Polyox solutions, PAM (N-110) solutions, PAM (A-20P) solutions, CMC-4H solutions, CMC-BS solutions, and HEC solutions were made. The initial bubble radius was chosen to be $R_0 = 1 \sim 0.01$ mm, the pressure ratio of the inside and outside of the bubble was $q = 0.01$, and the ratio of specific heats for the gas inside the bubble was $\gamma = 1.4$. The pressure far from the bubble wall $p_\infty = 101.3$ kPa.

5.1. Comparison of bubble radius-time histories in various polymer solutions

Figure 4 shows a comparison of the bubble radius-time histories in a PEO solution, a Polyox solution, a PAM (A-20P) solution, a CMC-4H solution, a HEC solution, and water, where the polymer concentration is $c = 1.0\%$. As shown in the figure, PEO is the polymer additive which makes the bubble oscillation damp very strongly. Secondly, PAM (A-20P) and CMC-4H are more effective in the suppression of bubble motion. Although the damping of the oscillation of bubbles in the Polyox solution and HEC solution are larger than in the case of water, the effects are less than the other polymers. When $c = 0.5\%$, the damping of the curve of the bubble radius-time in a PAM (N-110) solution was largest. Also, the effects of the polymer concentration and the size of the initial bubble radius on the behavior of the bubble were numerically examined. Consequently, it was found that the larger the polymer concentration and the initial bubble radius, the larger the damping in time of the bubble radius.

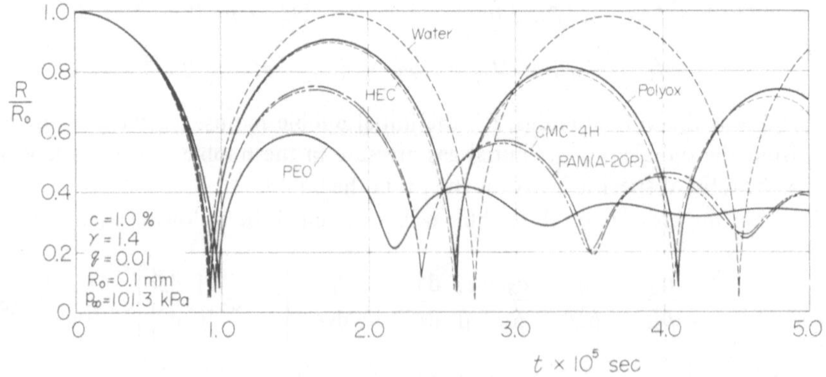

Figure 4. Comparison of bubble radius-time histories in various polymer solutions.

In the investigations on the cavitation inhibition by the additives, such as the drag reducing polymers, the difference from the case of water has been mainly discussed so far, and a detailed examination based on the kind of the polymer will be desired in the future. It may be also expected, by the result of this study, that PEO and PAM have more suppressive action on the motion of cavitation bubbles than the other polymers.

5.2. Maximum impulse pressures

The maximum impulse pressure p_{max} occurs on the bubble wall when the bubble reaches its minimum radius.

Figures 5 (a) and (b) show the relations of $p_{max} - c$ in various polymer solutions for the cases where $R_0 = 0.1$ and 0.01 mm, respectively. In the concentration limitation of $c \lesssim 0.5\%$, polyacrylamide, particularly PAM (N-110), is the most effective polymer to decrease p_{max}. Second, the polymer effects in the PEO solution and CMC-4H solution are larger. The decreasing rates of p_{max} in the Polyox solution, CMC-BS solution and HEC solution are less than the other polymer solutions. In the limitation of $c \gtrsim 1.0\%$, the value of p_{max} in the PEO solution is least, and further CMC-4H, CMC-BS, Polyox, and HEC are put in order. Either way, it may be expected from the present result that an addition of polymer into water will suppress the cavitation damage. Further, it will be anticipated that the kind of polymer and the concentration of the solution have a great influence on the cavitation damage in polymer solutions.

6. Conclusion

Rheology of polymer solutions, used often in cavitation experiments, were experimentally analysed, and a rheological model was proposed. By conducting the measurement of the surface tensions for various polymer solutions, the effects of the polymer additives on the surface tension were found.

By using the derived rheological equation, the equation of motion of a spherical gas bubble and the pressure equation were obtained. According to the numerical calculations for the bubbles in various polymer solutions, the effects of polymer kind and polymer concentration on the bubble radius-time histories and the maximum impulse pressures were made clear.

Acknowledgements

The authors wish to express their hearty thanks to Prof. Emeritus F. Numachi of Tohoku University for his useful suggestions in this paper. The authors are also grateful to Mr. N. Miura and Miss S. Takahashi for their assistance in making the manuscript.

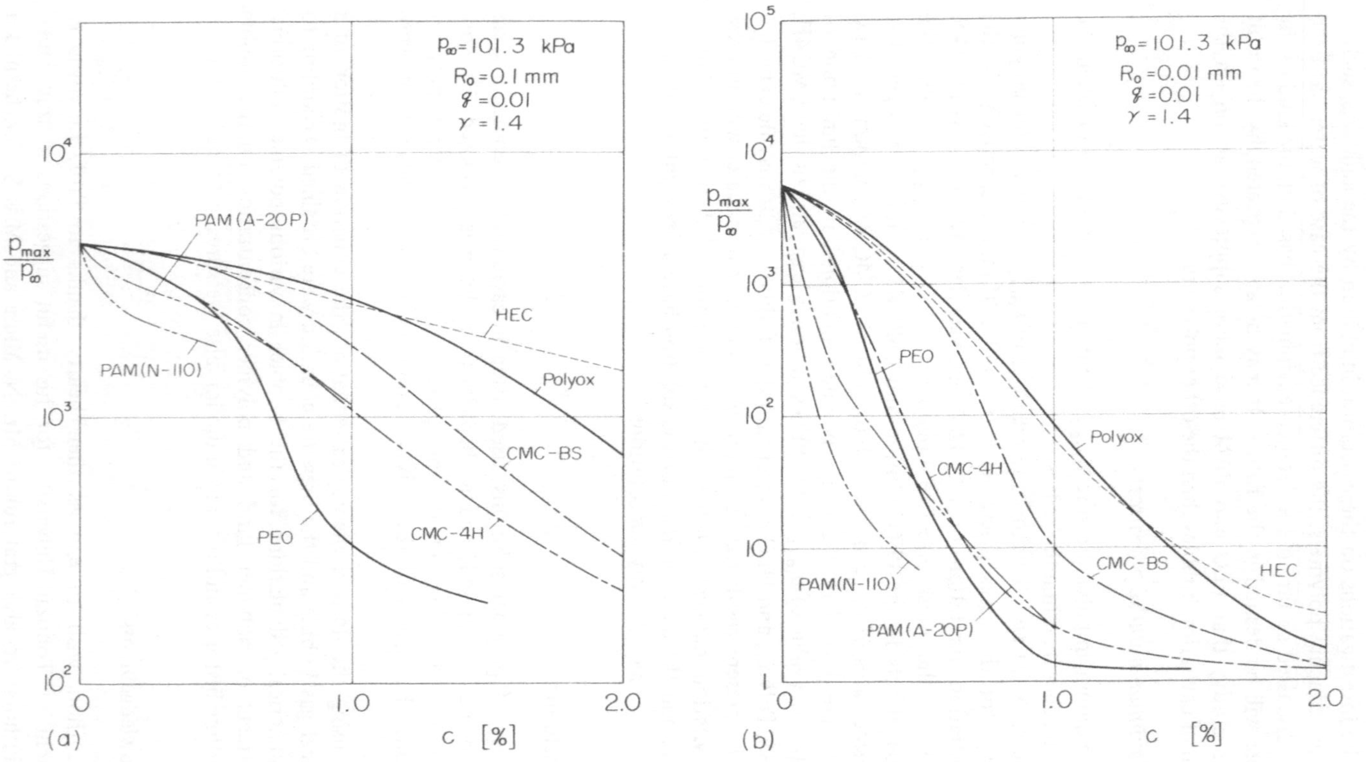

Figure 5. Comparison of $p_{max} - c$ curves in various polymer solutions.

References

1. Baker CB, Arndt REA and Holl JW (1973) Inst Sci Engng Appl Res Lab The Penn State Univ Rep TM73-257: 1–30.
2. Bird RB, Stewart WE and Lightfoot NE (1960) Transport phenomena. N Y: Wiley & Sons.
3. Ellis AT, Waugh JG and Ting RY (1970) J Basic Engng Trans ASME 92: 459–466.
4. Hoyt JW (1967) Naval Ord Test Start Rep TP 4364.
5. Hoyt JW (1976) J Fluids Engng Trans ASME 98: 106–112.
6. Murai H, Watanabe H and Katagiri K (1980) IAHR 10th Symp 1: 65–76.
7. Ōba R, Itō Y and Uranishi K (1978) J Fluids Engng Trans ASME 100: 493–499.
8. Shima A and Tsujino T (1977) Chem Engng Sci 31: 863–869.
9. Shima A and Tsujino T (1981) Chem Engng Sci 36: 931–935.
10. Shima A and Tsujino T (1978) J Appl Mech Trans ASME 45: 37–42.
11. Tanasawa I and Yang W-J (1970) J Appl Phys 41: 4526–4531.
12. Ting RY (1978) Phys Fluids 21: 898–901.
13. Yang W-J and Yeh H-C (1966) AIChE J 12: 927–931.
14. Yang W-J and Lawson ML (1974) J Appl Phys 45: 754–758.

PART FIVE

Collective phenomena in bubbly liquids

Mathematical modelling of bubbly liquid motion and hydrodynamical effects in wave propagation phenomenon

R.I. NIGMATULIN

Institute of Mechanics of the M.V. Lomonosov Moscow State University, Moscow V-234, USSR

1

Certain effects occurring in the propagation of shock waves in a liquid containing small gas or vapour bubbles are considered. The state-of-the-art in this area up to 1966 has been reviewed by Batchelor [2], and up to 1972 by van Wijngaarden [33].

The wave processes are investigated on vertical shock tubes shown schematically in Figure 1. The tube consists of a high pressure chamber (1) (HPC) and a low pressure chamber (2) (LPC) separated by a diaphragm (3). Gas is pumped into the HPC to create high pressure and some liquid is filled into the LPC to a level slightly below the diaphragm. Bubbles of a given radius a are injected from below into the liquid to produce an approximately monodispersed mixture. The radius of bubbles in different experiments varied from 0.2 to 2 mm, and their volume content, α_2, as measured by the increase in the liquid column height, varied from 0.01 to 0.1. After the rupture of the diaphragm, a shock wave travels downwards through the bubble mixture in the LPC. Wave evolution in the mixture is recorded by quick-response pressure transducers (4) fixed on the LPC wall at several heights. Typical oscillograms published in [7, 15, 17, 18, 22, 28, 33] are shown in Figure 2.

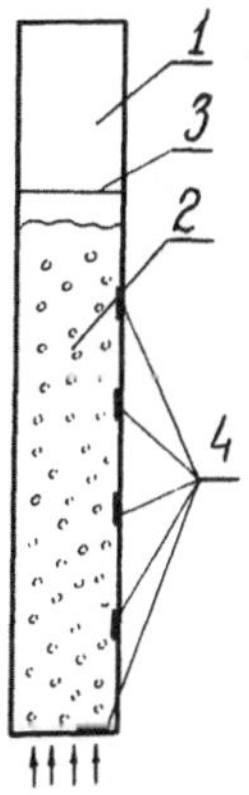

Figure 1.

Applied Scientific Research 38: 267–289 (1982) 0003–6994/82/0383–0267 $03.45.

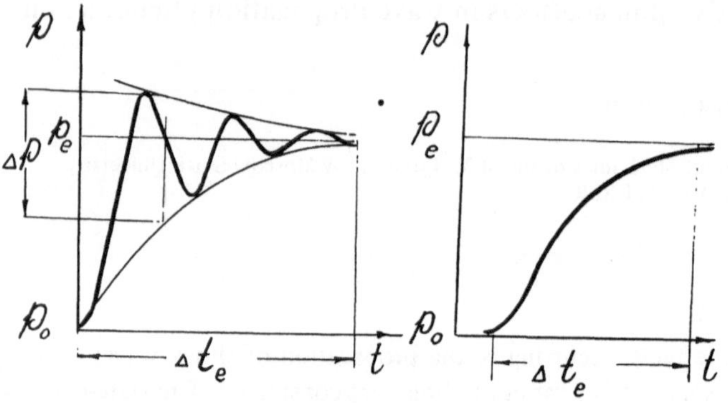

Figure 2.

The main feature of these shock waves in liquids containing small bubbles of insoluble and non-condensable (inert) gas is that under certain conditions they are of oscillatory type due to the pulsations of the bubble volume resulting from the radial inertia of the liquid and elasticity of the gas.

2

In studying these waves we shall consider a system of equations, which takes into account the non-equilibrium of phases in velocities, pressures and temperatures, to describe one-dimensional flows of a monodispersed bubble mixture. At first we shall confine ourselves to the case where there are no phase transitions or fragmentation of bubbles.

Details of the derivation of these equations and a review of the topic are given in [26]. The equations of conservation of masses of phases under the assumption that the liquid is incompressible, and the equation for the number of bubbles can be expressed as follows:

$$\frac{\partial \alpha_1}{\partial x} + \frac{\partial \alpha_i v_i}{\partial x} = 0, \qquad \frac{\partial n}{\partial t} + \frac{\partial n v_2}{\partial x} = 0,$$

$$\rho_i^0 = \text{const}, \qquad \frac{\rho_2^0}{\rho_{20}^0} = \left(\frac{a_0}{a}\right)^3, \qquad \alpha_2 = 1 - \alpha_1. \tag{2.1}$$

The subscripts 1, 2 show that the parameters pertain to liquid and gas, respectively; while the subscript 0 denotes that the values of the parameters in the initial state; ρ_i^0, α_i, v_i are the true densities, volume contents of phases and their macroscopic velocities along the X-axis; n is the number of bubbles in a unit volume of the mixture. As regards the assumption that the liquid is incompressible (ρ_1^0 = const) we can show that if

$$\alpha_2 \gg \alpha_C = p/\rho_1^0 C_1^2,$$

where C_1 is the velocity of sound in a liquid without bubbles, compression of mxiture takes place solely due to the compression of the gas, while the liquid remains virtually incompressible. Under the usual conditions, we have $\alpha_C \sim 10^{-4}$. Therefore, for a volume content of bubbles $\alpha_2 \gtrsim 0.01$, the assumption that the liquid phase is incompressible is quite justified.

Assuming that the bubbles are always spherical in shape, and that their volume content is small ($\alpha_C \ll \alpha_2 \ll 1$), we obtain the following expressions for the volume contents of the phases:

$$\alpha_2 = \tfrac{4}{3}\pi a^3 n, \qquad \alpha_1 = 1 - \alpha_2 \approx 1. \tag{2.2}$$

We shall describe the size of bubbles along their streamlines by the equation of combined deformation in the form of the classical Rayleigh-Lamb equation derived under the assumption that the liquid is incompressible, that the volume content of bubbles is small, their spherical shape is conserved, and also taking into account the radial inertia and viscosity of the liquid:

$$\frac{\mathrm{d}_2 a}{\mathrm{d}t} = w_1,$$

$$a\frac{\mathrm{d}_2 w_1}{\mathrm{d}t} + \frac{3}{2}w_1^2 - \frac{v_{12}^2}{4} + \frac{4\mu_1}{\rho_1^0 a} w_1 = \frac{p_2 - 2\Sigma/a - p_1}{\rho_1^0}. \tag{2.3}$$

Here w_1 is the radial velocity of the liquid at the bubble wall; $v_{12} = v_1 - v_2$ is the velocity of the flow round a bubble; p_i ($i = 1, 2$) are the average pressures of phases*; Σ, μ_1 are the coefficients of surface tension and the viscosity of the liquid. We shall also use the total derivative along the streamline of the i-th phase:

$$\frac{\mathrm{d}_i}{\mathrm{d}t} = \frac{\partial}{\partial t} + v_i\frac{\partial}{\partial x}.$$

The equations of motion or of momentum of phase solved for accelerations with an account of interphase force due to virtual mass are of the form [26, 27]:

$$\alpha_1\rho_1^0\frac{\mathrm{d}_1 v_1}{\mathrm{d}t} = -\frac{\partial p_1}{\partial x} + \rho_1^0\alpha_1 g,$$

*The corrections due to the fact that a bubble is not single in the mixture and the volume content of bubbles in finite have been ignored (see [26, 27]), which, for $\alpha_2 \ll 1$, is fully justified for our purposes.

$$\frac{d_2 v_2}{dt} = -\frac{3}{\rho_1^0 \alpha_1}\frac{\partial p_1}{\partial x} + \frac{3}{2\pi a^3 \rho_1^0} f_\mu + \frac{3\alpha_1^2}{a} w_1 v_{12} + g, \tag{2.4}$$

where g is the intensity of external mass force, e.g., acceleration due to gravity; f_μ is the force acting on bubbles due to the viscosity of the liquid. We shall not give the expressions for this force (see [26, 31]), because, as will be shown below, they are unimportant in the analysis of wave processes. Here the action of interphase force nf_μ due to bubbles can be neglected in the equation of motion of the liquid. This is justified because the mass of bubbles is negligibly small as compared with the mass of the liquid ($\rho_2^0\alpha_2 \ll \rho_1^0\alpha_1$).

The pressure of the gas in bubbles p_2 is described by the equation of heat influx to the gas. As the pressure (unlike the temperature) in a bubble is uniform, this equation reduces to the form:

$$\frac{d_2 p_2}{dt} = \frac{3(\gamma - 1)}{4\pi a^2} q_2 - \frac{3\gamma p_2}{a} w_2, \tag{2.5}$$

where γ is the adiabatic exponent of the gas; w_2 is the radial velocity of the gas on the bubble wall; q_2 is the heat flux from the liquid per bubble. Determination of q_2 is discussed below. The second term on the right-hand side describes the work of the internal pressure forces, and in the absence of phase transitions, we have

$$w_1 = w_2 = \frac{d_2 a}{dt}. \tag{2.6}$$

It should be noted that the negligible mass of bubbles cannot perceptibly change the liquid temperature T_1, so we can assume $T_1 = T_0 = \text{const}$.

Sometimes, a simplified scheme of a polytropic gas is used to determine the gas pressure p_2

$$\frac{p_2}{p_{20}} = \left(\frac{a_0}{a}\right)^{3\kappa} = \left(\frac{\rho_2^0}{\rho_{20}^0}\right)^{\kappa}, \tag{2.7}$$

where $1 \leqslant \kappa \leqslant \gamma$ is the polytropic exponent.

If the expression for the heat influx q_2 to a bubble and that for the viscid force f_μ acting on a bubble are known in terms of the parameters introduced for the mixture, the equations (2.1)–(2.6) become a closed system.

3

In order to study the volume deformations of a bubble, it is worth considering the solutions of the Rayleigh-Lamb equations (2.3) and the gas pressure equation (2.5) under certain specific laws of liquid pressure variations along the streamline of a test bubble which are typical of shock waves*.

As was already mentioned, the interphase heat exchange described by q_2 ought to be specified in the differential equation for the gas pressure. For this purpose, in the general case we have to solve the nonstationary equation of heat conductivity in a compressible gas and incompressible liquid with a moving boundary $a(t)$ separating the gas and the liquid. This boundary is determined in the course of solution. The formulation of this problem and its solutions are given in [26] with references to original papers, as well as in [25] and [30].

For a gas bubble, in the absence of phase transitions, the external problem of heat conductivity which is related to the solution of the equation of heat conductivity in a liquid is unimportant, because the temperatures of the gas and of the liquid may be taken to be a certain constant T_0 for moderate actions on the bubble wall. For a fixed action $p_1(t)$, this process is determined by the following parameters in the initial state: a_0, p_0, ρ_{20}^0, γ, λ_2 (thermal conductivity coefficient of the gas), c_{p2} (specific heat capacity at constant pressure), Σ_0, μ_1 which form four independent dimensionless parameters:

$$\gamma, \qquad \nu_* = \frac{\nu_2}{a_0\sqrt{p_0/\rho_1^0}}, \qquad \Sigma_* = \frac{\Sigma_0}{a_0 p_0},$$

$$\mu_* = \frac{\mu_1}{a_0\sqrt{p_0/\rho_1^0}} \qquad (\nu_2 = \lambda_2/\rho_{20}^0 c_{p2}), \tag{3.1}$$

where ν_2 is the thermal diffusivity of the gas. These four dimensionless parameters (3.1) which determine the properties of the gas seem to be rather too many to a simple problem of spherically symmetric motion of an incompressible liquid round a bubble filled with insoluble and noncondensable gas. Here lies the main difficulty in studying the processes occuring in two-phase media characterized by a large number of thermo-physical parameters of phases.

For $p = p_0 =$ constant, first consider small-amplitude free oscillations of a gas bubble of the type:

$$a(t) = a_0\left[1 + A_0 \exp\left(-\frac{\Lambda\omega_r t}{2\pi}\right)\cdot \sin\omega_r t\right] \tag{3.2}$$

$$(A_0 \ll 1).$$

This problem has been solved by Chapman and Plesset [5] (see also [26]).

*In the system of equations (2.1)–(2.6) describing the joint motion of phases, the change in the bubble volume and translatory motion of the liquid are interrelated precisely through the pressure p_1.

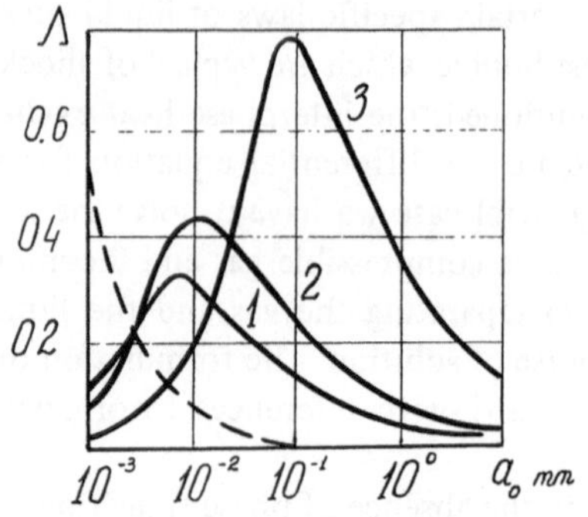

Figure 3.

Thus, the following expressions have been obtained for the frequency ω_r of free oscillations and for the logarithmic decrement Λ for large Peclet numbers $\mathrm{Pe}_2 \gg 1$ when a thin temperature boundary layer exists in the gas near the bubble wall:

$$\omega_r = \frac{1}{a_0}\sqrt{\frac{3\gamma p_0}{\rho_1^0}}, \qquad \Lambda = \Lambda^{(\mu)} + \Lambda^{(T)}$$

$$\left(\mathrm{Pe}_2 = \frac{2a_0}{\nu_2}\sqrt{\frac{3\gamma p_0}{\rho_1^0}}\right),$$

$$\Lambda^{(\mu)} = \frac{4\pi\mu_*}{\sqrt{3\gamma}} = \frac{4\pi\mu_1}{\rho_1^0 a_0^2 \omega_r},$$

$$\Lambda^{(T)} = \frac{3\pi(\gamma-1)}{\sqrt{\mathrm{Pe}_2}} = \frac{3\pi(\gamma-1)}{\sqrt[4]{12\gamma}}\sqrt{\nu_*}. \tag{3.3}$$

In Figure 3 the logarithmic decrements caused by liquid viscosity $\Lambda^{(\mu)}$ and by gas thermal conductivity $\Lambda^{(T)}$ are shown as a function of the size of bubbles of carbon dioxide, or air, of helium in water at $T_0 = 300\,\mathrm{K}$. The broken line corresponds to $\Lambda^{(\mu)}$, while the curves 1, 2 and 3 to $\Lambda^{(T)}$ for carbon dioxide, air and helium, respectively. Obviously, for $a_0 > 10^{-2}$ mm (this includes the range of bubble radius 0.2–2.0 mm considered in this paper) the pulsations of bubble volume in liquids of a viscosity of the order of the viscosity of water are mainly attenuated by the thermal diffusivity of the gas ν_2 rather than by the liquid viscosity μ_1, i.e. thermal dissipation is predominant. Dissipation due to the liquid viscosity may be predominant only in highly viscous liquids. For example, for air bubbles in pure glycerine (whose viscosity is 10^3 times that of water) for $a_0 = 1$ mm we have $\Lambda^{(T)}/\Lambda^{(\mu)} = 0.17$.

Heat dissipation of kinetic energy is related to the irreversibility or non-polytropicity of the processes taking place in a gas, e.g. during compression, where the gas temperature is higher than the liquid temperature T_0, gas emits more heat into the liquid than the liquid returns heat during expansion where the gas temperature is lower than the liquid temperature T_0.

From curves 2 and 3 in Figure 3, which show the logarithmic decrements for air and helium bubbles in water, we find that under similar conditions the radial or volume oscillations of helium bubbles would damp faster than those of the air bubbles. Although this fact, which follows from a rigorous solution [5] of the mathematical problem repeatedly discussed in the literature, is rather unusual, nevertheless, it is very essential in studying wave propagation in bubbly liquids.

Since heat dissipation is predominant under radial oscillations of bubbles, we shall consider simplified models of equations which approximately take account of heat dissipation.

4

The simplest is the model with effective or increased viscosity which is based on the Rayleigh-Lamb equation (2.3) and the polytropic gas equation (2.7). In this model to make allowance for heat dissipation, instead of solving the temperature problem and determining the interphase heat transfer q_2, in the Rayleigh-Lamb equation the viscosity μ_1 is to be replaced by the effective or increased viscosity μ which gives the same logarithmic decrement (say, under free oscillations) as the exact solution (3.3):

$$\mu = \mu_1 + \mu^{(T)},$$

$$\frac{4\sqrt{2}}{3(\gamma-1)}\,\frac{\mu^{(T)}}{\rho_1^0} = a_0\sqrt{\nu_2\omega_r} = (a_0\nu_2)^{1/2}\left(\frac{3\gamma p_0}{\rho_1^0}\right)^{1/4}. \tag{4.1}$$

This model or its generalization to frequencies ω other than the frequency ω_r of free oscillations may only give an adequate description for oscillations close to free oscillations or in those areas where the oscillation frequency is known. In wave analysis this model can be used only as a first approximation because the oscillation frequency is not known and modes different from oscillatory ones may appear in shock waves.

5

Another approximate model of gas behaviour is the two-temperature model based on the differential equation of heat influx (2.5) in which the heat transfer intensity q_2 is defined by a finite ratio:

$$q_2 = 4\pi a^2 \lambda_2 \,\mathrm{Nu}_2\, \frac{T_0 - T_2}{2a},$$

$$T_2 = \frac{p_2}{(\gamma-1)c_{p2}\rho_2^0} = T_0 \frac{p_2}{p_{20}}\left(\frac{a}{a_0}\right)^3, \tag{5.1}$$

where finite expressions are given for the heat transfer parameter or Nusselt number Nu_2. This model takes a direct account of the irreversibility of heat processes. Owing to the nonstationary nature of thermal conductivity processes in a gas, the Nusselt number Nu_2 depends not only on the state of the medium, but on the prehistory of the process as well; in other words, on the process as a whole. If the nature of the process, its amplitude and time characteristics (oscillation period, characteristic time of compression or expansion) are known beforehand, the values of Nusselt number can be specified, with the help of which the average (say, over one oscillation period) heat balance in the gas can be correctly described. In particular, for oscillatory modes typical of wave perturbations and modes of uniform compression or expansion of bubbles with a characteristic period Δt_0, the following estimate for the Nusselt number

$$\mathrm{Nu}_2 \approx \frac{2a}{(\nu_2 \Delta t_0)^{1/2}} \tag{5.2}$$

is given in [23, 24]. This estimate is based on the fact that in time Δt_0 a heat wave from the bubble wall penetrates to a depth of $\delta \sim (\nu_2 \Delta t_0)^{1/2}$ and this depth gives the typical thickness of the temperature boundary layer in the gas near the bubble wall where the temperature difference $T_0 - T_2$ occurs.

As a result of further development of these ideas made with due regard for the temperature fields in the course of analytical and numerical solutions of the nonstationary temperature problem in a bubble, the following approximate formula has been obtained for the instantaneous intensity of interphase heat transfer in terms of the current values of the maxture parameters (a, w_1, T_2, T_0) and the physical parameters of the gas (γ, ν_2) without taking recourse to *a priori* evaluation of the nature of the process and its characteristic time Δt_0:

$$\mathrm{Nu}_2 = \begin{cases} \sqrt{\mathrm{Pe}_2}, & \mathrm{Pe}_2 > 100, \\ 10, & \mathrm{Pe}_2 < 100, \end{cases}$$

$$\mathrm{Pe}_2 = 12(\gamma-1)\frac{T_0}{|T_0 - T_2|}\frac{a|w_1|}{\nu_2}. \tag{5.3}$$

Note that the singularity that arises at $T_0 = T_2$ can be avoided, because from (5.1) and (5.2) it follows that $q_2 \sim (|T_0 - T_2|)^{1/2}$. Here we shall not discuss the formula (5.3), but only mention that it correctly predicts the asymptotic behaviour of the Nusselt number and, consequently, the behaviour of q_2 for $\mathrm{Pe}_2 \gg 1$ and $\mathrm{Pe}_2 \ll 1$ in the small perturbation stage ($\Delta a \ll a_0$), if compression or expansion of a bubble takes place according to the exponential law:

$$a = a_0(1 + A_0 \exp \epsilon t).$$

For other regimes the validity of the expression (5.3) still remains to be verified. For oscillatory regimes, a comparison with the numerical solutions of exact equations has shown that on average it satisfactorily describes the heat balance in a bubble over one oscillation period.

6

Theoretical investigation of shock waves of finite intensity began [1, 6, 15, 23, 24, 29, 32] with the studies on the structure of a stationary shock wave propagating with a constant velocity $\mathscr{D}_0$ in a monodispersed medium without suffering any change in its structure. An analysis based on the system of equations (2.1)–(2.5), (5.1) with due regard for the heat effects is made in [1, 23, 24]. In a coordinate frame associated with a stationary wave, these equations are reduced to a system of ordinary differential equations describing continuous transition (of a structure) from an initial equilibrium state 0 to another equilibrium state e for the case where $p_e > p_0$ (p_e is the pressure in the state e, i.e. behind the shock wave).

The solution of this system suggests the following: first, the width of a shock wave $\mathscr{L} = \mathscr{D}_0 \Delta t_e$ (see Figure 2) or the zone of transition from 0 to e (relaxation zone) for a bubble of radius $a_0 \sim 1$ mm is about 1 m. Therefore, in experiments conducted with a length of the LPC $H \sim 1$ m, it is not possible to verify the results obtained in stationary analysis (e.g. the number of oscillations in a shock wave, whether a shock wave is monotone or oscillatory, etc.) because the waves in the LPC are obviously nonstationary, i.e. their structure varies in the course of propagation.

Second, unlike the waves in suspensions in gas where the two-velocity effects (effects of motion of carrier and dispersed phases with different translatory velocities v_1 and v_2) play the decisive part, the two-velocity effects are unimportant in the wave dynamics of bubbly liquids because the relative velocity* $v_{12} = v_1 - v_2$ is very small as compared with the phase velocities v_1 and v_2.

In one-velocity ($v_1 = v_2 = v$) and two-velocity models [29, 32] the equation of polytropic gas (2.7) is used, and the viscosity coefficient instead of the effective viscosity of the liquid is applied in (2.3), i.e. in both the models heat dissipation is disregarded. It is therefore clear that viscous dissipation in radial motion with a velocity w and viscous dissipation due to the flow round a bubble with a velocity v_{12} are quantities of the same order as the velocities w and v_{12} are of the same order. Calculations made in these two models have shown that in the two-velocity model the amplitude of oscillations is less and they damp faster than in the one-velocity model,

*Although this relative motion is responsible for the fragmentation of bubbles (see below), and may intensify heat exchange.

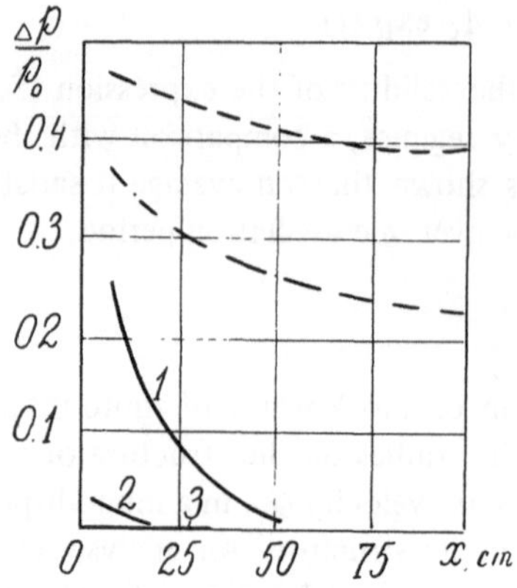

Figure 4.

which, in the opinion of the authors of [29, 32], indicate the the two-velocity effects, i.e. the translatory motion of bubbles relative to the liquid, ought to be taken into account.

In fact, both these two types of viscous dissipation are negligibly small as compared with the heat dissipation which was ignored in [29, 32]. Figure 4 shows the curves which we plotted for the variation in the pressure amplitude Δp (see Figure 2; in monotone and non-oscillatory waves, we have $\Delta p = 0$) along a stationary wave (to be more precise, along its relaxation zone) of intensity $p_e/p_0 = 1.32$ in 1:1 solution of glycerine in water containing air bubbles ($a_0 = 1.4$ mm, $\alpha_{20} = 0.025$, $p_0 = 0.9$ bar). The broken curve was plotted in the one-velocity single-temperature model with an isothermic gas ($\kappa = 1$, or $Nu_2 = \infty$), the dot-dash curve in the two-velocity model with an isotherm gas, and the solid lines in a two-temperature model for $Nu_2 = 3000$ (plot 1), 300 (plot 2), and for $Nu_2 < 100$ (plot 3). These plots are not affected whether the two-velocity effects are taken into account or not. Thus, though the two-velocity effects do appear in the calculations made in one-temperature or any other polytropic model when heat dissipation is disregarded, these effects would not perceptibly influence the wave structure*, if heat dissipation is taken into account.

Interestingly, the stationary waves calculated with due regard for the real temperature effects (Figure 4) should not exhibit oscillations (plot 3, $\Delta p = 0$), while the calculations made in unreal one-temperature model erroneously lead to essentially oscillatory structure of these waves (broken and dot-dash plots).

*The relative motion of a liquid and bubbles may, besides being responsible for the viscosity effects discussed above, be the reason for intensification of heat transfer in a bubble and loss of spherical shape and its ultimate manifestation, fragmentation of bubbles. But there does exist, however, a large range of regime parameters where these effects do not appear. On the other hand, there is a range of regime parameters where these effects may be decisive (see below).

The fact that the considered (Figure 4) wave observed in experiments [28] was oscillatory can be explained as follows: in experiments the waves could not assume a steady state because the LPC tubes were of insufficient length.

However, we should bear in mind that the structure and characteristics of a shock wave do not lie between the corresponding values of adiabatic and isothermic regimes.

The structure of stationary shock waves was studied in [1] with an account of the temperature fields in bubbles without using the approximate formulas for the Nusselt number. Calculations [1] confirm the conclusions drawn and the suitability of the two-temperature model for describing interphase heat transfer.

7

Since the width of shock waves (or their relaxation zone) in mixtures containing bubbles of radius $a \sim 1$ mm is ~ 1 m, i.e. is comparable with the shock tube length, the waves observed in these tubes were nonstationary, though the boundary conditions at the place where these waves were generated (the upper top of the bubble mixture) were stationary. Therefore, a nonstationary theory is necessary to study these waves. As mentioned above, V.E. Nakoryakov and others have studied the evolution of weak nonstationary perturbations with the help of the BKdV equation. Below we shall discuss a general approach [10, 11] to numerical modelling of one-dimensional unsteady waves (including reflections from various boundaries) in bubbly liquids with an incompressible carrier phase in terms of the one-velocity two-temperature model (2.1)–(2.6), (5.1).

It is easier to solve the problem of nonstationary motion of one-velocity medium in Lagrange variables (r, t), where r is the distance of a particle from the origin at $t = 0$. The parameters at $t = 0$ will ba affixed with the subscript 0.

Using the Lagrange variables (r, t), from (2.1) and the first equation of (2.4), we get

$$\frac{\partial v}{\partial r} = \frac{3\rho_0}{\rho}\frac{\alpha_2 w_1}{a}, \qquad \frac{\partial v}{\partial t} = -\frac{1}{\rho_0}\frac{\partial p}{\partial r} + g$$

$$(\rho_0 = \rho_1^0 \alpha_{10}). \tag{7.1}$$

The first equation shows that a mixture containing an incompressible carrier phase can suffer longitudinal compression solely due to the volume compression of bubbles. The second equation is the equation of momentum of the mixture.

Differentiating the first equation with respect to t, while the second with respect to r, and then equating the resulting expressions, by virtue of the Rayleigh-Lamb equations (2.3), we get

$$\frac{\partial^2 p}{\partial r^2} - \frac{1}{\rho_0}\frac{d\rho_0}{dr}\frac{\partial p}{\partial r} = -\frac{3\rho_0^2\alpha_2}{\rho a^2} \times$$

$$\left[(1-12\alpha_2)\frac{w_1^2}{2} - \frac{4\mu_1 w_1}{\rho_1^0 a} + \frac{p_2 - p - 2\Sigma/a}{\rho_1^0}\right]. \tag{7.2}$$

From this second order equation containing only r-derivatives, we can determine the pressure distribution at any instant, if the distributions of α_2, a, w_1, p_2 at this instant are given. The influence of the bubbles on the pressure distribution is expressed through the right-hand side of equation (7.2).

To solve this equation we have to specify the boundary conditions at the tube end-faces $r_* = (0, H)$. Thus, the distribution at any instant is determined from the boundary value problem for equation (7.2). Two types of boundary conditions can be distinguished: boundary conditions of the first kind in which the pressure is specified at the boundary:

$$r = r_*, \qquad p = p(t) \tag{7.3}$$

and the boundary condition of the second kind in which the velocity $V(t)$ of the 'piston' (in particular, $V(t) = 0$ for the stationary wall) is defined at the boundary. Thus, according to the momentum equation (second equation of (7.1)), we obtain

$$r = r_*, \qquad \frac{\partial p}{\partial r} = \rho_0(r_*)\left[g - \frac{dV}{dt}\right]. \tag{7.4}$$

As regards the initial condition, we should mention that the initial conditions should determine the distributions of α_2, a, w_1, and p_2 with respect to r at $t = 0$. The distributions of pressure p and of velocity v at any instant, including that at the initial instant $t = 0$, are not independent. The pressure distribution is determined from the boundary value problem stated above, while the velocity distribution from the first equation of (7.1). The time variations of α_2, a, w_1, and p_2 are determined from the differential equations which follow from (2.1)–(2.5) and (5.1):

$$\frac{\partial \alpha_2}{\partial t} = \frac{3\alpha_1\alpha_2 w_1}{a}, \qquad \frac{\partial a}{\partial t} = w_1 = w_2,$$

$$a\frac{\partial w_1}{\partial t} = \frac{p_2 - p - 2\Sigma/a}{\rho_1^0} - \frac{4\mu_1 w_1}{\rho_1^0 a} - \frac{3w_1^2}{2},$$

$$\frac{\partial p_2}{\partial t} = \frac{3(\gamma-1)}{2a^2}\lambda_2\,\mathrm{Nu}_2(T_0 - T_2) - \frac{3\gamma p_2 w_2}{a},$$

$$T_2 = T_0\frac{p_2}{p_{20}}\left(\frac{a_0}{a}\right)^3. \tag{7.5}$$

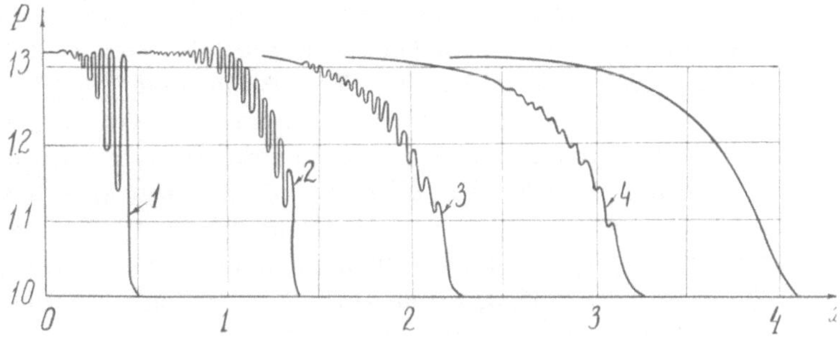

Figure 5.

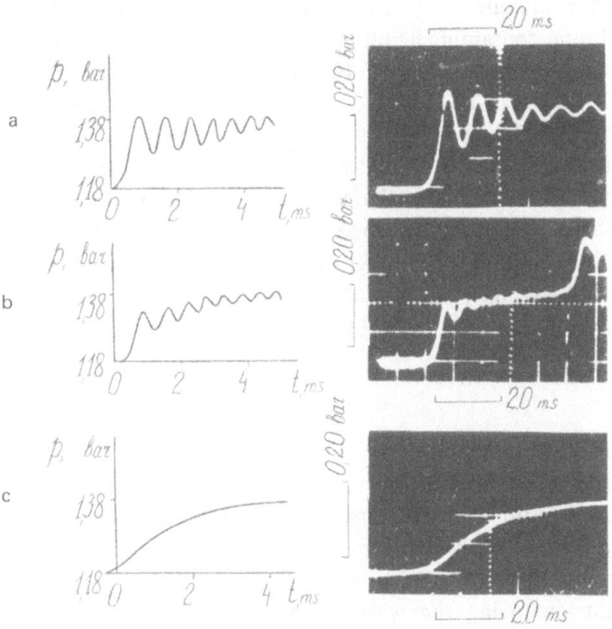

Figure 6.

The solutions obtained for different variants are given in [10, 11]. Figure 5 shows the evolution of a shock wave calculated with the parameters specified in Section 4 (Figure 4) where we discussed the case of limiting stationary structure when the perturbations at $r = 0$ and unperturbed state at $r = \infty$ are stationary. The figures 1, 2, 3, 4 and 5 on the plots correspond to the instants 6.4, 19.2, 32.0, 46.8 and 59.0 ms, respectively. Obviously, the wave is oscillatory near the place of generation ($r < 2$ m) and thereafter it becomes monotonic.

From Sections 3–6, it follows that the evolution of the wave structure and attenuation of short pulses in a bubbly mixture should depend on the type of the gas in the bubbles; more precisely, on its thermal diffusivity ν_2. The oscillations in a liquid containing helium bubbles, in particular, should die down faster than in a liquid with air bubbles. Figure 6 shows the calculated and experimental* 'oscillograms' (in equal scales) for the pressure in shock waves of constant intensity propagating in a mixture consisting of a given liquid ($\rho_1^0 = 1160\,\text{kg/cm}^3$, $\mu_1/\rho_1^0 = 0.9 \times 10^{-5}\,\text{m}^2/\text{s}$, $p_0 = 1.18\,\text{bar}$) with bubbles of given volume content ($\alpha_{20} = 0.94 \times 10^{-2}$) and size ($a_0 = 1.0$–$1.1\,\text{mm}$). The difference lies only in the gas contained in the bubbles: (a) carbon dioxide ($\nu_2 = 1.8 \times 10^{-5}\,\text{m}^2/\text{s}$); (b) air ($\nu_2 = 2.2 \times 10^{-5}\,\text{m}^2/\text{s}$); (c) helium ($\nu_2 = 18 \times 10^{-5}\,\text{m}^2/\text{s}$). The oscillograms were recorded at a fixed depth of $L = 1.6\,\text{m}$. Evidently, the nature of the gas exerts a strong influence on the structure of shock waves in a bubbly mixture. Shock waves in liquids containing carbon dioxide or air bubbles are, in particular, obviously oscillatory, while in a liquid containing helium bubbles, monotonic. The calculated curves are in satisfactory agreement with the experimental oscillograms.

Let us mention an interesting fact. High-speed filming has shown that the shape of the bubbles in shock waves is far from being spherical [17]; nevertheless, the theory based on the Rayleigh-Lamb equation derived for radial motion round bubbles under the assumption that the bubbles preserve their spherical shape, well describes the evolution of waves even in the absence of fragmentation of bubbles. Though the bubbles lose their spherical shape, yet it seems that the Rayleigh-Lamb equation correctly describes the main feature, viz., the change in the volume of the bubbles.

8

The loss of spherical shape of bubbles ultimately results in their fragmentation. Fragmentation essentially changes the structure of waves in a bubbly medium. As a result of intensive shattering taking place in sufficiently strong waves, as a rule, at the wave front small bubbles exist in the wave relaxation zone. This substantially shortens the pulsation period and the cooling time of bubbles, thus reducing the width of the wave relaxation zone. As a consequence, the equilibrium model, i.e. an ideal compressible liquid with the equation of state $p(\rho)$ defined beforehand, may be sufficient for describing the mixture. If, in addition to the compressibility of the carrier phase, the capillary pressure ($2\Sigma/a \ll p$) is also ignored, this equation of state takes a simple form:

$$p(\rho) = \frac{\alpha_{20} p_0}{\alpha_{10}} \frac{\rho}{\rho_1^0 - \rho}. \tag{8.1}$$

In this model the equilibrium shock adiabatic curve can for sufficiently small bubbles be given as:

*Experimental results are drawn from [18].

$$\mathscr{D}^2 = \frac{p_e}{\rho_1^0 \alpha_{20} \alpha_{10}}, \qquad v_e = \frac{p_e - p_0}{\sqrt{p_e}} \sqrt{\frac{\alpha_{20}}{\rho_1^0 \alpha_{10}}}, \tag{8.2}$$

where $\mathscr{D}$ and v_e are the velocities of the jump and of the medium behind the jump relative to the medium in front of the jump.

Owing to the low velocity of sound, a liquid with minute bubbles may be used in modelling high velocity flow of compressible media [12].

If the bubbles are very small in size and the liquid is highly viscous, so that $a(p\rho_1^0)^{1/2}/\mu_1 \ll 1$, the pressures of phases are equalized by the viscous relaxation, if radial inertia and heat can be disregarded. Consequently, the Rayleigh-Lamb equation and the equation of gas pressure become simplified as follows:

$$\frac{4\mu_1 w_1}{a} = p_2 - p_1 - \frac{2\Sigma}{a}, \qquad \frac{p_2}{p_{20}} = \left(\frac{a_0}{a}\right)^3. \tag{8.3}$$

In this case, taking into account the compressibility of the carrier liquid, we can show that the mixture behaviour is described by the system of equations for a visco-elastic liquid with frozen sound velocity $C^{(f)}$ and volume velocity ζ:

$$\frac{1}{\rho}\frac{\mathrm{d}\rho}{\mathrm{d}t} = -\frac{\partial v}{\partial x}, \qquad \rho\frac{\mathrm{d}v}{\mathrm{d}t} = -\frac{\partial p}{\partial x},$$

$$\frac{1}{\rho (C^{(f)})^2}\frac{\mathrm{d}p}{\mathrm{d}t} = -\frac{\partial v}{\partial x} - \frac{p - p_e^0}{\zeta}$$

$$\left(C^{(f)} = \frac{C_1}{\alpha_1}, \qquad \zeta = \frac{4\mu_1}{\alpha_2}\right),$$

$$\frac{\mathrm{d}a}{\mathrm{d}t} = \frac{a}{4\alpha_1 \mu_1}(p - p_2^0), \qquad p_2^0 = p_{20}\left(\frac{a_0}{a}\right)^3 - \frac{2\Sigma}{a},$$

$$\alpha_2 = \alpha_{20}\frac{a^3}{a_0^3}\frac{\rho}{\rho_0}. \tag{8.4}$$

The properties of such a medium depend on the physical properties of the liquid (ρ_1^0, C_1, μ_1), volume content, α_2, and radius, a, of bubbles. G.M. Lyakhov [20] studied nonstationary flows of bubbly mixture in the framework of this model.

9

Since fragmentation can radically influence the behaviour of waves, it is necessary to investigate the conditions which promote fragmentation, the parameters of phases which aid fragmentation, and in particular, whether the properties of the gas in bubbles affect the critical intensity of the shock waves above which the wave destroys the bubbles. If we consider the flow round a bubble or a drop with a velocity v_{12}, quantitative analysis shows that the condition for the annihilation of bubbles is that the Weber number should be greater than a certain value:

$$\mathrm{We} = \frac{\rho_* v_{12}^2 a}{\Sigma} \geqslant \mathrm{We}_* \sim 1, \tag{9.1}$$

where ρ_* is the density. At first glance, this density seems to be the density of the flowing phase, viz., the gas density $\rho_* = \rho_2^0$ in the case of drops, and the liquid density $\rho_* = \rho_1^0$ in the case of bubbles. The coefficient of surface tension Σ depends on the liquid phase (gaseous phase does not virtually affect it), and on its temperature T_a at the interphase. As already mentioned in Section 3, unlike the temperature of the main bulk of the gas, the liquid temperature T_a does not practically vary ($T_a = T_0$). Very strong shock waves ($p_e/p_0 > 10$) are necessary so as to raise the liquid temperature at the bubble wall due to the rise in the temperature of the gas in a bubble. Next, the velocity of the flow round bubbles $v_{12} = v_1 - v_2$ should not also depend on the gas (see (2.4)) as the inertia of a bubble is almost determined by the virtual mass of the liquid which far exceeds the mass of a bubble ($\rho_1^0 \gg \rho_2^0$).

Thus, we are compelled to conclude that the gas properties have no influence on the conditions which promote fragmentation of bubbles. Experiments [8] however gave the opposite result. At a constant initial pressure of $p_0 = 1$ bar and for bubble size of $a_0 = 2$ mm, air or nitrogen bubbles are fragmented by a wave of intensity 3 bar. A wave of intensity $p_e \approx 15$ bars almost shatters nitrogen bubbles to fine 'dust' particles ($a \lesssim 0.1$ mm). Helium bubbles do not suffer fragmentation under the action of a wave of intensity $p_e = 4$ bar, neither hydrogen bubbles even by a wave of intensity $p_e = 50$ bar. Thus, helium and hydrogen bubbles have greater resistance to the action of shock waves than air or nitrogen bubbles.

In order to understand this effect, we have to take recourse to more subtle analysis than what we have done above, for instance, examine the behaviour and stability of a liquid sphere of density ρ_2^0 in a stream of another liquid of density ρ_1^0. But even the linear problem of the development of small perturbations on a sphere has so far not been resolved. Therefore, first consider the stability of the initial plane boundary between two non-viscous liquids of densities ρ_1^0 and ρ_2^0 ($\rho_1^0 \gg \rho_2^0$) and surface tension Σ, moving with a relative velocity v_{12} along this boundary. Furthermore, assume that the interface has an acceleration g normal to this boundary. Linear analysis [3] shows that, if $\rho_1^0 \gg \rho_2^0$, the amplitude δ of the harmonic oscillations of wavelength b chances with time as

$$\delta = \delta_0 \exp It,$$

$$I(b) = \left[4\pi^2 \frac{\rho_2^0 v_{12}^2}{\rho_1^0 b^2} + 2\pi \frac{g}{b} - 8\pi^3 \frac{\Sigma}{\rho_1^0 b^3}\right]^{1/2}. \tag{9.2}$$

The first term gives the destabilizing effect (corresponds to the Kelvin-Helmholtz instability) due to dynamic pressure drops, the second term (corresponds to the Rayleigh-Taylor instability) can either be a stabilizing or a destabilizing factor, depending on the direction of the acceleration, and finally, the third term is the stabilizing factor due to the surface tension. Assuming that the perturbations on a spherical interface in the form of a large circle of radius a in a plane normal to the velocity of the flow varies according to equation (9.2), and retaining only the Kelvin-Helmholtz instability and since a sphere can be destroyed only by increasing perturbations ($I(b) > 0$) of wavelengths $b < 2a$, i.e. if $I(2b) > 0$, we obtain the condition for fragmentation:

$$\mathrm{We} = \frac{2a\rho_2^0 v_{12}^2}{\Sigma} > \mathrm{We}_* \sim 2\pi. \tag{9.3}$$

In this scheme we can only estimate the order of We_*. However, this analysis of interaction between dynamic and capillary forces reveals the influence of the gas density (though small as compared with the liquid density) on the fragmentation of bubbles and explains the experimentally observed greater strength of hydrogen and helium bubbles than that of air bubbles.

10

So far we have been discussing shock waves in a liquid containing bubbles of non-condensable and insoluble gas where there are no phase transitions. What would be the effects, if the bubble interior were filled with the vapor of the carrier liquid and the initial mixture were in equilibrium, i.e. at saturation temperature.

Weak perturbations ($\Delta p/p_0 \ll 1$), as they propagate, fade out rapidly (see [16]) due to large dispersion when the velocity of high-frequency perturbations is much larger than the velocity of the low-frequency ones, because the equilibrium velocity of sound (the velocity of low-frequency perturbations) in such a mixture is very small (see [19]). The explanation is that the vapor pressure does not increase during compression of a bubble due to condensation of the 'superfluous' vapor.

If the initial action is sufficiently strong ($p^{(1)}/p_0 > 1.1$), experiments [4] conducted with liquid nitrogen containing gaseous nitrogen bubbles have shown that the nature of wave evolution changes. Figure 7 shows an oscillogram recorded under the following conditions $\alpha_{20} \sim 10^{-1}$, $a_0 \sim 1$–2 mm, $p_0 \sim 1$ bar. An initial wave $\mathscr{D}^{(1)}$ of pressure $p^{(1)}$ propagates deep into the boiling bubbly nitrogen followed by a wave packet $\mathscr{D}^{(2)}$ of intense damping pressure oscillations at a frequency of $\sim 3 \times 10^3$ Hertz and amplitude of

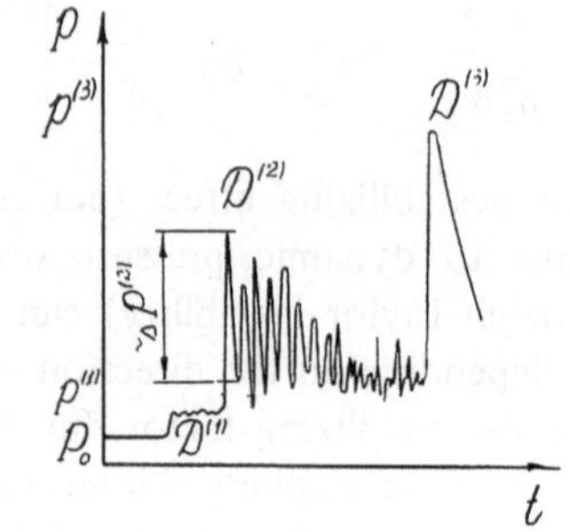

Figure 7.

$\Delta p^{(2)} \sim 5$–20 bar. This amplitude increases with the increasing $p^{(1)}$. Next, a shock wave reflected from the lower endface of the LPC propagates. The pressure in this wave is much higher than the pressure behind a reflected wave generated by a similar initial perturbation in water-air system. If the pressure of the initial wave is equal to $p^{(1)} = 3$ bar, the amplitude of the secondary oscillations is equal to $\Delta p^{(2)} = 13$ bar and the pressure behind the reflected wave recorded at the tube end face is $p^{(3)} = 120$ bar. The reflected wave is quickly damped out by the strong rarefaction wave following it.

The waves in a liquid containing bubbles filled with its own vapor are evidently amplified due to phase transitions. Similar effects were observed [9] in a liquid with bubbles of carbon dioxide which dissolves in water more readily than other gases. Amplification effects were also observed in [13]. But the wave amplification mechanism is indeed a complicated problem.

The mechanism of amplification of waves in liquids containing bubbles with their own vapor or a readily soluble gas is apparently connected with the rapid condensation collapse of bubbles. It is therefore desirable to study such a collapse with one 'test' bubble as we did in Section 3 for a gas bubble. Take a spherical shell of radius R around the test bubble. This radius depends on the initial volume content of vapor: $(a/R)^3 = \alpha_{20}$. Besides the law governing the collapse $a(t)$, we shall also study the variation of the mean liquid pressure $\langle p \rangle_1$ in the spherical shell:

$$\langle p \rangle_1 = \frac{3}{4\pi(R^3 - a^3)} \int_a^R 4\pi r^2 p'(r)\, \mathrm{d}r, \tag{10.1}$$

where the distribution $p'(r, t)$ is determined from the Cauchy-Lagrange integral [14, 26] for a given pressure $p_\infty(t)$ at a large distance from the bubble and the law of variation of bubble radius $a(t)$ which is described by the Rayleigh-Lamb equation:

$$\frac{p'(r,t)}{\rho_1^0} = \frac{p_\infty(t)}{\rho_1^0} + \frac{1}{r}\frac{\mathrm{d}A(t)}{\mathrm{d}t} - \frac{A^2(t)}{2r^4} \qquad (A(t) = w_1 a^2),$$

$$a\frac{dw_1}{dt}+\frac{3}{2}w_1^2+\frac{4\mu_1 w_1}{\rho_1^0 a} = \frac{p_2-p_\infty-2\Sigma/a}{\rho_1^0},$$

$$\frac{da}{dt}\approx w_1. \tag{10.2}$$

If $p_2 = \text{const}$, $p_\infty = \text{const}$ and if the capillary and viscous effects can be disregarded, we obtain accelerated inertial collapse or the Rayleigh regime:

$$\frac{da}{dt}\approx w_1 = -\sqrt{\frac{2(p_\infty-p_2)}{\rho_1^0}\left[1-\left(\frac{a_0}{a}\right)^3\right]}. \tag{10.3}$$

Simple calculations [4] show that as $a \to 0$, the mean liquid pressure in the shell tends to infinity:

$$\langle p'\rangle_1 = 0.5\sqrt[3]{\alpha_{20}}\, p_\infty (a_0/a)^{2/3}. \tag{10.4}$$

The real process of collapse of vapor bubbles is limited by the finite thermal conductivity of the liquid*, so the liquid cannot give out that amount of heat which is needed for the realization of the condition $p_2 = \text{const}$ that is necessary for (10.3) and (10.4) to be satisfied. Therefore at some stage the pressure p_2 will increase, thereby hampering the collapse and slowing down the increase of $\langle p'\rangle_1$. To investigate whether the heat conducting properties of liquids (liquid nitrogen, in particular) can significantly aid in increasing the mean liquid pressure $\langle p'\rangle_1$ during collapse of a vapor bubble, the system of equations of spherically symmetric motion with nonstationary thermal conductivity in vapor and in liquid under the boundary conditions specified at the interface $a(t)$, which is determined in the course of solution, was numerically solved.

By the way of example, Figure 8 shows the diagram for the time variation of $\langle p'\rangle_1$ for a nitrogen bubble in liquid nitrogen under the following initial conditions: $a_0 = 1$ mm, $\alpha_{20} = 0.05$, $p_0 = 1$ bar, $T_0 = T_s(p_0) = 77$ K, where T_s is the saturation temperature, $p_\infty = 5$ bar. From the diagram it is seen that the mean pressure reaches 13 bars, which is perceptibly greater than the pressure $p_\infty = 5$ bar initiating the collapse process. But, unlike in the Rayleigh regime (10.3) shown in Figure 8 by the broken plot with a vertical asymptote $t = t_R = 0.91\, a_0 \sqrt{\rho_1^0/(p_\infty - p_0)}$, several pulsations occur due to the finite thermal conductivity of the liquid. The collapse and annhilation of a bubble may be boosted up by fragmentation which accelerates condensation because of increased interface surface.

*Unlike the heat processes for a gas bubble, which depend on the thermal diffusivity of the gas, the heat processes and phase transitions for a vapor bubble are limited by the thermal diffusivity of the liquid.

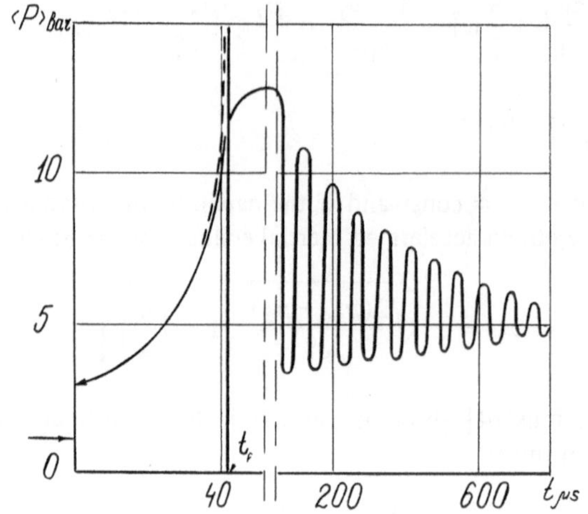

Figure 8

11

Thus, in bubbly mixtures such situations may arise in which bubbles may collapse and vanish due to condensation of vapor or dissolution of gas, and the two-phase mixture may turn into a one-phase one. The situation becomes still more complicated if at the moment of collapse the kinetic energy of the radial motion per unit mass of the liquid $k = \frac{3}{2}\alpha_2 w_1^2$ is finite (this is the case in the Rayleigh regime). After the collapse, this energy should transform into some other form which the one-phase liquid can possess. If the final stage of bubble collapse takes place very fast in a very narrow mixture zone, we can assume that this stage proceeds as a 'condensation jump', i.e. instantaneously. The instantaneous collapse is aided by the loss of spherical shape and fragmentation which speeds up condensation or dissolution due to increased interface surface. Therefore, the problem of bubbly liquids in which bubbles may vanish should include the determination of the volume or zones $U^{(1)}$ and $U^{(2)}$ where one-phase and two-phase liquids exist and the determination of the interfaces $F^{(12)}$ separating these zones. Moreover, boundary conditions similar to those at the discontinuity surfaces should be specified at these interfaces.

Now consider these boundary conditions in a coordinate system in which the surface $F^{(12)}$ is at rest. Denote the two-phase state (with bubbles) of the medium before the wave by the subscript f, and the state behind the wave (one-phase liquid) by the subscript e. Neglecting the mass, momentum and energy of bubbles as compared with these corresponding parameters of the liquid, we can express the laws of conversion of mass, momentum and energy in the following form:

$$\alpha_{1f}\rho_{1f}^0 v_f = \rho_{1e} v_e, \qquad \alpha_{1f}\rho_{1f}^0 v_f^2 = \rho_{1e}^0 v_e^2 + p_e \qquad (p \approx p_1),$$

$$\alpha_{1f}\rho_{1f}^0 v_f[\tfrac{1}{2}v_f^2 + k_f + u_{1f}] + p_f v_f = \rho_{1e}^0 v_e[\tfrac{1}{2}v_e^2 + u_{1e}] + p_e v_e, \tag{11.1}$$

where u_{1f}, u_{1e} are the specific internal energies of the liquid before and behind the wave, respectively. These energies, just like the pressure, contain elastic (if the compressibility of the liquid is taken into account) and heat components. At moderate temperatures and pressure drops, for which $|\rho_1^0 - \rho_{10}^0| \ll \rho_{10}^0$, we can assume the following acoustic equation of state:

$$p - p_0 = C_1^2(\rho_1^0 - \rho_{10}^0), \tag{11.2}$$

where C_1 is the velocity of sound in a liquid without bubbles.

12

Owing to highly complicated computation algorithms, in a number of problems we may encounter great difficulties in determining the zones $U^{(1)}$, $U^{(2)}$ and their interface $F^{(12)}$ with unknown laws of motion, when the equations of one-phase liquid in $U^{(1)}$ and the equations of two-phase liquid in $U^{(2)}$ and the jump equations on the interface $F^{(12)}$ are given. In such cases it is preferable to use a unified system of equations for the whole flow field, that can describe a nondegenerate continuous transition of the two-phase liquid into one-phase liquid. A similar situation exists in the gas dynamics of an ideal gas in which it is convenient to change the equations of continuous motion by introducing a viscosity called the pseudo-viscosity which is many times greater than the physical viscosity of the gas. Thus, the boundary conditions at the jump or shock wave become superfluous, and the shock wave becomes a narrow zone of sharp (but continuous) compression with high gradients.

In conclusion we may mention that the special feature of bubbly mixtures that determines their wave properties is the presence of radial inertia of liquid and elasticity of vapor. Their existence depends on the interphase heat and mass exchange processes, compressibility and viscosity of the liquid and fragmentation of bubbles.

Each of these factors plays a different part, depending on the pressure p_0, radius a_0, volume content α_{20} of bubbles and physical properties of the phases ($\rho_1^0, \mu_1, \lambda_2, c_1, C_1, \Sigma, \rho_2^0, \gamma, c_{p2}, \lambda_2$ and l).

The wave properties of liquids containing bubbles filled with a non-condensable and insoluble gas are paradoxical in that the wide variation in such physical properties of the liquid phase (which occupies almost the whole bulk of the mixture) as thermal conductivity λ_1, specific heat c_1, viscosity μ_1 (for instance, caused by the variation in the glycerine content in the solution) does not virtually influence the structure and attenuation of shock waves. Viscosity of bubbly mixture plays a perceptible part only when its value tends to the viscosity of pure glycerine. On the contrary, variation in the properties of the gas (which occupies a very small volume in the mixture and its mass content is negligibly small) in bubbles (such as the thermal conductivity λ_2, specific heat c_{p2}, density ρ_2^0, adiabatic exponent γ) exerts considerable influence on the structure and attenuation of shock waves.

Furthermore, the nature of the gas, more explicitly, its density ρ_2^0, though very small as compared with the liquid density ρ_1^0, strongly influences fragmentation which may result in essential reduction of the bubble size and may radically change the part played by relaxation processes.

Finally, if bubbles are filled with condensable vapor or a readily soluble gas (once again, it is the influence of the gas), small perturbations fade out quickly and attenuate at a relatively fast rate, while stronger perturbations may increase at some stage of the process.

References

1. Aidagulov RR, Khabeev NS and Shagapov VSh (1977) Structure of a shock wave in a liquid with gas bubbles with reference to non-stationary interphase heat transfer. Prikl Mekh i Tekh Fiz 3: 334–340.
2. Batchelor GK (1967) Compression waves in a suspension of gas bubbles in liquid. Fluid Dynamics Transactions, Vol 4, Warszawa.
3. Birkhoff G (1960) Hydrodynamics. Princeton, New Jersey: Princeton University Press.
4. Borisov AA, Gel'fand BE, Gubaidullin AA, Gubin SA, Gubanov AV, Ivandaev AI, Nigmatullin RI, Filin NV, Timofeev EI and Khabeev NS (1977) Amplification of shock waves in a liquid with vapor bubbles. In Kutateladze SS (ed), Nonlinear wave processes in two-phase media. Novosibirsk.
5. Chapman RB and Plesset MS (1971) Thermal effects in the free oscillations of gas bubbles. Trans ASME, ser D. J Basic Eng 93 (3).
6. Crespo A (1969) Sound and shock waves in a liquid containing bubbles. Phys Fluids 12: 2274–2282.
7. Gel'fand BE, Gubin SA, Kogarko BS and Kogarko SM (1973) The study of compression waves in a mixture of a liquid and gas bubbles. Doklady AN SSSR 213 (5).
8. Gel'fand BE, Gubin SA, Nigmatulin RI and Timofeev EI (1977) The influence of gas density on fragmentation of bubbles by shock waves. Doklady AN SSSR 235 (2): 292–294.
9. Gel'fand BE, Stepanov VV, Timofeev EI and Tsyganov SA (1978) Amplification of shock waves in non-equilibrium system consisting of a liquid and bubbles of soluble gas. Doklady AN SSSR 239 (9).
10. Gubaidullin AA, Ivandaev AI and Nigmatulin RI (1976) Non-stationary waves in a liquid with gas bubbles. Doklady AN SSSR 226 (6).
11. Gubaidullin AA, Ivandaev AI and Nigmatulin RI (1978) Nonsteady shock waves in gas-liquid mixtures of bubble structure. Prikl Mekh i Tekh Fiz 2: 204–210.
12. Jakimov YuL, Eroshin VA and Romanenko NI (1978) Modelling of the motion of a body in water involving compressibility. In Grigoryan SS (ed) Some problems of mechanics of continua. Inst Mekh, Moscow University Press.
13. Kedrinsky VK (1968) Propagation of perturbations in a liquid with gas bubbles. Prik Mekh i Tekh Fiz 4.
14. Knapp RT, Daily JW and Hammitt FG (1970) Cavitation. New York: McGraw-Hill Book Company.
15. Kutateladze SS, Burdukov AP, Kuznetsov VV, Nakoryakov VE, Pokusaev BG and Schreiber IP (1972) On the structure of weak shock waves in gas-liquid medium. Doklady AN SSSR 207 (2).
16. Kutateladze SS, Nakoryakov VE and Pokusaev BG (1979) Experimental investigation

of wave processes in gas- and vapour-liquid media. In Two-phase momentum, heat and mass transfer in chemical process and energy engineering sust., Vol I, New York.

17. Kuznetsov VV, Nakoryakov VE, Pokusaev BG and Schrieber IR (1977) Experimental study of propagation of perturbations in a liquid with gas bubbles. In Nonlinear wave processes in two-phase media. Novosibirsk.
18. Kuznetsov VV, Nakoryakov VE, Pokusaev BG and Schreiber IR (1978) Propagation of perturbations in gas-liquid mixtures. J Fluid Mech 85 (I): 85–96.
19. Landau LD and Lifshitz EM (1959) Fluid mechanics. Oxford: Pergamon Press.
20. Lyakhov GM and Okhitin VN (1977) Plane waves in nonlinear viscous multicomponent media. Prikl Mekh i Tekh Fiz 2: 241–248.
21. Nakoryakov VE, Sobolev VV and Schreiber IR (1972) Long wave perturbation in a gas-liquid medium. Izv AN SSSR, Mekhan Zhid i Gaza 5: 763–769.
22. Nakoryakov VE, Pokusaev BG, Schreiber IR, Kuznetsov VV and Malykh NV (1975) Experimental study of shock waves in liquid with gas bubbles. In Wave process in two-phase systems. Novosibirsk.
23. Nigmatulin RI, Khabeev NS and Shagapov VSh (1974) On shock waves in a liquid with gas bubbles. Doklady AN SSSR 214 (4).
24. Nigmatulin RI and Shagapov VSh (1974) The structure of shock waves in a liquid with gas bubbles. Izv AN SSSR, Mekhan Zhid i Gaza 6: 890–899.
25. Nigmatulin RI and Khabeev NS (1978) Dynamics and heat-mass transfer between vapor-gas bubbles and a liquid. In Grigoryan SS (ed) Some problems of mechanics of continua. The Institute of Mechanics, Moscow University Press.
26. Nigmatulin RI (1978) Fundamentals of mechanics of heterogeneous media. Moscow: Nauka.
27. Nigmatulin RI (1979) Spatial averaging in the mechanics of heterogeneous and dispersed systems. Int J Multiphase Flow 5: 353–385.
28. Noordzij L (1973) Shock waves in mixtures of liquids and air bubbles. Thesis Twente Technological University, Eusclede.
29. Noordzij L and Wijngaarden L van (1974) Relaxation effects caused by relative motion on shock waves in gas-bubbles/liquid mixtures. J Fluid Mech 66 (I): 115–144.
30. Plesset MS and Prosperetti A Bubble dynamics and cavitation. Ann Rev Fluid Mech 9: 145–186.
31. Wijngaarden L van (1968) On the equations of motion for mixtures of fluid and gas bubbles. J Fluid Mech 33 (3): 465–474.
32. Wijngaarden L van (1970) On the structure of shock waves in liquid-bubble mixture. Appl Sci Res 22 5: 366–381.
33. Wijngaarden L van (1972) One-dimensional flow of liquids containing small gas bubbles. Ann Rev Fluid Mech 4: 369–396.

Acoustic nonlinearity of bubbly liquids

LEIF BJØRNØ

Technical University of Denmark, Building 352, DK-2800 Lyngby, Denmark

Abstract. Sources contributing to the acoustic nonlinearity of a gas/liquid mixture are discussed and calculations of a coefficient β_{eff} for the second-order nonlinearity of the mixture are performed based on source strength density functions for parametric acoustic arrays and on formation with source distance of the second harmonic to a monochromatic wave propagating through the mixture. Some procedures for experimental determination of β_{eff} are suggested, and it is concluded on basis of the calculations that the dynamic bubble nonlinearity will yield the predominant contribution to the acoustic nonlinearity of the mixture, in particular at resonance.

1. Introduction

It is well known that the presence of bubbles in liquids modifies the acoustical properties of the medium by the introduction of dispersion, of increased attenuation, i.e. absorption and scattering, and of acoustic nonlinearity. The effects of the increased acoustic nonlinearity in bubbly liquids, i.e. the formation of harmonics and intermodulation products for instance, have during the past been suggested to be used for several practical applications of which only very few have turned out to be feasible. It has been suggested [12] that the change in the second-order properties of a medium due to the presence of microbubbles could be used for detection of gas bubbles in animal blood and tissue, and experimental and theoretical studies have been performed [5, 6, 11] aiming at an increase in the virtual source strength by parametric acoustic arrays through introduction of microbubbles produced by acoustic cavitation or by forcing air through a microporous filter into the interaction region for the primary waves. In spite of a considerable increase in the source strength density function found due to increase in nonlinearity of the medium, the introduction of bubbles into the interaction region of the parametric array was found to terminate the array and to destroy its high directivity, resulting in a beam broadening and in some cases leading to omnidirectionality of the secondary radiation due to scattering by the bubbles. The use of bubbles for virtual source strength improvement must therefore be considered prohibitive in parametric arrays. Some recent studies of acoustic nonlinearity of tissue performed at the Technical University of Denmark in an attempt to be able to characterize tissue by its second-order properties have again focussed the interest around acoustic nonlinearity of bubbles in liquids due to the presence of microbubbles in tissue in vivo, and in vitro, in particular [10].

The present study is particularly aiming at determining through an engineering approach the magnitude of the bulk acoustic nonlinearity of

Applied Scientific Research 38: 291–296 (1982) 0003–6994/82/0383–0291 \$00.90.

bubbly liquids, by producing numerical values (β_{eff}) for the second-order nonlinearity of the medium, and at determining the dependence of these values on bubble parameters.

Theoretical considerations

The nonlinear distortion of a finite-amplitude wave propagating through single-phase fluids can be characterized by Earnshaw's expression [7] for the velocity of propagation of a constant phase in the wave:

$$\left(\frac{dx}{dt}\right)_{u\,=\,\text{constant}} = c_0 + \left(1 + \frac{B}{2A}\right)u = c_0 + \beta u, \tag{1}$$

where u and c_0 denote the local particle velocity and the velocity of propagation of infinitesimal amplitude waves in the fluid, respectively. A and B are forming a dimensionless second-order nonlinearity ratio, and they originate from a Taylor series development of the pressure-density relation of the fluid for constant entropy [2, 3]. As shown in [3] the ratio between the contribution from the nonlinear equation of state of the fluid and the convective contribution to the wave distortion course is $B/2A$, which shows that for liquids the equation of state contribution prevails, while for gases the convective contribution will be predominant.

In two-phase fluids, e.g. bubbly liquids, the bulk acoustic nonlinearity receives contributions from: (a) nonlinearity of the equation of state of the liquid; (b) nonlinearity of the equation of state of the gas; (c) the dynamical nonlinearity of the bubbles; (d) the convective influence; (e) nonlinearity of the equation for the density of the liquid/gas mixture.

The contributions (d) and (e) can be neglected in the present study due to their dependence on initial wave amplitude (d) and their extremely small magnitude (e) [14].

a. *The parametric array approach*

For a single-phase fluid the difference-frequency pressure p_s is determined by [13].

$$\Box p_s = \nabla^2 p_s - \frac{1}{c_0^2}\frac{\partial p_s}{\partial t^2} = \rho_0\left(\frac{\partial q}{\partial t}\right), \tag{2}$$

where

$$q = \frac{\beta}{\rho_0^2 c_0^4}\frac{\partial}{\partial t}(p_i^2)$$

is the source strength density. ρ_0 and p_i denote the ambient density of the fluid and the pressure function of the primary waves, respectively.

For a two-phase fluid, i.e. a gas/liquid phase, (2) may be rewritten using the approach given in [5] as:

$$\Box p_s \approx \rho_l \left\{ \frac{b_l}{2a_l p_l^2} + \delta \frac{b_g}{2a_g p_g^2} \right\} \frac{\partial^2}{\partial t^2} (p_i^2), \tag{3}$$

where $a = \rho c^2 / p$ and $b = (2A/p)\beta$, and with indices l and g denoting qualities of the liquid and of the gas, respectively. δ is the volume concentration of the gas. The ratio between the source strength densities of the gas/liquid mixture q_t and the pure liquid q_l may now be written as:

$$\frac{q_t}{q_l} = \left[1 + \delta \left(\frac{\rho_l c_l^2}{\rho_g c_g^2} \right)^2 \right] = \frac{\beta_{\text{eff}}}{\beta_l} \cdot \frac{\rho_l^2 c_l^4}{\rho_t^2 c_t^4},$$

from which the effective second-order nonlinearity coefficient β_{eff} may be derived as:

$$\beta_{\text{eff}} = \beta_l \left(\frac{\rho_t^2 c_t^4}{\rho_l^2 c_l^4} \right) \left[1 + \delta \left(\frac{\rho_l c_l^2}{\rho_g c_g^2} \right)^2 \right]. \tag{4}$$

For air bubbles in water at 20°C and 10^5 Pa, (4) gives for various values of δ:

δ	10^{-5}	10^{-4}	10^{-3}	10^{-2}	10^{-1}
β_{eff}	$3.1 \cdot 10^5$	$3.1 \cdot 10^4$	$3.1 \cdot 10^3$	$3.1 \cdot 10^2$	$3.1 \cdot 10^1$,

which, as may be shown easily, is not in agreement with Ballou's rule [1] for single-phase fluids, which states $B/A \propto 1/c_t$.

Introduction of frequency and attenuation influence by considering the bubbles as being damped harmonic oscillators exposed to forced vibration will for monochromatic waves by some modification of the approach given in [11] lead to:

$$\beta_{\text{eff}} = \frac{\beta_l}{\rho_l^2 c_l^4} (\rho_t^2 c_t^4) \cdot 8.677 \cdot 10^7 \cdot \delta \cdot \frac{R_r}{L}$$

$$\left[\frac{1}{(1-\Omega^2)^2 + C\Omega^2} \right] \left[0.287 + \left(1 - \frac{\Omega^2}{2.4} \right) \right], \tag{5}$$

where R_r is the Rayleigh distance for a transducer operating at the angular frequency ω, Ω is the dimensionless ratio between ω and the resonance frequency of the particular bubble size considered, and L is the distance through the liquid from the source to the field of gas/liquid mixture of concentration δ. The damping constant C, receiving contributions from viscosity, heat conductivity and acoustic radiation, can be calculated from the logarithmic decrement of the bubble oscillation.

In resonance, i.e. $\Omega = 1$, and for various values of C, β_{eff} may be calculated for $\delta = 10^{-5}$, bubble radius $R_0 = 50\,\mu\text{m}$, $R_r = 0.20$ m and $L = 1$ m for air bubbles in water at 20°C and 10^5 Pa, giving:

C	0.01	0.062	0.1	1.0	10
β_{eff}	$1.35 \cdot 10^7$	$3.5 \cdot 10^5$	$1.35 \cdot 10^5$	$1.35 \cdot 10^3$	$1.35 \cdot 10^1$,

where the damping constant value $C = 0.062$ was calculated based on air-water data at 20°C and 10^5 Pa.

Far from resonance, i.e. $\Omega = 0.1$, (5) leads to:

C	0.01	0.062	0.1	1.0	10
β_{eff}	$1.9 \cdot 10^2$	$1.9 \cdot 10^2$	$1.9 \cdot 10^2$	$1.86 \cdot 10^2$	$0.93 \cdot 10^2$,

These values should be compared to values of the nonlinearity coefficient β_{eff} found for air bubbles in water at 20°C and 10^5 Pa by modifying and recalculating the expression for S_ω [8] to give the acoustic nonlinearity of the bubbly liquid. These values for β_{eff} are for $\delta = 10^{-5}$ and for monochromatic waves (the expressions in [8] were originally developed for the study of the sum and difference-frequency components of two sinusoidal waves of frequencies ω_1 and ω_2):

Resonance ($\omega_1 = \omega_0$)	Non-resonance ($\omega_1/\omega_0 = 0.1$)
$\beta_{eff} = 1.16 \cdot 10^5$	$\beta_{eff} = 267$

which are in fair agreement with the results obtained for $C = 0.062$.

b. *Other procedures for determination of* β_{eff}

Also the formation of the second-harmonic amplitude with distance from the source for wave propagation through a strongly nonlinear medium may be used for a calculation of values of β_{eff}.

Starting from the nonlinear expressions for the density of bubbly liquids and for bubble motion and determining the second-harmonic amplitude through a perturbation procedure like the one used in [14], a rather involved expression can be derived for β_{eff} as a function of frequency, bubble radius and liquid and gas qualities. For air bubbles of radius 50 μm in water at 20°C and 10^5 Pa and for $\delta = 10^{-5}$, the following values of β_{eff} have been calculated:

Resonance ($\omega_1 = \omega_0$)	Non-resonance ($\omega_1/\omega_0 = 0.1$)
$\beta_{eff} = 4.6 \cdot 10^5$	$\beta_{eff} = 250$

which show a good agreement with the values determined through modified expressions for source strength density functions by parametric arrays.

These calculations show, that for $\omega_1 \ll \omega_0$ (ω_1 and ω_0 are angular frequencies for excitation and bubble resonance, respectively) the dynamic bubble nonlinearity may be neglected and the dominant role in contributing nonlinearity is taken over by the nonlinearity of the equation of state of the gas. For comparison it shall be mentioned, that at 20°C and 10^5 Pa pure water and pure air individually possess the following second-order nonlinearity coefficients:

$$\text{water: } \beta_l = 3.48 \qquad \text{and} \qquad \text{air: } \beta_g = 1.2.$$

For experimental determination of the second-harmonic amplitude p_2 as a function of distance x from the source for wave propagation through bubbly liquids, the following expression [9] may be used for qualitative and order-of-magnitude studies at frequencies ω far from resonance in the propagation region close to the source:

$$\frac{p_2}{p_1} = R\left[N(1-R) - \left(\frac{N^3}{4}\right)(1-R)^3(3+R) + \frac{N^5}{16}(1-R)^5(22+3R-5R^2-3R^3-R^4)\right], \tag{6}$$

where $R = \exp(-2\alpha_t x)$ and

$$N = \beta_{\mathrm{eff}}\left(\frac{\omega p_0}{4\alpha_t \rho_t c_t^3}\right) = \frac{\Gamma_t}{4}.$$

An experimental determination of $p_2(x)$ may for known values of the attenuation coefficient α_t in the bubbly liquid permit a calculation of β_{eff}. In (6) p_0 is the ambient pressure, ρ_t and c_t denote the density and the sound velocity in the bubbly liquid, respectively. The quantity Γ_t is the so-called Gol'dberg number, expressing the dimensionless ratio between nonlinear and dissipative effects, and which is inverse proportional to the thickness of shocks in the medium [3, 4]. The production of bubbles covering a band of resonance frequencies will probably lead to some reduction in the measured β_{eff} compared to the calculated, which is based on constant and only one resonance frequency.

Far from resonance a thermodynamical procedure originally suggested by Beyer [2] and later used for an experimental determination of second-order nonlinearity ratios of various liquids and gases [2, 3] and of water-filled marine sediments [4] may be applicable. This procedure is based on the expression:

$$\beta_{\mathrm{eff}} = 1 + \rho_t c_t\left[\left(\frac{\partial c_t}{\partial p}\right)_T\right]_{\rho=\rho_t} + \left(\frac{c_t T \kappa}{c_p}\right)\left[\left(\frac{\partial c_t}{\partial T}\right)_p\right]_{\rho=\rho_t}, \tag{7}$$

where κ is the isobaric compressibility and c_p is the specific heat at constant pressure for the gas/liquid mixture at the absolute temperature T.

The crucial factors to be determined are the relations between the velocity of sound in the mixture c_t and the temperature T and pressure p in order to be able to form the derivatives given in (7).

Conclusions

In spite of the lack for the moment of experimental evidence for the magnitude of the second-order nonlinearity coefficient β_{eff} for bubbly liquids, calculations have suggested that the predominant contribution to acoustic nonlinearity will come from the dynamic bubble nonlinearity, in particular at resonance, while only minor contributions will derive from the nonlinearity of the equations of state of the gas and the liquid, respectively.

References

1. Beyer RT (1974) Nonlinear Acoustics. Naval Ship Systems Command. Department of the Navy.
2. Beyer RT (1960) Parameter of nonlinearity of fluids. J Acoust Soc Amer 32: 719–723.
3. Bjørnø L (1976) Nonlinear Acoustics. In Stephens RWB and Leventhall HG (eds) Acoustics and Vibration Progress, Vol 2, pp 101–198. London: Chapman and Hall.
4. Bjørnø L (1977) Finite-amplitude wave propagation through water-saturated marine sediments. Acustica 38 (4): 195–200.
5. Clynch RC and Rolleigh RL (1974) Measurement of enhanced nonlinear radiation in the presence of microbubbles. Applied Research Laboratories. University of Texas at Austin. ARL-TM-74-17.
6. Clynch JR and Dittman CW (1976) Bubble-enhanced nonlinear sound generation. J Acoust Soc Amer 59 (Suppl no 1): S88.
7. Earnshaw S (1860) On the mathematical theory of sound. Phil Trans Royal Soc 150: 133–143.
8. Fenlon FH and Wonn JW (1980) On the amplification of modulated acoustic waves in gas-liquid mixtures. In Cavitation and Inhomogeneities in Underwater Acoustics. Proceedings of the 1st International Conference, Göttingen 1979, pp 141–150. Springer Verlag.
9. Keck, W and Beyer RT (1960) Frequency spectrum of finite amplitude ultrasonic waves in liquids. Phys Fluids 3: 346–351.
10. Lewin PA and Bjørnø L (1981) Acoustic amplitude thresholds for rectified diffusion in gaseous microbubbles in biological tissue. J Acoust Soc Amer 69 (3): 846–852.
11. Lockwood JC and Smith DP (1975) Difference frequency generation by forced-air bubbles. J Acoust Soc Amer 57 (Suppl no 1): 573–574.
12. Welsby VG and Safar MH (1969) Acoustic nonlinearity due to micro-bubbles in water. Acustica 22: 177–182.
13. Westervelt PJ (1963) Parametric acoustic array. J Acoust Soc Amer. 35: 535–537.
14. Zabolotskaya EA (1976) Acoustic second-harmonic generation in a liquid containing uniformly distributed air bubbles. Soviet Phys-Acoust 21 (6): 569–571.

Existence and properties of flow structure waves in two-phase bubbly flows

J.A. BOURÉ and Y. MERCADIER*

Département des Réacteurs à Eau, Service des Transferts Thermiques,
Centre d'Etudes Nucléaires de Grenoble, Grenoble, France

Abstract. Structure waves occur in two-phase flows because one phase drifts with respect to the other, the drift flux being primarily a function of the flow structure. The wave properties provide information on the closure laws required in engineering models. Experiments made with an air-water bubbly mixture flowing in a vertical annular test section are reported. Void fluctuations involving structure disturbances were detected by capacitance measurements, the effect of individual bubbles being always negligible. Only low frequency disturbances were present, high frequency disturbances being strongly damped. Within the low frequency range, the wave velocity is independent of the frequency, and the damping is small. The wave velocity is always comprised between the average liquid velocity and the average gas velocity.

Résumé. Des ondes de structure apparaissent dans les écoulements diphasiques parce que les phases n'ont pas la même vitesse moyenne, le flux de glissement correspondant étant essentiellement fonction de la structure de l'écoulement. Les propriétés des ondes apportent des informations sur les lois de fermetures requises par les modèles pratiques. On présente les résultats d'expériences effectuées avec un écoulement eau-air à bulles dans une section annulaire verticale. Les fluctuations de taux de vide dues aux perturbations de structure sont détectées par mesure de capacité (l'effet d'une bulle unique est toujours négligeable). Seules les perturbations de basse fréquence peuvent subsister, les perturbations de haute fréquence étant fortement amorties. Dans la gamme basse fréquence, la vitesse des ondes est indépendante de la fréquence et l'amortissement est faible. La vitesse des ondes est toujours comprise entre la vitesse moyenne du liquide et la vitesse moyenne du gaz.

1. Introduction; Motivation of the study

Propagation phenomena play an important role in two-phase flows whenever the two-phase mixture undergoes rapid changes. Such is the case during transients, but also as long as the flow is not fully developed. In another light, since the closure laws used in two-phase flow models have a strong influence on the propagation properties of the models, the knowledge on propagation phenomena may be used to check the validity of the closure laws. Working out an acceptable set of closure laws is currently an essential problem in the development of advanced engineering models for two-phase flows (models used for instance, in nuclear safety studies) [1].

In engineering applications, there is no choice but to ignore small scale phenomena (such as individual bubble behaviors), and the interest is

*Present address: Département de Génie Mécanique, Université de Sherbrooke, Sherbrooke, Quebec, Canada.

Applied Scientific Research 38: 297–303 (1982) 0003–6994/82/0384–0297 $01.05.

restricted to collective phenomena. For bubbly flows, the corresponding assumption is that bubble sizes and distances between neighboring bubbles are small with respect to the length scales of interest.

With usual engineering models consisting of algebraic and first order partial differential equations, the number of describable propagation phenomena is equal to the number of partial differential equations. Besides material transports, the two-phase flow literature deals with two kinds of propagation phenomena: pressure waves, which have been the subject of numerous studies, and are not considered hereunder, and continuity or kinematic waves, introduced independently in the early sixties by Wallis [6] and Zuber [8], and which surprisingly, have been the subject of very few studies [4, 5]. Small disturbance, one-dimensional kinematic waves in bubbly flows are the subject of this paper.

2. Kinematic waves and modeling

In many problems, some flux G and some concentration ρ are related through the balance equation (t, time; z, space coordinate):

$$\frac{\partial \rho}{\partial t} + \frac{\partial G}{\partial z} = S, \tag{1}$$

S being a source term.

Kinematic waves occur [7] whenever G is an algebraic function of ρ, which provides a closure law

$$G = G(\rho) \tag{2}$$

for equation (1). Combination of equations (1) and (2) yields a wave equation

$$\frac{\partial \rho}{\partial t} + G'_\rho \frac{\partial \rho}{\partial z} = S, \quad \text{where} \quad G'_\rho = \frac{\mathrm{d}G}{\mathrm{d}\rho}. \tag{3}$$

The wave conveys concentration disturbances. Its velocity G'_ρ, and its other properties (damping) are strongly influenced by the closure laws for G and for S.

The foregoing concept may be applied to two phase flows: In one-dimensional fully developed two-phase flow, the average velocity difference (in other words, the drift flux of one phase with respect to the other) may be expected to be essentially a function of the void fraction through flow structure effects, and of the physical properties of the phases. A purely kinematic description of the flow is possible if the physical properties may be assumed to be constant and if the mass transfer between the two phases may be assumed to be an algebraic function of the kinematic variables (acceptable in two-component flows; unrealistic in one-component flows). The relevant equations are

(a) the mass balance equations

$$\frac{\partial(A\alpha_G\rho_G)}{\partial t} + \frac{\partial(A\alpha_G\rho_G W_G)}{\partial z} = M_G, \tag{4}$$

$$\frac{\partial(A\alpha_L\rho_L)}{\partial t} + \frac{\partial(A\alpha_L\rho_L W_L)}{\partial z} = -M_G; \tag{5}$$

(b) the closure law for the velocity difference

$$\Delta W \triangleq W_G - W_L = f(\alpha_G). \tag{6}$$

A, α_K, ρ_K, W_K, with $K = G$ or L, are respectively the cross section area, the volumetric concentration, the density and the velocity of phase K (the classical void fraction is α_G and $\alpha_L = 1 - \alpha_G$); M_G is the mass transfer term.

Two propagation phenomena are associated with the set (4–6). A first wave equation may be obtained by summing up equations (4) and (5), respectively divided by ρ_G and ρ_L. It is, introducing the volumetric fluxes per unit area $j_K \triangleq \alpha_K W_K$ and $j \triangleq j_G + j_L$:

$$\frac{\partial(Aj)}{\partial z} = M_G\left(\frac{1}{\rho_G} - \frac{1}{\rho_L}\right) - A'_t \quad \text{with} \quad A'_t = \frac{\partial A}{\partial t}. \tag{7}$$

Equation (7) expresses the propagation, at infinite velocity, of volumetric flux disturbances. A second wave equation may be obtained by forming $\frac{\alpha_L}{A\rho_G}$ (equation 4) $-\frac{\alpha_G}{A\rho_L}$ (equation 5), taking (equation 6) into account:

$$\frac{\partial\alpha_G}{\partial t} + \left[\alpha_L W_G + \alpha_G W_L + \alpha_G\alpha_L \frac{\mathrm{d}(\Delta W)}{\mathrm{d}\alpha_G}\right]\frac{\partial\alpha_G}{\partial z}$$

$$= \frac{M_G}{A}\left(\frac{\alpha_L}{\rho_G} + \frac{\alpha_G}{\rho_L}\right) - \alpha_G\alpha_L(\Delta W)\frac{A'_z}{A}. \tag{8}$$

Equation (8) expresses the propagation, at velocity

$$C = \alpha_L W_G + \alpha_G W_L + \alpha_G\alpha_L \frac{\mathrm{d}\Delta W}{\mathrm{d}\alpha_G} \tag{9}$$

of void fraction, i.e. of flow structure, disturbances. Note that, in equation (9) the subscripts are crossed. Expressions equivalent to the left-hand side of equation (8) were first given by Wallis and Zuber. They introduced the drift velocity of the gas phase $W_G - j$.

Since $W_G - j = \alpha_L\,\Delta W$, equation (9) may be written

$$C = W_G + \alpha_G \frac{\mathrm{d}(W_G - j)}{\mathrm{d}\alpha_G} = \left[\frac{\partial(\alpha_G W_G - j)}{\partial\alpha_G}\right]_j$$

Again, the wave properties depend on the closure laws, and their knowledge may be used to obtain information on the closure laws.

In two-phase flow occurrences of practical interest, the above description is often insufficient. In particular:

the densities ρ_K are functions of pressures P_K and temperatures T_K which introduces compressibility effects, possible coupling between flow structure waves and pressure waves, and the necessity to write all the balance equations (mass, momentum, energy);

the fully-developed value of ΔW is also a function of P_K and T_K, through the physical properties of the phases. Moreover, the flow is seldom fully-developed (importance of so-called nonequilibrium effects);

the mass transfer term M_G, as well as the other interfacial terms which appear in the balance equations, is an unknown function of the other variables.

The problem is therefore much more complicated. However, it is reasonable to assume that the concept of flow structure wave is still valid. This is confirmed by the experiments presented hereunder.

3. Experiments

An experimental set-up was designed to enable detection and study of void fraction waves. It is described in [3]. The test section is a vertical annulus 2 m long, the diameter of the central rod being 32 mm and the inner diameter of the outer tube being 70 mm. Air bubbles are generated at the bottom of the test section. The experiments are performed at room temperature and the top of the test section is open to the atmosphere. The water flow rate Q_L has been varied from slightly negative values ($-0.5\ m^3/h$, countercurrent flow) to positive values (up to $15\ m^3/h$ corresponding to a single phase velocity of 1.37 m/s) through zero. The air flow rate has been varied to cover the whole range of bubbly flows (void fraction varying approximatley from 1% to 22%, depending on the water flow rate).

In most experimental runs, the set up was operating in so-called steady state, void fluctuations occurring naturally (Figure 1). However periodic disturbances of the air flow rate were imposed in a complementary series of experiments.

The void fraction is calculated from the pressure gradient, using the steady-state mass and momentum balance equations and experimental correlations for wall and interfacial friction. The average velocities of the phases are computed from the flow rates and void fraction. As expected, the pressure gradient, the void fraction and the velocities are practically constant along the test section, except near the inlet where the flow is developing.

The void fluctuations (Figure 1) are detected by capacitance measurements: the test section is equipped with nine annular capacitors, the distance between two successive capacitors being 20 cm. The capacitor electrodes are

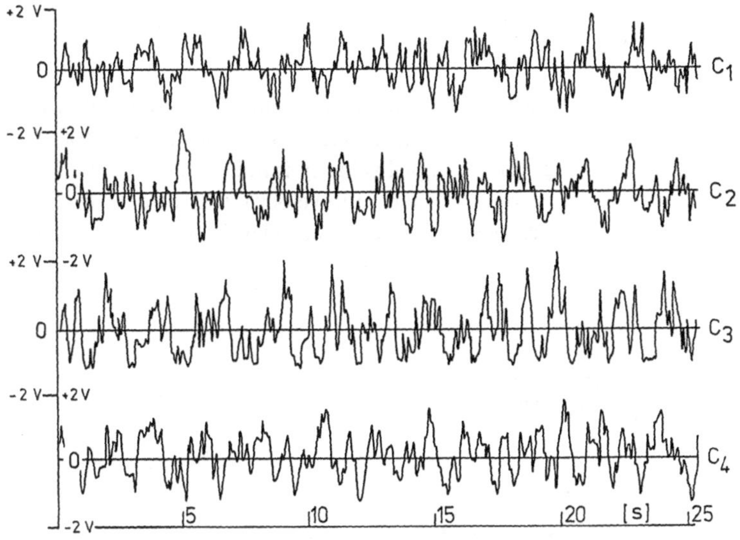

Figure 1. Typical capacitance signals ($\alpha_G = 0.135$).

part of the test section central rod and outer tube, and they do not disturbe the flow. Seven capacitors are 5 cm long. Their dielectric volume is large and the effect of a single bubble on the capacitance is well below the sensitivity threshold of the measurement. The absence of any effect of capacitor length was checked by using the two uppermost capacitors, which are only 2.5 cm long.

4. Data processing

The data (Figure 1) are processed using statistical analysis techniques [2], the two phase mixture between any pair of capacitors being regarded as a linear filter. $x(t)$ and $y(t)$ being the fluctuating parts of the signals of two capacitors, and τ and ν being respectively a time interval and a frequency, the auto- and cross-correlation functions $C_{xx}(\tau)$ and $C_{xy}(\tau)$ are evaluated, as well as their Fourier transforms $S_{xx}(\nu)$ and $S_{xy}(\nu)$, which are respectively the spectral density function and the cross-spectral density function. $S_{xx}(\nu)$ is a real function and $S_{xy}(\nu)$ a complex function of the frequency ν:

$$S_{xy}(\nu) = R\,[S_{xy}(\nu)] + iI\,[S_{xy}(\nu)]\,. \tag{10}$$

The spectral density function gives information on the range of frequencies present in the signal. For each of these frequencies, the phase factor:

$$\phi(\nu) = \mathrm{tg}^{-1}\,\frac{I\,[S_{xy}(\nu)]}{R\,[S_{xy}(\nu)]} \tag{11}$$

and the gain factor

$$G(\nu) = \frac{|S_{xy}(\nu)|}{S_{xx}(\nu)} \tag{12}$$

enable evaluation of respectively the velocity and the damping of the signal. These evaluations are obtained with an error margin which, especially for the damping, may be large. If the signal velocity is independent of the frequency (which turns out to be the case here) the velocity may be determined with a better accuracy from the peak of the cross-correlation function (Figure 2).

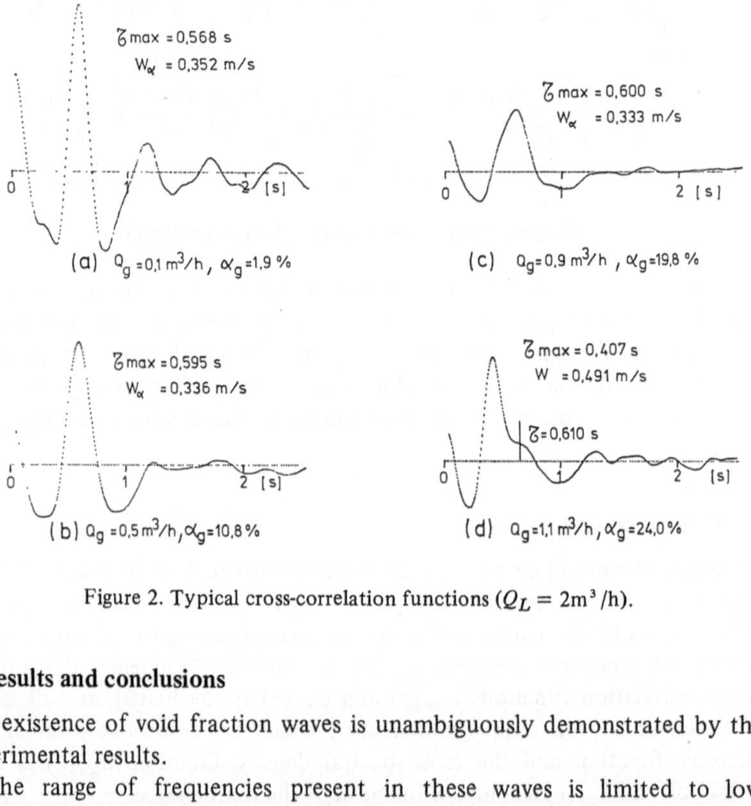

Figure 2. Typical cross-correlation functions ($Q_L = 2\text{m}^3/\text{h}$).

5. Results and conclusions

The existence of void fraction waves is unambiguously demonstrated by the experimental results.

The range of frequencies present in these waves is limited to low frequencies, depending on the flow rate, the practical cut-off frequency varying from a few hertz at low flow rate to 10–15 hertz at large flow rates. The very large damping of high frequency waves is confirmed in the runs made with imposed periodic disturbances.

Within most of the above low fequency range, the damping is small and the void fraction waves are non-dispersive (their velocity W_α does not depend on frequency). Typical results are given in Figure 2. Figure 2d corresponds to a case when the flow conditions were close to the transition between bubble flow and slug flow; this may be the explanation for the appearance of an inflexion point for $\tau = 0.61$ s.

The wave velocities are plotted in Figure 3. It is emphasized that the wave velocity W_α is clearly distinct, for all runs, from both W_G and W_L. It is always comprised between these two values.

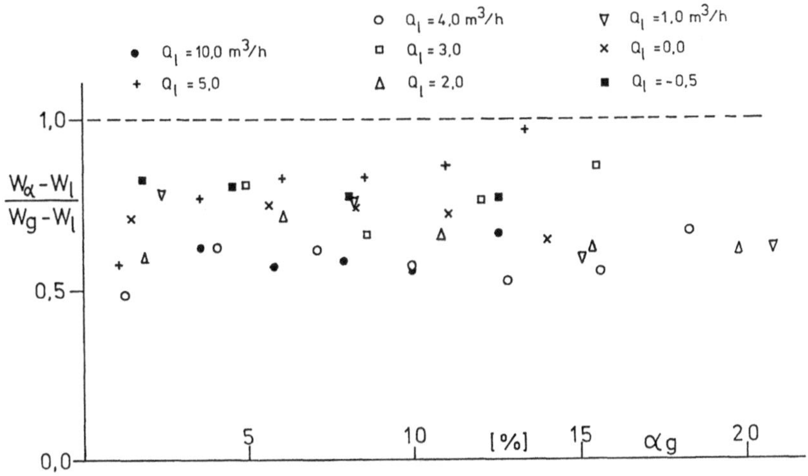

Figure 3. Void fraction wave velocities.

References

1. Bouré JA (1980) Multiphase flow modeling; I. Fundamentals; II. Development of practical models. Lecture notes, Seminar on multiphase processes in LMFBR safety analysis, Ispra.
2. Bendat JS and Piersol AG (1971) Random data: Analysis and measurement procedures. New-York: Wiley-Interscience.
3. Mercadier Y (1981) Contribution à l'étude des propagations de perturbations de taux de vide dans les écoulements diphasiques eau-air à bulles. Thèse de Docteur-Ingénieur, Université scientifique et médicale et Institut national polytechnique, Grenoble.
4. Nassos GP and Bankoff SG (1966) Propagation of density disturbances in an air water flow. Proc 3rd Int Conf on Heat Transfer, IV, pp 234–246. New-York: AIChE.
5. Van Schaik JCH (1979) On the propagation of void fraction disturbances in bubble flows. CENG report TT 600.
6. Wallis GB (1961) Two phase flow phenomena associated with the boiling of liquids. Ph D Thesis, Cambridge University.
7. Whitham GB (1974) Linear and non linear waves. New-York: Wiley & Sons.
8. Zuber N (1961) Steady state, transient response, operating limits and continuity waves in two-phase flow systems. General Electric 61GL215.

Some aspects of dynamics of bubbly liquids

D.Y. HSIEH

Division of Applied Mathematics, Brown University, Providence, RI 02912, USA

Abstract. The formulation of dynamics of a liquid containing gas bubbles or vapor bubbles is presented. It is applied to sound waves and one-dimensional flow problems.

1. Introduction

An attempt is made to present a systematic and practical analysis of the dynamics of a bubbly liquid. First the basic dynamical equations are formulated. Then these basic equations are applied to sound waves and other relatively simple flow problems.

Keeping in mind the existing general formal theories of mixtures, we formulate the basic dynamic equations from a rather heuristic approach as those taken by Wijngaarden [2]. Besides defining fluid properties like density ρ, pressure p, velocity $\mathbf{u}$, and self frictional force density $\mathbf{F}$ for each individual phase, we define also the local number density $n(\mathbf{x}, t)$ and average radius $R(\mathbf{x}, t)$ of the bubbles as a continuum variable in the fluid at the position $\mathbf{x}$ and time t. The number of bubbles, when there is no creation, collapse or coalescence of bubbles, are conserved. The change of R is to be related to pressures of two phases by the Rayleigh-Plesset equation. A mutual friction between the relative motion of two fluids can be introduced modeled after the Stokes's law. For gas bubbles, the mass content of each individual bubble is constant. We classify the bubble species by the mass content. Then the basic equation can also be formulated for bubbly liquids with many species of bubbles.

For vapor bubbles, the mass content in each bubble will vary because of evaporation and condensation. There will be mass source term S_m in the continuity equations and thermal source terms S_T in the heat equation of the liquid. These source terms are interrelated and can be derived from the study of the dynamics of individual vapor bubbles, e.g., the Plesset-Zwick theory [3].

2. Dynamic equations of a liquid containing gas bubbles

We shall use subscripts f and g to designate for quantities $\rho, p, \mathbf{u}$ and $\mathbf{F}$ of the liquid and gas phases respectively. Let m be the constant mass content of each bubble, β the volume occupied by the gas in a unit volume of the fluid mixture and $\mathbf{b}$ the external body force per unit mass. Then the basic equations are:

Applied Scientific Research 38: 305–312 (1982) 0003–6994/82/0384–0305 $01.20.

$$\beta = \frac{4\pi}{3} R^3 n, \tag{1}$$

$$m = \frac{4\pi}{3} R^3 \rho_g, \tag{2}$$

$$\frac{\partial n}{\partial t} + \nabla \cdot (n\mathbf{u}_g) = 0, \tag{3}$$

$$\frac{\partial}{\partial t} (\beta\rho_g) + \nabla \cdot (\beta\rho_g \mathbf{u}_g) = 0, \tag{4}$$

$$\frac{\partial}{\partial t} [(1-\beta)\rho_f] + \nabla \cdot [(1-\beta)\rho_f \mathbf{u}_f] = 0, \tag{5}$$

$$\rho_f \left[\frac{\partial \mathbf{u}_f}{\partial t} + (\mathbf{u}_f \cdot \nabla)\mathbf{u}_f\right] = -\nabla p_f + \rho_f \mathbf{b} + \mathbf{F}_f - \frac{\mathbf{F}_{fg}}{(1-\beta)}, \tag{6}$$

$$\rho_g \left[\frac{\partial \mathbf{u}_g}{\partial t} + (\mathbf{u}_g \cdot \nabla)\mathbf{u}_g\right] = -\nabla p_g + \rho_g \mathbf{b} + \mathbf{F}_g + \frac{\mathbf{F}_{fg}}{\beta}, \tag{7}$$

$$p_g - p_f = P\{R\}. \tag{8}$$

In these expressions, we take $\mathbf{F}_f$ and $\mathbf{F}_g$ to be given by the Newtonian viscous forces. Then

$$\mathbf{F}_f = \eta_f [\nabla^2 \mathbf{u}_f + \tfrac{1}{3}\nabla(\nabla \cdot \mathbf{u}_f)], \tag{9}$$

$$\mathbf{F}_g = \eta_g [\nabla^2 \mathbf{u}_g + \tfrac{1}{3}\nabla(\nabla \cdot \mathbf{u}_g)], \tag{10}$$

where η_f and η_g are the viscosity coefficients for liquid and gas respectively.

Modeling after Stokes law, we may take the mutual friction

$$\mathbf{F}_{fg} = D(\mathbf{u}_f - \mathbf{u}_g), \tag{11}$$

with

$$D = 2\pi\eta_f \left(\frac{2\eta_f + 3\eta_g}{\eta_f + \eta_g}\right) Rn. \tag{12}$$

$P\{R\}$ can be taken to be given by the Rayleigh-Plesset equation. Hence

$$P\{R\} = \rho_f R \frac{\partial^2 R}{\partial t^2} + \tfrac{3}{2}\rho_f \left(\frac{\partial R}{\partial t}\right)^2 + \frac{4\eta_f}{R}\left(\frac{\partial R}{\partial t}\right) + \frac{2\sigma}{R}, \tag{13}$$

where σ is the surface tension coefficient.

When the effect of heat transfer is not important, we also have the following equations of state for the two phases

$$p_g R^{3\gamma} = \text{constant}, \tag{14}$$

and

$$p_f = p_f(\rho_f). \tag{15}$$

The above formulation can be readily generalized to deal with dynamics of a liquid containing more than one species of gas bubbles.

3. Dynamic equations of a liquid containing vapor bubbles

Due to the process of evaporation and condensation, the mass contents of vapor bubbles are variable, and the thermal process can not be ignored. Equations (1), (3), (6)–(13) are still valid. Equations (4) and (5) become now:

$$\frac{\partial}{\partial t}(\beta\rho_g) + \nabla\cdot(\beta\rho_g\mathbf{u}_g) = S_m, \tag{16}$$

$$\frac{\partial}{\partial t}[(1-\beta)\rho_f] + \nabla\cdot[(1-\beta)\rho_f\mathbf{u}_f] = -S_m. \tag{17}$$

Equations of state (14) and (15) are changed to

$$p_g = NT_g\rho_g, \tag{18}$$

and

$$p_f = p_f(\rho_f, T), \tag{19}$$

where T is the local temperature of the liquid and T_g the temperature of the bubble. The local phase equilibrium relation will also require that

$$p_g = p_v(T_g). \tag{20}$$

The temperature T is to be governed by the following equation of heat transfer:

$$\frac{\partial T}{\partial t} + (\mathbf{u}_f\cdot\nabla)T = D_T\nabla^2 T - S_T, \tag{21}$$

where D_T is the coefficient of thermal diffusion. The mass source term S_m and thermal source term S_T are related by

$$S_m = \frac{\kappa}{LD_T}S_T, \tag{22}$$

where L is the latent heat of evaporation per unit mass and κ is the coefficient of thermal conduction in the liquid. The expression of S_T can be derived from the Plesset-Zwick theory. Denote the non-dimensional variable

$$\theta = \frac{3LD_T\rho_g}{2\pi\kappa(T-T_g)}, \tag{23}$$

then we have

$$S_T = \begin{cases} \dfrac{3\beta D_T(T-T_g)}{R^2}, & \text{for} \quad \theta > 1 \\ \dfrac{2\pi\beta\kappa\,(T-T_g)^2}{L\rho_g R^2}, & \text{for} \quad \theta < 1. \end{cases} \tag{24}$$

4. Sound waves in bubbly liquid

We now apply the general formulation to small amplitude wave propagation in a bubbly liquid at rest. We shall use subscript 0 to denote equilibrium quantities. For sinusoidal waves proportional to exp $[i(\mathbf{k}\cdot\mathbf{x}-\omega t)]$, the characteristic equation relating ω to k is a fifth degree equation in ω. The full expression of the characteristic equation will be presented elsewhere. We consider here the special case when the self-force terms and the surface tension term are negligible. Let us denote

$$c_f^2 = \left(\frac{dp_f}{d\rho_f}\right)_0, \qquad c_g^2 = \left(\frac{dp_g}{d\rho_g}\right)_0, \tag{25}$$

$$\omega_f = kc_f, \qquad \omega_D = \frac{D_0}{\beta_0(1-\beta_0)\rho_{g_0}}, \qquad q = (kR_0)^2, \tag{26}$$

$$\epsilon_1 = \frac{\rho_{g_0}}{\rho_{f_0}}, \qquad \epsilon_2 = \frac{c_g^2}{c_f^2}. \tag{27}$$

Since, in general we have $\epsilon_1 \ll 1$ and $\epsilon_2 \ll 1$, the characteristic equation is given approximately:

$$\omega^5 - \left[\frac{3\beta_0}{q(1-\beta_0)} + 1\right]\omega_f^2\omega^3 + \frac{3\beta_0\epsilon_2}{q(1-\beta_0)}\omega_f^4\omega$$
$$+ i\omega_D\left[(1-\beta_0)\omega^4 - \left(1+\frac{3\beta_0}{q}\right)\omega_f^2\omega^2 + \frac{3\epsilon_1\epsilon_2}{(1-\beta_2)q}\omega_f^4\right] = 0 \tag{28}$$

Let us consider the following limiting cases:
(i) $\omega_D \ll \omega$, i.e. when the mutual friction is insignificant.
The approximate solutions are:

$$\omega^2 = \omega_1^2 = \omega_f^2\left[1+\frac{3\beta_0}{q(1-\beta_0)}\right], \tag{29}$$

$$\omega^2 = \omega_2^2 = \epsilon_2\omega_f^2\left[1+\frac{q(1-\beta_0)}{3\beta_0}\right]^{-1}. \tag{30}$$

If $q \gg 1$, then $\omega_1^2 = c_f^2 k^2$ while if $q \ll 1$, then $\omega_2^2 = c_g^2 k^2$. These modes represent sound propagation in uncoupled liquids and gas phases. The fifth mode from (28) is a nonpropagating mode. We may substitute $\omega = \omega_1 - i\delta_1$ or $\omega = \omega_2 - i\delta_2$ to estimate the damping coefficients. We found

$$\delta_1 \approx -\frac{\beta_0}{2}\left(\frac{\omega_f}{\omega_1}\right)^4 \omega_D, \tag{31}$$

and

$$\delta_2 \approx \left(1 + \frac{3\beta_0}{q}\right)\omega_D \Bigg/ 2\left[1 + \frac{3\beta_0}{q(1-\beta_0)}\right]. \tag{32}$$

(ii) $\omega_D \gg \omega$, i.e. when the mutual friction is important.
The approximate frequencies and damping coefficients are

$$\omega^2 = \omega_3^2 = \left(1 + \frac{3\beta_0}{q}\right)\omega_f^2 \Big/ (1-\beta_0), \tag{33}$$

$$\omega^2 = \omega_4^2 = \epsilon_1\epsilon_2\omega_f^2 \Big/ \beta_0(1-\beta_0)\left(1 + \frac{q}{3\beta_0}\right). \tag{34}$$

$$\delta_3 = -\beta_0\omega_f^2/2(1-\beta_0)^2\omega_D, \tag{35}$$

and

$$\delta_4 = \epsilon_2\omega_f^2 \Big/ 2(1-\beta_0)\left(1 + \frac{q}{3\beta_0}\right)\omega_D. \tag{36}$$

The dampings associated with ω_1, and ω_3 are negative. These modes are the ones corresponding to sound waves in the liquid phase only. Their instability can perhaps be interpreted to mean that these modes are not likely to be excited in a bubbly mixture.

When $q \ll 1$, we can obtain the familiar relation:

$$\omega_4^2 = k^2 c_m^2,$$

where

$$c_m = \left[\frac{1}{\beta_0(1-\beta_0)}\,\frac{\rho_{g_0}}{\rho_{f_0}}\right]^{1/2} c_g, \tag{37}$$

which is usually an order of magnitude smaller than c_g.

5. Damping of sound wave due to vaporization

For sound wave propagation in a liquid containing vapor bubbles, the general frequency relation is considerably more complex than that containing only gas bubbles. We shall only consider the particular mode associated with ω_4 discussed in Section 4.

Denote

$$\Gamma = \left(\frac{dp_v}{dT_g}\right)_0 \frac{T_{g_0}}{p_{g_0}}, \qquad c_\theta^2 = \left(\frac{\Gamma}{\Gamma-1}\right)\left(\frac{p_{g_0}}{\rho_{g_0}}\right), \tag{38}$$

and for $\theta < 1$:

$$S_{m_0} = \frac{2\pi\beta_0\kappa^2(T_0 - T_{g_0})^2}{D_T\rho_{g_0}R_0^2L^2}, \qquad \alpha_2 = \frac{2T_{g_0}}{T_0 - T_{g_0}},$$

$$\alpha_3 = 1, \tag{39}$$

while for $\theta > 1$:

$$S_{m_0} = \frac{3\kappa\beta_0(T_0 - T_{g_0})}{R_0^2L}, \qquad \alpha_2 = \frac{T_{g_0}}{T_0 - T_{g_0}},$$

$$\alpha_3 = 0. \tag{40}$$

Then when $(S_{m_0}/\omega\rho_{g_0})$ and (D_Tk^2/ω) are small, we obtain approximately:

$$\omega = \omega_4' - i\delta_4', \tag{41}$$

where

$$\omega_4' = \left[\frac{\rho_{g_0}}{\rho_{f_0}\beta_0(1-\beta_0)}\right]^{1/2} kc_\theta, \tag{42}$$

and

$$\delta_4' = \frac{1}{2}\left(\frac{S_{m_0}}{\rho_{g_0}}\right)\left(\frac{\alpha_2}{\Gamma-1} + \alpha_3\right). \tag{43}$$

The damping δ_4' is small when the latent heat L is large, and this is the damping coefficient due solely to the thermal effects.

6. One-dimensional steady flows

For one-dimensional steady problem, we have $\mathbf{u}_f = (u_f(x), 0, 0)$, $\mathbf{u}_g = (u_g(x), 0, 0)$, $\mathbf{b} = (b, 0, 0)$ and all other quantities are functions of x only. Take $\gamma = 1$, which implies isothermal behavior and is justifiable from single bubble dynamics [1]. We shall also neglect the self-frictional forces $\mathbf{F}_f$ and $\mathbf{F}_g$, since the dominant friction is the mutual friction. Furthermore, we shall assume the liquid is incompressible. After some manipulation, the governing equations in Section 2 can be reduced to:

$$\frac{\mathrm{d}R}{\mathrm{d}x} = \frac{1}{A_5}\{A_1[\beta - \beta_e(R)] + A_3 b\}, \tag{44}$$

$$\frac{\mathrm{d}\beta}{\mathrm{d}x} = \frac{1}{A_5}\{A_2[\beta - \beta_e(R)] + A_4 b\}, \tag{45}$$

where

$$A_1 = \eta\left[\frac{\rho_{g_0} u_{g_0}^2}{(1-\beta)R_0^2}\left(\frac{\beta_0 R}{\beta R_0}\right)^2 - \frac{\rho_f u_{f_0}^2}{(1-\beta)R^2}\left(\frac{1-\beta_0}{1-\beta}\right)^2\right] A_6, \tag{46}$$

$$A_2 = -\eta\left[\frac{3p_{g_0}}{(1-\beta)R^3}\left(\frac{R_0}{R}\right)^3 - \frac{2\sigma}{R^4} - \frac{3\rho_{g_0} u_{g_0}^2 \beta_0^2}{\beta(1-\beta)R_0^3}\right] A_6, \tag{47}$$

$$A_3 = -\rho_f \rho_{g_0}\left[\frac{u_{g_0}^2}{\beta_0}\left(\frac{\beta_0 R}{\beta R_0}\right)^3 + \frac{u_{f_0}^2}{(1-\beta_0)}\left(\frac{R_0}{R}\right)^3\left(\frac{1-\beta_0}{1-\beta}\right)^3\right], \tag{48}$$

$$A_4 = \frac{3p_{g_0}}{R}\left(\frac{R_0}{R}\right)^3\left[\rho_f - \rho_{g_0}\left(\frac{R_0}{R}\right)^3\right] + \frac{2\sigma\rho_{g_0}}{R^2}\left(\frac{R_0}{R}\right)^3$$

$$- \frac{3\rho_f \rho_{g_0} u_{g_0}^2}{R_0}\left(\frac{\beta_0 R}{\beta R_0}\right)^2, \tag{49}$$

$$A_5 = \left(\frac{3p_{g_0}}{R} - \frac{2\sigma R}{R_0^3}\right)\frac{\rho_{g_0} u_{g_0}^2 \beta_0^2}{\beta^3} + \left[\frac{3p_{g_0}}{R}\left(\frac{R_0}{R}\right)^3\right.$$

$$\left. - \frac{3\rho_{g_0} u_{g_0}^2}{R_0}\left(\frac{\beta_0 R}{\beta R_0}\right)^2\right] \times \frac{\rho_f u_{f_0}^2}{(1-\beta)}\left(\frac{1-\beta_0}{1-\beta}\right)^2, \tag{50}$$

$$A_6 = \frac{\beta_0 u_{g_0}}{\beta(1-\beta)\beta_e(R)}\left(\frac{R}{R_0}\right)^3, \tag{51}$$

$$\eta = \frac{3}{2}\left(\frac{2\eta_f + 3\eta_g}{\eta_f + \eta_g}\right)\eta_f, \tag{52}$$

and

$$\beta_e(R) = \left(\frac{R}{R_0}\right)^3\left[\left(\frac{R}{R_0}\right)^3 + \left(\frac{u_{f_0}}{u_{g_0}}\right)\left(\frac{1-\beta_0}{\beta_0}\right)\right]^{-1}. \tag{53}$$

In these expressions, the subscript 0 refers to properties of some reference state.

The equilibrium point of the system (44) and (45), i.e. the uniform flow state, is given by

$$b = 0, \tag{54}$$

and

$$\beta = \beta_e(R). \tag{55}$$

Equation (55) is actually the condition that $u_f = u_g$. The stability of the uniform flow state can be obtained from (44) and (45). Let us denote

$$\lambda = \left[\frac{1}{A_5}\left\{A_2 - \frac{3A_1}{R}\,\beta_e(1-\beta_e)\right\}\right]_{\beta=\beta_e(R)}. \tag{56}$$

Then the uniform state is stable if

$$\lambda < 0. \tag{57}$$

When $b = 0$, the uniform flow states can be connected by a discontinuity. Let the state to the left of the discontinuity be denoted by the subscript 0, and let the state to the right be denoted by the subscript 1. Then it can be shown that

$$\ln\left(\frac{R_1}{R_0}\right) - \frac{1}{3}\left(\frac{\rho_{g_0}}{\rho_f}\right)\left[\left(1 - \frac{R_0^3}{R_1^3}\right) - \frac{2\sigma}{p_{g_0}}\left(\frac{1}{R_0} - \frac{1}{R_1}\right)\right] = 0. \tag{58}$$

Thus with R_0 specified, R_1 is determined irrespective of u_0 or u_1. For the state 0, since $\beta_e = \beta_0$, the criterion of its stability can be explicitly determined from (56).

When $b \neq 0$, no uniform flow state can be found. Numerical solutions of the system (44) and (45) and detailed discussions are to be presented elsewhere.

References

1. Plesset MS and Hsieh DY (1960) Theory of gas bubble dynamics in oscillating pressure field. Phys Fluids 3: 882–892.
2. Wijngaarden L van (1972) One-dimensional flow of liquids containing small gas bubbles. Ann Rev Fluid Mech 4: 369–396.
3. Zwick SA and Plesset MS (1955) On the dynamics of small vapor bubbles in liquids. J Math and Phys 33: 308–330.

Energy considerations on the collapse of cavity clusters

K.A. MØRCH

Laboratory of Applied Physics I, Technical University of Denmark,
DK-2800 Lyngby, Denmark

Abstract. A general equation for the energy balance in a liquid containing a cluster of cavities is used to derive the collapse equations for specific cluster configurations. These equations contain as parameters the volume fraction of cavities β and a coefficient γ for the energy transfer into the cluster due to collapse of the individual cavities. The influence of γ and β and of the spatial distribution of β is calculated. Experimental studies of the plane, one-dimensional collapse of a cavity layer are compared with the theoretical results. In this connection the influence of temperature on the cavity cluster formation and collapse is considered.

1. Introduction

In theoretical cavitation research the attention has primarily been directed at the collapse of a single cavity, but normally, cavitation leads to the formation of clusters containing large numbers of cavities. In these the cavities collapse successively from the outer boundary of the cluster, so that those collapsing first strengthen the collapse of the next ones.

Momentum considerations were used in [1, 3, 5] for analysis of the collapse of a cluster, but the influence of the pressure wave emitted from each of the cavities at collapse [2] was neglected. Meanwhile, in an energy balance equation for the cluster and the surrounding liquid these waves are readily included.

2. Collapse equations for cavity clusters

We consider a cluster of cavities which is bounded by an incompressible liquid, and possibly by solid surfaces. Initially the system is supposed to be in equilibrium at the vapour pressure $p_1 \simeq 0$, but increase of the far field pressure in the liquid to the constant value $p_\infty \gg p_1$ initiates collapse of the cluster from its free boundary, which moves towards the cluster centre as the cavities collapse. Just after the pressure increase the energy of the system is purely potential, but as the cluster volume Ω_c shrinks, this energy is gradually converted first into kinetic energy of the liquid in the region Ω_l outside Ω_c, then, as the individual cavities collapse, into wave energy which is radiated partly into the cluster, where it is absorbed and contributes to its further collapse, partly away from the cluster. The latter part is generally lost. As a consequence, the change of the sum of the potential and the kinetic energies of the system due to a small volume change of the cluster $\mathrm{d}\Omega_c$ is equal to the fraction $(1-\gamma)$ of the mean potential energy of the cavities in $\mathrm{d}\Omega_c$ (at the local pressure in the liquid surrounding the individual cavity) which is lost by radiation (and dissipation)

Applied Scientific Research 38: 313–321 (1982) 0003–6994/82/0384–0313 \$01.35.

$$\mathrm{d}\Bigg(\underbrace{\int_{\Omega_c} p_\infty \beta \,\mathrm{d}\Omega_c}_{\text{pot. energy of cluster}} + \underbrace{\int_{\Omega_l} \tfrac{1}{2}\rho v^2 \,\mathrm{d}\Omega_l}_{\text{kin. energy of liquid}}\Bigg) = (1-\gamma)\underbrace{\frac{P}{2}\beta \,\mathrm{d}\Omega_c}_{\text{mean pot. energy of collapsing cavities}}. \qquad (1)$$

Here ρ and v are the density and the velocity, respectively, in the liquid, while P is the pressure in the liquid at the cluster boundary (capital letters are used for the variables at this position). The sound velocity in the two-phase medium and the pressure ratio across the cluster boundary give

$$P = V_{\mathrm{sh}}^2\, \rho\beta(1-\beta), \qquad (2)$$

where V_{sh} is the velocity of the cluster boundary, and β is measured just inside this boundary. If the cluster configuration is specified (1) and (2) determine the collapse equation of the cluster.

For a *spherically symmetrical cavity cluster* of initial radius R_0 and instantaneous radius R in an infinite liquid we have $V_{\mathrm{sh}} = \mathrm{d}R/\mathrm{d}t = \dot{R}$ and $v = (\beta\dot{R})R^2/r^2$, which give the collapse equation

$$R\ddot{R} + \left(\frac{3}{2} - \frac{1}{2}(1-\gamma)(1-\beta) + \frac{R}{\beta}\frac{\mathrm{d}\beta}{\mathrm{d}R}\right)\dot{R}^2 = -\frac{p_\infty}{\rho\beta}. \qquad (3)$$

It should be noted that for $\mathrm{d}\beta/\mathrm{d}R = 0, \gamma = 1$ (3) is analogous to the spherical collapse equation for a single cavity, while for $\mathrm{d}\beta/\mathrm{d}R = 0, \gamma = 0$, it reduces to the form in [1, 3]. By spherical collapse of the individual cavities about half the collapse energy is radiated into the cluster, while at nonspherical collapse a larger fraction is preserved, and so $0.5 < \gamma < 1$. In Figure 1 the collapse of the cluster is calculated for $\gamma = 0 \wedge 1$ at $\beta = \beta_1 \wedge \beta_1(1-(R/R_0)^2)$, in which $\beta_1 = \text{constant} \ll 1$. In Figure 2 the corresponding pressures P at the cluster boundary are shown. The collapse time as well as the pressure P are significantly affected by the $\beta(R)$-function, while γ primarily affects P and thus the violence with which the individual cavities collapse. Theoretically the collapse velocity becomes infinite as $R \to 0$, and P grows unrestrictedly, but the model ceases to be valid when R becomes of order as the cavity distance. For small β-values the pressure P is independent of β, but for large ones (2) causes P to decrease.

Qualitatively similar results are obtained for the *cylindrical collapse of a cluster* of initial radius R_0, instantaneous radius R between two coaxial cylindrical rods of radius r_c and separated a distance h (an ultrasonic horn and a stationary specimen) as considered in [1]. Here the collapse equation becomes

$$\left(R\ddot{R} + \dot{R}^2\left(1 + \frac{R}{\beta}\frac{\mathrm{d}\beta}{\mathrm{d}R}\right)\right)\left[\ln\frac{r_c}{R} + \pi\delta\ln\frac{1+2\delta}{2\delta}\right]$$

$$-\frac{\dot{R}^2}{2}(1 + (1-\gamma)(1-\beta)) = -\frac{p_\infty}{\rho\beta}, \qquad (4)$$

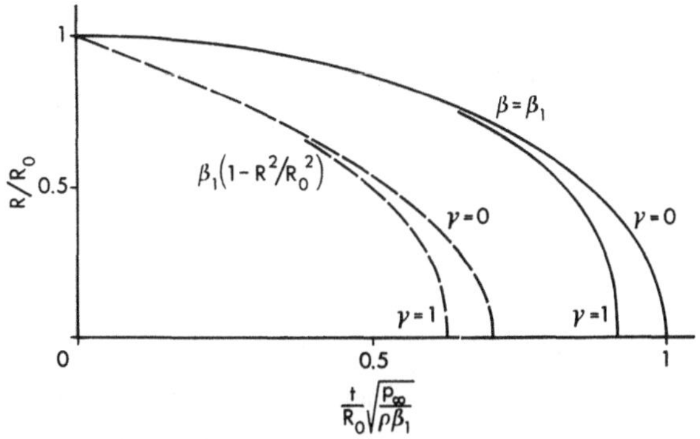

Figure 1. The cluster radius R vs time t during the collapse of a spherical cavity cluster for $\gamma = 0 \wedge 1$ at $\beta = \beta_1 \wedge \beta_1(1-(R/R_0)^2) \ll 1$.

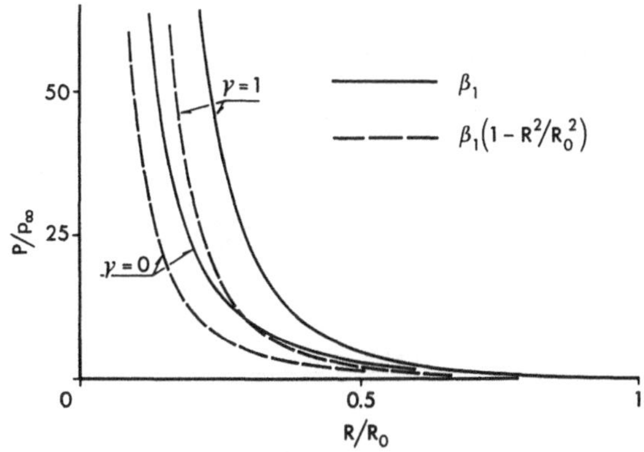

Figure 2. The pressure P at the cluster boundary vs cluster radius R corresponding to Figure 1.

where $\delta = h/\pi^2 r_c$. The focussing effect is weaker than in the spherical case and therefore the collapse time is 20–25% longer, and at corresponding cluster radii the pressure at the cluster boundary are significantly lower. However, in this case too a change of γ from 0 to 1 means a pressure increase by a factor 2 or more.

The most simple type of collapse is the *one-dimensional collapse of a plane layer of cavities.* Here the system is chosen to be bounded by a stationary solid surface at $y = 0$. The layer of cavities (the two-phase medium) initially occupies the region $0 \leqslant y < Y_0$, and pure liquid the region $Y_0 < y \leqslant L_0$, and during its collapse the cavity layer occupies $0 \leqslant y < Y$ and the liquid $Y < y \leqslant L$. At pressure increase from $p_1 \simeq 0$ to p_∞ at $y = L$ the cavity layer collapses from its free boundary at $y = Y$ which then moves at the velocity $V_{sh} = \dot{Y}$ towards the solid surface, and the liquid layer above moves with the uniform velocity $v = \beta\dot{Y}$. Then (1) gives

$$\ddot{Y}(L-Y)-\left[\left(1-\frac{\gamma}{2}\right)(1-\beta)-\frac{L-Y}{\beta}\frac{\mathrm{d}\beta}{\mathrm{d}Y}\right]\dot{Y}^2 = -\frac{p_\infty}{\rho\beta}, \qquad (5)$$

from which $Y(t)$, and with (2) also $P(t)$ can be calculated.

At the moment of total collapse the layer of liquid impacts the solid surface at the velocity $v = (\beta\dot{Y})_{Y=0}$ and an impact pressure p_w is produced

$$p_w = \rho c(\beta\dot{Y})_{Y=0}\,\frac{\rho_s c_s}{\rho_s c_s + \rho c}\,, \qquad (6)$$

in which c is the sound velocity and subscript s refers to the solid. This pressure propagates to the surface, where the wave is reflected as a rarefaction wave and a new cavitation field is produced, again reducing the pressure at the solid surface to p_1 after a time $\Delta\tau_w = 2c(L)_{Y=0}$. A new collapse process is then initiated, and this repeats itself until the energy is lost by dissipation.

In Figure 3 solutions to (5) are given for $\beta = \beta_1 \ll 1$ and $L_0/Y_0 = 1 \wedge 2$, $\gamma = 0 \wedge 1$, and in Figure 4 the corresponding values of P are shown. It is

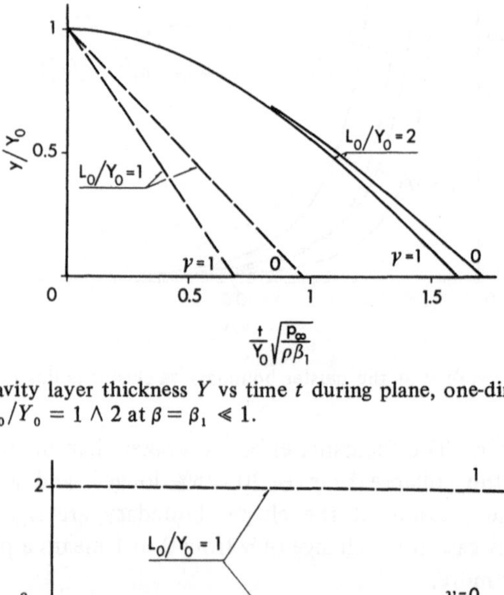

Figure 3. The cavity layer thickness Y vs time t during plane, one-dimensional collapse for $\gamma = 0 \wedge 1$, $L_0/Y_0 = 1 \wedge 2$ at $\beta = \beta_1 \ll 1$.

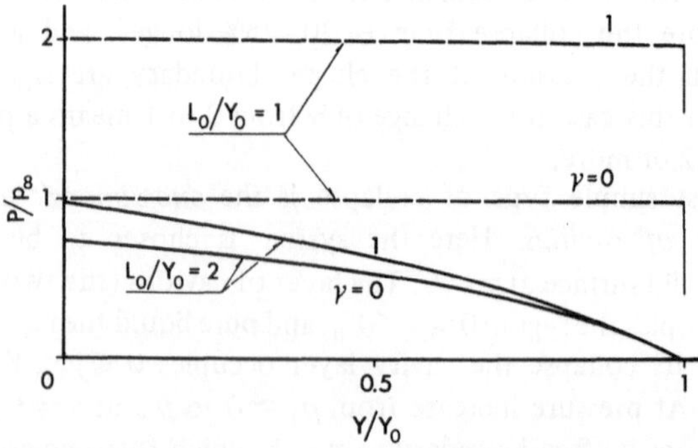

Figure 4. The pressure P at the cavity layer boundary vs thickness Y of the layer corresponding to Figure 3.

noted that the focussing effect is absent, and only for $L_0/Y_0 = 1$ the limiting value of the collapse velocity $\dot{Y} = \left[p_\infty/\rho\beta_1(1-\beta_1)\left(1-\frac{\gamma}{2}\right)\right]^{1/2}$ is obtained. If $L_0/Y_0 > 1$ the energy connected to the collapsing cavities is converted into kinetic energy of the liquid contained not only in the collapse zone but in all the liquid layer. Consequently both $\dot{Y}$ and P decrease with increasing L_0/Y_0. It is apparent from Figures 3 and 4 that then the influence of γ becomes small within the collapse time of the layer. The case $L_0/Y_0 = 1$, $\gamma = 0$ is equivalent to the case considered by [5] and with $\rho = 10^3$ kg/m^3, $c = 1500$ m/s, $p_\infty = 0.10$ MPa and $\rho_s c_s \gg \rho c$

$$\dot{Y} = \frac{10}{\sqrt{\beta_1(1-\beta_1)}}\ \text{m/s}, \quad P = 0.10\ \text{MPa},$$

$$p_w = 15\sqrt{\frac{\beta_1}{1-\beta_1}}\ \text{MPa}.$$

It is noted that P is independent of β_1, while p_w depends on β_1.

It was assumed above that the solid surface is stationary, but if it moves during the collapse of the cavity layer work is done on the system and the potential energy of the layer becomes a function of its motion, and the liquid in the cavity layer also acquires a kinetic energy. In case that the latter is negligible (5) is valid when β is known at the collapse zone and if the Y- and L-coordinates are measured from the position of final collapse ($Y = 0$) in a reference system moving with the velocity of the liquid at inception. Then p_w is determined from the motion of the liquid layer relative to the solid surface when $Y = 0$.

3. Experiments

Experimentally a plane, one-dimensional cavity layer is studied using a tube containing water. A vertical, flat bottomed plexiglas tube with rectangular cross section, supported by a spring is chosen, and cavitation is produced by vertical impact of a mass on the open upper end of the tube. At the impact the tube is given a pulse acceleration and tensile forces arise in the lower part of the liquid column, where cavities develop, Figure 5a. Right at the bottom they quickly grow into a separation zone in which the pressure is the vapour pressure and β is close to 1, while the cavities above collapse, and so the tube itself and the liquid column move essentially independently, Figure 5b. On this assumption the development and collapse of the separation zone can be calculated from (5) when the motion of the bottom is given and the initial thickness of the separation layer can be estimated. In Figure 6a such a calculation is compared with the experimental results partly shown in

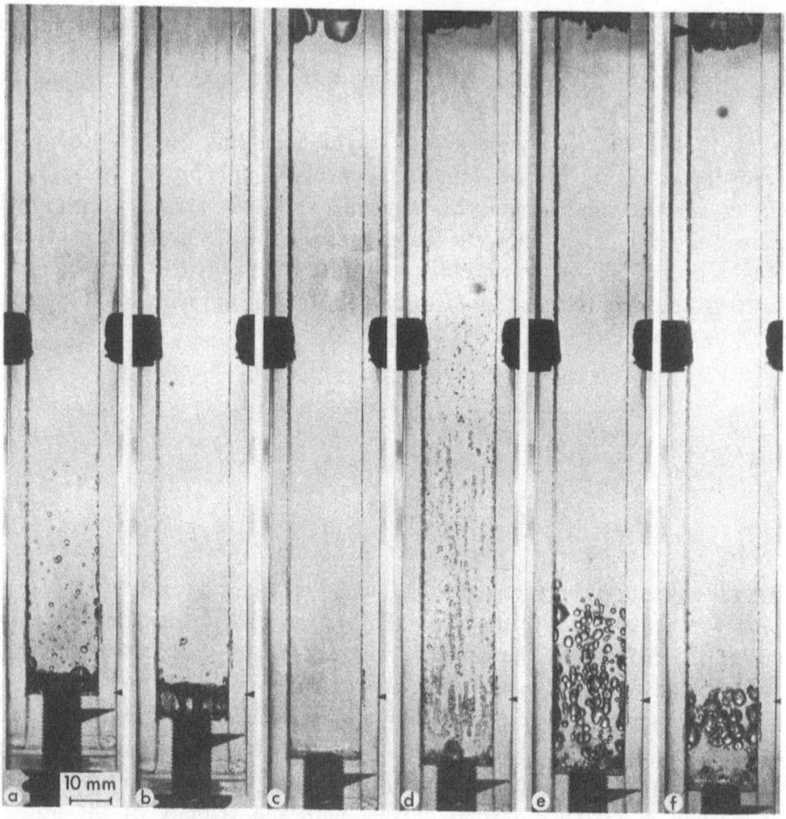

Figure 5. The one-dimensional collapse of cavity layers in the tube, photographed in successive experiments at different time delays Δt after the moment of impact of the mass. (a)–(c) collapse of the separation zone for $\Delta t = 1.3$ ms, 3.3 ms and 7.8 ms and (d)–(f) the successive collapse of the rarefaction-wave-induced cavity layer for $\Delta t =$ 8.8 ms, 10.8 ms and 12.8 ms. Temperature: 18°C.

Figures 5a–c. At the moment when the separation layer collapses (Figures 5c) the column of water impacts at the bottom, and the pressure rises to the impact pressure. For Figure 6a calculation from (6) gives $p_w \simeq 2.4$ MPa, which fits reasonably with measurements as shown in Figure 7, where the first pulse in each graph is due to the separation layer collapse.

Reflection of the impact pressure wave as a rarefaction wave at the surface leads to the formation of a cavitation field in the whole liquid column ($L_0/Y_0 = 1$). The collapse of this cavity layer starts from the surface just after its inception. During the collapse the β-distribution develops inside the cavity layer due to the motion of the bottom of the tube relative to the initial cavity layer, and the cavities grow markedly by time especially in its lower part, Figures 5d–f. Deceleration of the tube eventually causes cluster collapse also from below, and the point of final collapse of the layer is located

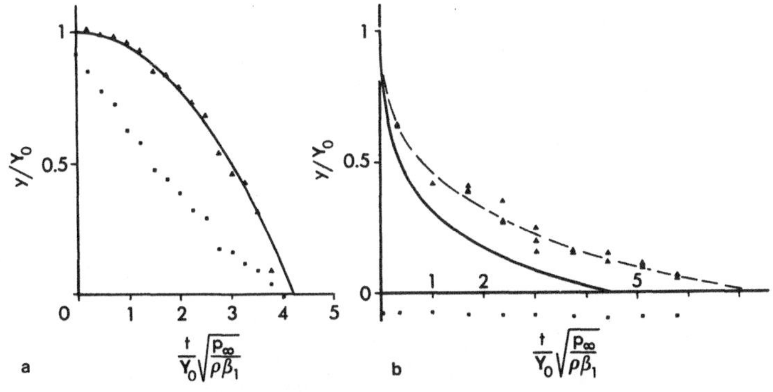

Figure 6. (a) Collapse of the separation zone. $Y_0 = 19.7\,\text{mm}$, $L_0 = 198\,\text{mm}$, $\beta \simeq 1$, $\gamma = 1$. (b) Collapse of the rarefaction-wave-induced cavity layer. $\beta = 0.20\ \mathrm{e}^{-8.0\,Y/Y_0}$, $\gamma = 0.5$, $Y_0 = L_0 = 167\,\text{mm}$.
Experimental position of: ▲ cavity layer boundary Y, ■ bottom of tube. ——— Y calculated from (5).

off the bottom. The photographs show that the cavities collapse within a narrow zone at the boundary Y of the cavity layer and measurements from these give $\beta \simeq \beta_1 \mathrm{e}^{\alpha Y/Y_0}$ just inside this boundary. For Figures 5d–f $\alpha \simeq -8.0$, $\beta_1 \simeq 0.20$. In Figure 6b the experimental results for the cavity layer collapse are plotted for comparison with calculations from (5) using a reference system in which the column of liquid is stationary before the cavity layer inception. The kinetic energy of the medium in the cavity layer produced by the motion of the bottom is neglected, and the discrepancy between theory and experiment is ascribed to this approximation. The calculated values of the impact pressure p_w and its duration $\Delta\tau_w$ are 2 MPa and 0.54 ms, respectively. Measurements of the pressure-time dependence at the bottom, Figures 7a, b show that the pressure drops to zero at impact of the mass, and that it jumps to p_w (reasonably in accordance with the above calculation) during successive impacts of the water column. A notable dependence of the value of p_w and of the pule shape on the temperature is observed at measurements in the interval 6°C to 40°C. With identical impact conditions of the mass it is found that the tube displacement at collapse of the separation zone is larger at high temperatures than at low ones, and the impact pressure is higher. This indicates a temperature dependence of the cavitation inception.

Photographs of cavity layers formed in water which is close to equilibrium of dissolved air content show that the number of cavities increases and their size decreases with increase of temperature, Figures 8a, b. This supports the above conclusion. Deviations from equilibrium of air content also influence the cavity formation, sub-saturation giving a small number of large cavities, super-saturation a large number of small cavities.

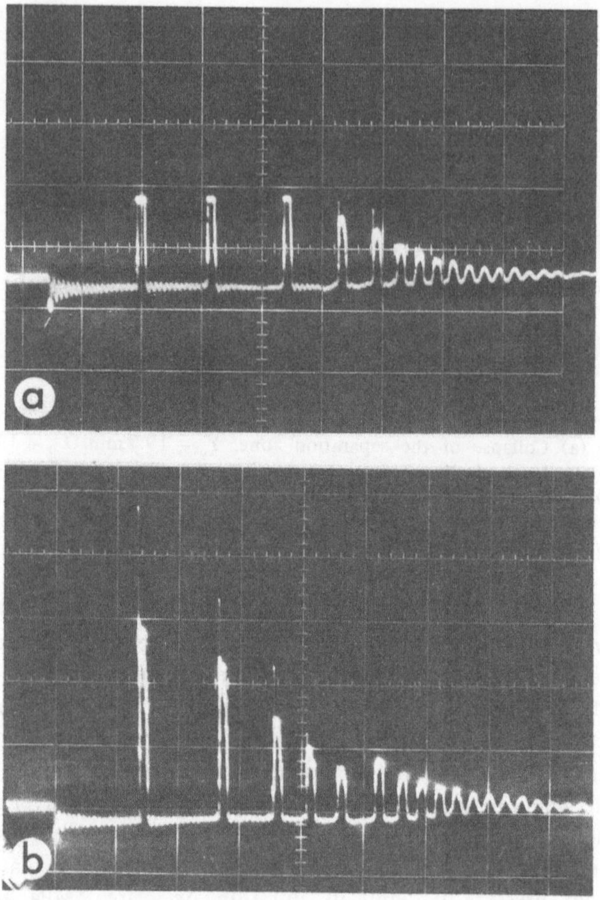

Figure 7. The pressure at the bottom of the tube. Initial liquid column 180 mm, $p_\infty = 10^5$ Pa. Water at the temperatures (a) 6.5°C, (b) 22.7°C. x-sweep 5 ms/div, Y: 0.6 MPa/div.

Conclusion

The collapse of a cavity cluster depends on the volume fraction of cavities β and its distribution in the cluster and on the coefficient of energy transfer γ, but the cluster configuration is decisive for the collapse process. In spherical and cylindrical collapses, energy focussing results in a collapse similar to that of a single cavity, while in plane collapses no focussing effect occurs. The theoretical results are satisfactorily supported by experiments on plane, one-dimensional collapse. These experiments have shown a significant influence of the temperature and of the air saturation on the collapse process, and it is concluded that these parameters influence the inception of the cluster and thus β as well as the size distribution of the cavities. The latter

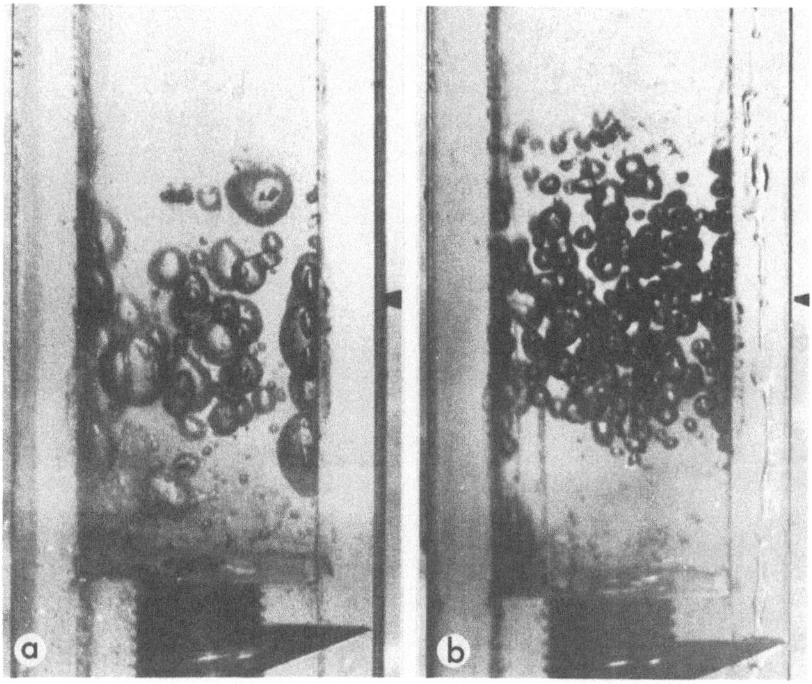

Figure 8. Collapsing cavity layer at (a) 8°C and (b) 40°C.

may influence γ, but investigations concerning this remains to be done. The effect of temperature increase in cold liquids is to intensify the collapse, and for focussed cluster collapses this means that the erosive effects of cavitation are enhanced [4].

Acknowledgements

The author wishes to thank warmly Dr. I. Hansson for valuable discussion and support and Mr. M. Møllgaard and Mr. L. Engelsen for assistance in the experimental work.

References

1. Hansson I and Mørch KA (1980) J Appl Phys 51: 4651–58 and (1981) 52: 1136.
2. Mørch KA (1977) Proc Acoustic Cavitation Meeting, 62–70 Institute of Acoustics, London.
3. Mørch KA (1980) Proc 1st Int Conf Cavitation and Inhomogeneities in Underwater Acoustics. Springer Series in Electrophysics 4: 95–100.
4. Plesset MS (1972) J Basic Eng Trans ASME 94: 559–566.
5. van Wijngaarden L (1964) Proc 11th Int Congress Appl Mech. Applied Mechanics, Springer, 854–861.

O(α)-accurate modelling of the virtual mass effect in a liquid-bubble dispersion

P. VAN BEEK

Delft University of Technology, Department of Mathematics, Julianalaan 132, Delft, The Netherlands

1. Introduction

Purpose of this paper is to investigate the behaviour of a randomly distributed dispersion of bubbles in an infinitely extending fluid under the following conditions: (a) the bubbles are spherical and of equal radius; (b) every bubble moves like a rigid sphere in an ideal potential flow, i.e. friction, heat transfer, volumetric oscillations of the bubble, etc. are neglected.

Under these assumptions we have to deal with continuity of mass and momentum, for both fluid and bubbles. Since the system, obtained by writing down the conservation equations on the level of separate bubbles, is intractable, one has to resort to modelling in terms of averaged variables.

Below a set of non-linear first order partial differential equations will be derived for the variables α (the void-fraction), u (the average fluid velocity) and v (the bubble velocity).

Although the imposed restrictions hardly admit any practical application (with exception perhaps for dispersions of spherical bubbles in an almost frictionless fluid which experiences a relatively strong acceleration), the analysis of the dispersion behaviour is interesting for several reasons.

1. It appears that the governing equations are non-hyperbolic and therefore ill-posed (in an initial-boundary value setting). This phenomenon sets in frequently in two-phase flow modelling and has been subject to discussion in recent years. In this paper we will show that ill-posedness is an inevitable consequence of the interaction between the bubbles and that it can be seen as the macroscopic expression of a tendency to bubble clustering. At the same time we make plausible that the ill-posed nature of the equations need not be disastrous.

2. Additional physical mechanisms can be incorporated either analytically or by means of experimentally determined correlations. For practical purposes we have to resort mainly to the second alternative. Correlations are mostly given as relations between variables and will not involve their derivatives. Incorporation of such correlations therefore not alters the nature of the differential system. From this point of view it is important to know the mathematical properties of the 'basic' differential system, especially if one has numerical calculations in mind.

Applied Scientific Research 38: 323–330 (1982) 0003–6994/82/0384–0323 $01.20.

2. The zero-order approximation

The most simple way to attack the problem is to neglect any interaction between the bubbles. Statistics do not come into play then and the analysis of the dispersion is reduced to that of a single bubble moving in pure fluid. The only problem is to express the acceleration $\dot{V}$ of a rigid sphere in terms of the local pure fluid velocity $U(t, r)$, which may be considered as given: it results simply from the solution of Laplace's equation.

In [1] we derived the following expression:

$$\dot{V}^\alpha = 3\left(\frac{\partial U^\alpha}{\partial t} + (U \cdot \nabla)U^\alpha\right) + 3\sum_{k=2}^{\infty} C_k a^{2k-2} \frac{\partial}{\partial x^\alpha}\left(\frac{\partial^{k-1} U^{\alpha_1}}{\partial x^{\alpha_2} \ldots \partial x^{\alpha_k}}\right)^2, \tag{1}$$

where

$$C_k = \frac{k}{(k+1)!\,(2k-1)\,(2k-3)\ldots 3.1}$$

and a is the bubble radius.

At this point the macroscopic assumption is introduced: we suppose that the length-scale L for variations in U is much larger than a and that the time-scale for U is $L/\|U\|$, where $\|U\|$ is some measure for the magnitude of U.

Then

$$\frac{\partial^{k-1} U^{\alpha_1}}{\partial x^{\alpha_2} \ldots x^{\alpha_k}} = \mathrm{O}\left(\frac{\|U\|}{L^{k-1}}\right)$$

and the magnitude of the series in (1) is an order a^2/L^2 smaller than that of the term $3\dfrac{\mathrm{d}U^\alpha}{\mathrm{d}t}$. We may replace (1) by

$$\dot{V}^\alpha = 3\frac{\mathrm{d}U^\alpha}{\mathrm{d}t} + \mathrm{O}\left(\frac{a^2}{L^2}\frac{\|U\|^2}{L}\right) \quad \text{with} \quad a/L \ll 1.$$

In Eulerian coordinates $V = v(t, r)$ and

$$\frac{\partial v}{\partial t} + (v \cdot \nabla)v = 3\left(\frac{\partial U}{\partial t} + (U \cdot \nabla)U\right) + \mathrm{O}\left(\frac{a^2}{L^2}\frac{\|U\|}{L}\right). \tag{2}$$

The description of the system is completed by the number density equation (a continuity equation for the gaseous phase). The equation can be derived in a formal way from the assumption that a point of the gaseous phase will continue to be a point of that phase. This leads to [3]

$$\frac{\partial \alpha}{\partial t} + \operatorname{div}(\alpha v) = 0. \tag{3}$$

Equation (3), in fact a geometrical equation, and equation (2) provide a 'trivial' approximation for the behaviour of the dispersion. Dropping the terms of $O\left(\frac{a^2}{L^2}\frac{\|U\|}{L}\right)$ in equation (2) both characteristic directions appear to be equal to v.

3. The O(α)-approximation

The interaction between the bubbles of the dispersion can be approximated hierarchally by two bubble interactions, three bubble interactions, etc. Two bubble interactions yield terms of a relative order of magnitude $O(\alpha)$ (to be added to the zero-order equations); the three bubble interactions yield $O(\alpha^2)$ terms, etc. For low void-fractions one may restrict to two-bubble interactions.

In the $O(\alpha)$-approximation the acceleration of a bubble B_1 at point r is determined by the total induced velocity $u_i(t, r/q_2)$, which is the sum of the pure fluid velocity $U(t, r)$ and the velocity $w_i(t, r/q_2)$ induced at r by a second bubble B_2, immersed in the pure fluid at point q_2. Replacement in (1) of the pure fluid velocity $U(t, r)$ by $u_i(t, r/q_2)$ gives an expression of the acceleration [1]:

$$\dot{V}^{\alpha}(t, r/q_2) = 3\frac{\mathrm{d}u_i(t, r/q_2)}{\mathrm{d}t} + 6\sum_{k=2}^{\infty} C_k a^{2k-2} D_r^{\alpha}(D_r^{\alpha_2}\cdots{}^{\alpha_k} u_i^{\alpha_1}(t, r/q_2))^2, \tag{4}$$

with

$$\frac{\mathrm{d}u_i(t, r/q_2)}{\mathrm{d}t} = \frac{\partial u_i(t, r/q_2)}{\partial t} + u_i(t, r/q_2)\cdot\nabla_r u_i(t, r/q_2).$$

The induced velocity $u_i(t, r/q_2)$ is the velocity that would exist at point r if B_1 were imagined to be absent (for a more precise definition see [1]).

The local fluid velocity, denoted by $u(t, r/q_2)$, is found by adding to $U(t, r)$ the velocity $w(t, r/q_2)$ caused by a single bubble immersed in pure fluid at q_2 [1]:

$$w(t, r/q_2) = \tfrac{1}{2}a^3 D_r^{\alpha}\frac{x^{\beta}-q_2^{\beta}}{|q_2 - r|^3}(U^{\beta}(q_2) - V^{\beta}(q_2)) + \tfrac{1}{3}a^5 D_r^{\alpha}\times \frac{(x^{\beta}-q_2^{\beta})(x^{\gamma}-q_2^{\gamma})}{|q_2 - r|^5}\frac{\partial U^{\beta}(q_2)}{\partial q_2^{\gamma}} + O\left(\frac{a^2}{L^2}\frac{a^5}{|q_2 - r|^5}\|U\|\right). \tag{5}$$

The induced velocity and the local fluid velocity are not equal. This is a consequence of what we call the reflection effect: B_1 induces a certain velocity at q_2, which causes a change in the velocity that B_2 induces at r, which in its turn changes the velocity that B_1 induces at q_2, etc.

Neglecting the reflection effect we may put $u_i(t, r/q_2) = u(t, r/q_2)$ and the average acceleration of B_1 is found by substituting (5) into (4) and averaging over all positions $|q_2 - r| \geqslant 2a$. However $w(t, r/q_2)$ behaves like $|q_2 - r|^{-3}$ for large $|q_2 - r|$, and so the integral $\int_{|q_2 - r| \geqslant 2a} \frac{du(t, r/q_2)}{dt} n(q_2) dq_2$ will not be convergent ($n(q_2)$ is the number density at q_2). The following observation solves the problem.

Omitting the reflection effect the average $\langle u_i \rangle (t, r)$ of the induced velocity differs from the average $u(t, r)$ of the local fluid velocity only as far as the region of accessibility of q_2 is concerned and therefore:

$$\langle u_i \rangle (t, r) = u(t, r) - \int_{a \leqslant |q_2 - r| \leqslant 2a} w(t, r/q_2)\, n(q_2)\, dq_2 .$$

Departing from this expression it can be shown [2] that

$$\begin{aligned} \left\langle \frac{du_i}{dt} \right\rangle &= \frac{\partial}{\partial t}((1-\alpha)u) + \operatorname{div}((1-\alpha)uu) \\ &\quad + \operatorname{div} \int_{|q_2 - r| \geqslant 2a} w(t, r/q_2) w(t, r/q_2) n(q_2)\, dq_2 \\ &\quad - \int_{a \leqslant |q_2 - r| \leqslant 2a} \frac{dw(t, r/q_2)}{dt} n(q_2)\, dq_2 . \end{aligned} \tag{6}$$

The second integral arises by expressing the average of the non-linear convection term as the product of average fluid velocities. Both integrals in (6) are evaluated by substituting expression (5) for $w(t, r/q_2)$. Note that the volume integral is convergent since $w(t, r/q_2) w(t, r/q_2)$ is proportional to $|q_2 - r|^{-6}$.

The series in (4) can be averaged directly: the integrand decays sufficiently fast towards infinity.

Conclusion: if the reflection effect is omitted the average acceleration can be written as the sum of $3\left(\frac{\partial}{\partial t}((1-\alpha)u) + \operatorname{div}((1-\alpha)uu)\right)$ and a number of terms which result from the evaluation of certain convergent integrals and which contain α, v and U and their spatial derivatives (for details see [2]). In these terms of magnitude $O(\alpha \|U\|^2/L)$ the pure fluid velocity U may be replaced by the average fluid velocity, since $u = U(1 + O(\alpha))$.

As for the reflection effect, it turns out that we may confine ourselves to the first reflection, i.e. the effect of the velocity induced by B_1 at q_2 on the acceleration of B_1 itself. The evaluation of the average contribution of this first reflection is complicated, but all integrands decay fast enough to yield convergent integrals.

We conclude this paragraph by merely stating the final result (relative error $O(\alpha^2) + O(\alpha a/L) + O(a^2/L^2)$):

$$\frac{\partial v}{\partial t} + (v \cdot \nabla)v - (3 + \kappa\alpha)\left(\frac{\partial u}{\partial t} + (u \cdot \nabla)u\right) + a_1(v-u)^2\nabla\alpha$$

$$+ a_2(v-u)(v-u) \cdot \nabla\alpha + \alpha b_1 \nabla(v-u)^2$$

$$+ \alpha b_2 \operatorname{div}((v-u)(v-u)) + \alpha c_1 (v-u) \cdot \nabla u = 0, \tag{7}$$

with

$$\kappa = 0.46$$
$$a_1 = 0.58$$
$$a_2 = -0.90$$
$$b_1 = 0.47$$
$$b_2 = -0.25$$
$$c_1 = 2.08$$

Unlike the pure fluid velocity the average fluid velocity does not satisfy the incompressibility condition. By differentiating the definition $(1-\alpha)w = \int_{|q_2 - r| \geqslant a} w(t, r/q_2)\, n(q_2)\, dq_2$ one obtains

$$\operatorname{div}((1-\alpha)w) = \int_{|q_2 - r| \geqslant a} \operatorname{div} w(t, r/q_2)\, n(q_2)\, dq_2$$

$$- \int_{|q_2 - r| = a} n \cdot w(t, r/q_2)\, n(q_2)\, dq_2 .$$

Since $\operatorname{div} w(t, r/q_2) = 0$ for all q_2 the volume integral vanishes. Evaluation of the surface integral yields

$$\operatorname{div} u + \operatorname{div}(\alpha(v-u)) = 0 \tag{8}$$

with a relative error of $O(\alpha^2) + O(\alpha a^2/L^2)$.

The number density equation (3) is based on the formal definitions and remains unchanged in any approximation.

Combination of (3) and (8) yields the form by which the fluid continuity equation is usually expressed:

$$\frac{\partial(1-\alpha)}{\partial t} + \operatorname{div}((1-\alpha)u) = 0.$$

Like (3) this equation can also be derived in a formal way and is therefore independent of the approximation made. By equation (8) one has a statistical interpretation of the fluid continuity equation.

In three dimensions equations (3), (7) and (8) constitute 5 partial differential equations for the 7 unknowns α, u and v. As (7) shows all spatial derivatives of u are needed to determine the acceleration. But only the sum of the derivatives $\frac{\partial u^1}{\partial x^1}, \frac{\partial u^2}{\partial x^2}$ and $\frac{\partial u^3}{\partial x^3}$ is fixed by equation (8). Our proposal is to provide additional relations for the spatial derivatives to close the system. In general direct evaluation of the u-derivatives is frustrated by non-convergent integrals. However since rot $u(t, r/q_2) = 0$ for all q_2 an expression for rot u can be derived:

$$\operatorname{rot} u - \tfrac{1}{2} \operatorname{rot}(\alpha(v-u)) = 0. \tag{9}$$

Equations (8) and (9) present an analogy to the pure fluid case where $\operatorname{div} U = 0$ and $\operatorname{rot} U = 0$, together with appropriate boundary conditions, determine U uniquely.

4. One-dimensional analysis of the differential equations

The following one-dimensional solution of the equations (3), (7), (8) and (9) can be given.

Suppose that at $t = 0$, α, v and U are functions of t and x^1 only and that v^2, v^3, U^2 and U^3 are identically zero in the whole flow field. For symmetry reasons $u^2 = u^3 = 0$ holds everywhere and u^1 is a function of t and x^1 only. Obviously the average acceleration in x^2 and x^3 directions will be zero and so the above conditions will hold for all $t > 0$. The following set of equations results:

$$\frac{\partial u}{\partial x} + \frac{\partial}{\partial x}(\alpha(v-u)) = 0, \tag{10}$$

$$\frac{\partial \alpha}{\partial t} + \frac{\partial}{\partial x}(\alpha v) = 0, \tag{11}$$

$$\frac{\partial v}{\partial t} + v\frac{\partial v}{\partial x} - (3+\kappa\alpha)\left(\frac{\partial u}{\partial t} + u\frac{\partial u}{\partial x}\right) + a(v-u)^2\frac{\partial \alpha}{\partial x} + 2\alpha b(v-u)\frac{\partial v}{\partial x} + \alpha c(v-u)\frac{\partial u}{\partial x} = 0 \tag{12}$$

with $a = a_1 + a_2$, $b = b_1 + b_2$, $c = c_1 - 2(b_1 + b_2)$.
Observe that (9) is identically satisfied.

An elementary calculation shows that the characteristic directions λ_1 and λ_2 are given by

$$\lambda_{1,2} = v + (3-b)\alpha(v-u) + (v-u)\sqrt{-\alpha(3-a) + O(\alpha^2)} + O(\alpha^2).$$

For $\alpha = 0$ the characteristic directions coincide with those of the zero-order approximation. For $a < 3$ (which is the case) the characteristic directions will be complex for α small enough. Let us try to analyse this phenomenon. Write $u = U + w$, then

$$\frac{\partial u}{\partial t} + u\frac{\partial u}{\partial x} = \frac{\partial U}{\partial t} + U\frac{\partial U}{\partial x} + \frac{\partial w}{\partial t} + w\frac{\partial U}{\partial x} + U\frac{\partial w}{\partial x} + w\frac{\partial w}{\partial x}.$$

In one dimension there is only one solution $U(t, x)$ which satisfies the incompressibility equation, viz. $U(t, x) = U(t)$. Since $w = O(\alpha \|U\|)$, we can write within $O(\alpha)$-accuracy

$$\frac{\partial u}{\partial t} + u\frac{\partial u}{\partial x} = \frac{\partial U}{\partial t} + u\frac{\partial w}{\partial x} + \frac{\partial w}{\partial t}$$

The main contribution to $\frac{\partial w}{\partial t}$ (the rate of change of w in a fixed point) will be that due to the change in average position of the bubble cloud relative to point x:

$$\frac{\partial w}{\partial t} = -v\frac{\partial w}{\partial x},$$

where $\frac{\partial w}{\partial x}$ is given by equation (10):

$$\frac{\partial w}{\partial x} = -\frac{\partial}{\partial x}(\alpha(v-u)).$$

So approximately

$$\frac{\partial u}{\partial t} + u\frac{\partial u}{\partial x} = \frac{\partial U}{\partial t} + (v-u)^2\frac{\partial \alpha}{\partial x} + \alpha(v-u)\frac{\partial v}{\partial x}$$

holds and

$$\dot{V} \simeq 3\frac{\partial U}{\partial t} + (3-a)(v-u)^2\frac{\partial \alpha}{\partial x} + (3-2b)(v-u)\,\alpha\frac{\partial v}{\partial x}. \quad (13)$$

Suppose that $\frac{\partial U}{\partial t} = 0$ and that $3 - a > 0$. Then both remaining terms in (13) are of the same order of magnitude and $\dot{V}$ will be positive if $\frac{\partial \alpha}{\partial x}$ is large enough. Likewise $\dot{V}$ will be negative if $\frac{\partial \alpha}{\partial x}$ is small enough. That means that for $\left|\frac{\partial \alpha}{\partial x}\right|$ large enough (how large depends on the magnitude of the third term in (13)) the acceleration of the bubble is in the direction of increasing α. As a consequence the α-gradient is increased, which on its turn tends to increase $\dot{V}$ further in the direction of increasing α. In other words the interaction between $\dot{V}$ and $\frac{\partial \alpha}{\partial x}$ is one of an unstable kind: there is a tendency for α to increase unboundedly in certain regions. Now $3 - a > 0$ is just the condition for the characteristics of (10), (11) and (12) to be complex. If $3 - a$ would be negative, the characteristics would be real and at the same time the argument above shows that α-curves would tend to flatten out. We conclude that complex characteristics are connected with a bubble clustering tendency, at least in absence of an acceleration of the pure fluid.

If $\frac{\partial U}{\partial t}$ is significant however, it will dominate the average bubble acceleration. Accordingly the argument given above breaks down and we conjecture the unstable behaviour to be absent or to be of minor importance. This aspect will be brought out more clearly for three-dimensional flows with spatial gradients in the pure fluid velocity. A characteristics analysis in three dimensions is under current research.

References

1. Beek P van (1981) O(α)-accurate modelling of the virtual mass effect in a liquid-bubble dispersion (to appear).
2. Beek P van (1979) On the motion of an expanding sphere in a non-steady solenoidal irrotational flow field. Report NA-24, Delft University of Technology, Dept of Mathematics.
3. Buyevich Yu A and Shchelchkova IN (1978) Flow of dense suspensions. Progr in Aerospace Sci 18: 121–150.

Bubble interactions in liquid/gas flows

L. VAN WIJNGAARDEN

Technological University Twente, Enschede, The Netherlands

Abstract. The system of equations, usually employed for unsteady liquid/gas flows, has complex characteristics. This as well as other facts have led to the search for a more accurate description of effects associated with relative motion. For liquid/bubble systems the fluctuations resulting from hydrodynamic interaction between the bubbles may be taken into account in the same way as particle interactions in the theory of viscous suspensions. This is illustrated for the pressure. In a description accurate up till the third power of the void fraction two-bubble interactions are of primary importance. Numerically obtained results for the relative motion in bubble pairs are presented and interpreted with help of simplified equations from which conclusions can be drawn in an analytic way.

1. Introduction

In recent years the numerical instabilities which arise in the computation of transients in two-phase flows, have stimulated research on the interaction between phases. Such interactions, inertial or of other nature, give rise to additional terms in the equations of motion. These terms might, in rendering real characteristics, prevent instabilities.

In our laboratory a study is in progress regarding the interaction effects in a mixture of massless spheres and a perfect liquid. Such a fluid reasonably approximates a bubbly flow under circumstances in which bubbles are small enough to be kept spherical by surface tension and in which surface active agents are absent in the liquid. The latter condition means that the flow around an individual bubble in the mixture can be with good accuracy approximated by a potential flow. The interaction effects produce in any point in the fluid fluctuating pressures and velocities. Just as in turbulence one is interested in mean values.

When a distinction between fluctuations and mean values is made and some type of averaging is carried out (ensemble averaging, volume averaging or otherwise) effects of the fluctuations on the stress in the fluid remain. These have been calculated by Voinov and Petrov [7] with the use of a cell method. Such a method however leads to results of unknown accuracy. Our approach is similar to that used by Batchelor and his associates, see e.g. [2] in the theory of suspensions dominated by viscous effects. As an illustration we deal here with the calculation of the average or bulk pressure in the inhomogeneous liquid.

Applied Scientific Research 38: 331–339 (1982) 0003–6994/82/0384–0331 $01.35.

2. Bulk pressure in bubbly liquid

Consider a large volume V of a suspension of N bubbles of zero mass in a perfect liquid. A point in the fluid, whether in a bubble or in liquid, is indicated with its position vector $\mathbf{x}$. We assume that the ensemble average of the velocity $\mathbf{u}$ over all possible configurations of the N bubbles is given as $\mathbf{U}_0$,

$$\frac{1}{N!}\int \mathbf{u}(\mathbf{x}, C_N)P(C_N)\,\mathrm{d}C_N = \mathbf{U}_0, \tag{1}$$

where $P(C_N)$ is the probability distribution of N bubbles, or

$$\langle \mathbf{u}\rangle (\mathbf{x}) = \mathbf{U}_0. \tag{2}$$

The volume flow $\mathbf{U}_0$ can be divided in a gas flow (assuming that the bubbles are filled with gas of negligible density) $\alpha \mathbf{U}_g$, α being the concentration of gas by volume, and a volume flow $(1-\alpha)\mathbf{U}_1$ of liquid,

$$\alpha \mathbf{U}_g + (1-\alpha)(1-\alpha)\mathbf{U}_l = \mathbf{U}_0. \tag{3}$$

Our aim eventually is to formulate equations of motion for the averaged quantities. If we carry out the ensemble averaging, the average pressure $\langle p\rangle$ makes its appearance. Assuming that it is permitted to replace ensemble averaging by volume averaging (for this statistical homogeneity is required) we have

$$\langle p\rangle = \frac{1}{V}\int_V p\,\mathrm{d}V = \frac{1}{V}\int_{V_l} p\,\mathrm{d}V + \sum \frac{1}{V}\int_{V_B} p\,\mathrm{d}V.$$

In this equation, V_l denotes the volume occupied by liquid and V_B the volume of one of the N identical bubbles. Next we define $\langle p\rangle_l$ as the pressure averaged over the liquid alone. Then

$$\begin{aligned}\langle p\rangle &= (1-\alpha)\langle p\rangle_l + \sum \frac{1}{V}\int_{V_B} p\,\mathrm{d}V \\ &= \langle p\rangle_l + \sum \frac{1}{V}\int_{V_B} \{p - \langle p\rangle_l\}\,\mathrm{d}V.\end{aligned} \tag{4}$$

Upon introducing the number density $n = N/V$ and the quantity S given by

$$S = \int_{V_B} (p - \langle p\rangle_l)\,\mathrm{d}V, \tag{5}$$

we write (4) as

$$\langle p\rangle = \langle p\rangle_l + n\langle S\rangle. \tag{6}$$

The quantity S in (5) is similar to the 'particle stress' discussed in [1]. In particular it was shown there that (5) may be written as

$$S = \frac{1}{3} \int_{A_B} (p - \langle p \rangle_l)\, \mathbf{r} \cdot \mathrm{d}\mathbf{A}, \tag{7}$$

the integration being over a surface A_B which lies just at the liquid side of the interface between liquid and gas.

When $\alpha \ll 1$ we may in a first approximation assume each bubble to be alone in the liquid. Far from the bubble, the pressure is $\langle p \rangle$ and the velocity $\mathbf{U}_0$. With bubble velocity $\mathbf{U}_g$, the flow is presented by the potential

$$\phi = \mathbf{U}_0 \cdot \mathbf{r} + \frac{(\mathbf{U}_0 - \mathbf{U}_g) \cdot a^3\, \mathbf{r}}{2r^3}, \tag{8}$$

where a is the radius of a bubble and $\mathbf{r}$ gives the position of a point with respect to the centre of the bubble. Upon calculation of p with Bernoulli's Theorem ($\mathbf{U}_0$ may depend on time) and upon carrying out of the integration in (7), we find for S,

$$S \approx S_0 = -\frac{\pi a^3}{3} \{|\mathbf{U}_0 - \mathbf{U}_g|\}^2. \tag{9}$$

Accordingly we find for the bulk pressure, from (6),

$$\langle p \rangle = \langle p \rangle_l - \tfrac{1}{4}\, \alpha\rho \{|\mathbf{U}_l - \mathbf{U}_g|\}^2 + \mathrm{O}(\alpha^2), \tag{10}$$

ρ being the density of the liquid. (Voinov and Petrov [7] find $\frac{1}{3}$ where $\frac{1}{4}$ occurs in [10].) After a similar calculation of the 'Reynolds stress', by calculation of $\langle \rho \mathbf{u}\mathbf{u} \rangle$ we obtain, see [10], to $\mathrm{O}(\alpha)^2$, for one-dimensional flow in x direction

$$\rho(1-\alpha)\left\{\frac{\partial U_l}{\partial t} + U_l \frac{\partial U_l}{\partial x}\right\} = -\frac{\partial}{\partial x} \langle p \rangle_l + \frac{1}{20} \frac{\partial}{\partial x} \{\alpha (U_g - U_l)^2\}, \tag{11}$$

which may be compared with the result found by Van Beek [8] in his contribution to this Symposium.

In the next order of approximation, we start with the exact expression for $\langle S \rangle$,

$$\langle S \rangle = \frac{1}{N!} \int S(\mathbf{x}_0, C_N) P(C_N/\mathbf{x}_0)\, \mathrm{d}C_N, \tag{12}$$

where S is given by (5) for a bubble with centre in $\mathbf{x}_0$, and where $P(C_N/\mathbf{x}_0)$ is the probability distribution for N bubbles, an additional one being in $\mathbf{x}_0$ (the so-called conditional probability). We now replace in each configuration the N bubbles by just one bubble situated in $\mathbf{x}_1$ and with $P(\mathbf{x}_1/\mathbf{x}_0)$ as the probability of finding a bubble centred at $\mathbf{x}_1$, when there is one centred in $\mathbf{x}_0$, we have

$$\langle S \rangle = S_0 + \int (S - S_0)(\mathbf{x}_0, \mathbf{x}_1) P(\mathbf{x}_1/\mathbf{x}_0)\, \mathrm{d}^3\mathbf{x}_1 . \qquad (13)$$

The integration in (13) is over all possible positions of the second bubble.

In S, as given by (5) or (7), this time p is the pressure at distance $\mathbf{r}$ (see Figure 1) from the centre of one bubble, another one having its centre in $\mathbf{x}_1$. The potential for the flow involving two spheres can in the present context conveniently be expressed in terms of twin spherical expansions as used in [4] and [9]. When this is done and the associated pressure is introduced in (13) we are confronted with the fact that $(S - S_0)$ behaves at large values of $(\mathbf{x}_1 - \mathbf{x}_0)$ like $\{|\mathbf{x}_1 - \mathbf{x}_0|\}^{-3}$ and that therefore the integral is not uniformly convergent. This problem can be solved by using the normalization technique, reviewed in [2].

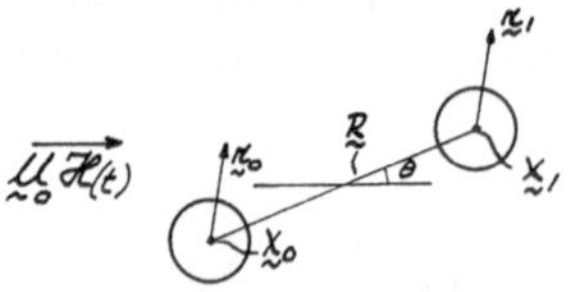

Figure 1. Two bubbles, in $\mathbf{x}_0$ and $\mathbf{x}_1$, immersed in a liquid which acquires a velocity $\mathbf{U}_0 H(t)$.

Another problem is the determination of $P(\mathbf{x}_1/\mathbf{x}_0)$. In contrast to inhomogeneous media with a fixed structure, like a porous bed, the probability distribution is in general affected by the flow itself. The question whether a stationary distribution is reached can only be considered after the problem of determining relative motion in a pair of bubbles has been solved. Both for the calculation of $(S - S_0)$ and for the investigation of $P(\mathbf{x}_1/\mathbf{x}_0)$ we turn our attention to relative motion between two bubbles.

3. Relative motion in a pair of bubbles

To be specific we consider pairs of bubbles in a bubbly flow which is at $t = 0$ instantaneously accelerated to a velocity $\mathbf{U}_0$. The equation of motion is for each bubble

$$\int_{A_0} p\, \mathrm{d}\mathbf{A} = 0, \qquad (14)$$

because the bubble mass is neglected. The pressure p can be derived, by using Bernoulli's Theorem, from the potential of the flow. This consists, if we consider a specific bubble with centre in $\mathbf{x}_0$, of a part which is regular in $\mathbf{x}_0$, with gradient $\mathbf{u}_R$ and a singular part. The latter can be represented by monopoles, dipoles and multipoles situated in points $\mathbf{x}_s$ within the bubble (for a spherical bubble $\mathbf{x}_s$ coincides with $\mathbf{x}_0$). The force on a bubble can be expressed in the strength $\mathbf{M}_q$ of these singularities and the derivatives in $\mathbf{x}_0$ of $\mathbf{u}_R$.

This has been done recently by Landweber and Miloh [6]. Here we disregard, for simplicity, changes of the volume τ of a bubble – inclusion offers no essentially new problems – and write using the result in [6] and denoting the velocity of a bubble with $\mathbf{v}$, the relation (14) as

$$\frac{\mathrm{d}}{\mathrm{d}t}\{\rho\tau\mathbf{v}-4\pi\mathbf{M}_1\}+\mathbf{F} = 0, \tag{15}$$

with

$$\mathbf{F} = -4\pi\rho \sum \mathbf{M}_q \frac{\partial^q}{\partial\mathbf{x}^q}(\mathbf{u}_R)_{\mathbf{x}=\mathbf{x}_s}. \tag{16}$$

The part of the force which is indicated with $\mathbf{F}$ is due to the 'velocity squared' term in Bernoulli's Theorem. $\mathbf{M}_q$ is the singularity of order q ($\mathbf{M}_1$ is a dipole, $\mathbf{M}_2$ a quadrupole etc.) and is multiplied in the expression for the force with the gradient of order q of $\mathbf{u}_R$ in $\mathbf{x}_0$. At time $t = 0^+$ only the terms in the braces in (15) are effective and the resulting velocity is, see [9], the same for each bubble and of magnitude $\mathbf{v} = 3\mathbf{U}_0 + O(\alpha)$. The $\mathbf{M}_q$ can be found from the potential mentioned above and equations for the velocities of the bubbles in $\mathbf{x}_0$ and in $\mathbf{x}_1$ can be found by applying (15) to each of them. These are complicated expressions because the $\mathbf{M}_q$ contain the unknown velocities of the bubbles. Note that because of the occurrence of $\mathbf{F}$ relative motion with velocity

$$\mathbf{V} = \frac{\mathrm{d}\mathbf{R}}{\mathrm{d}t} \tag{17}$$

developes for $t > 0$, $\mathbf{R}$ being the distance $\mathbf{x}_1 - \mathbf{x}_0$ between the two centres. Next by combination of these equations an equation for $\mathbf{V}$ can be constructed, which has to be solved numerically. This program has been carried out in [5] by Knibbe. Some trajectories $\mathbf{R} = \mathbf{R}(t)$ obtained from this by Biesheuvel are shown in Figure 2.

Since the analysis is quite complicated, it is difficult to understand and interprete the results on the basis of the full problem. A qualitative insight can be obtained by taking only the leading singularities, in terms of the parameter (a/R) into account. The leading term in $\partial/\partial x(\mathbf{u}_R)$ in (16) is the gradient of the velocity induced in $\mathbf{x}_0$ by the dipole in $\mathbf{x}_1$. The latter behaves as $\mathbf{U}_0 a^3/R^3$ and because the initial dipole strength is $\mathbf{U}_0 a^3$ we can neglect the variation in the dipole strength due to the relative motion. The leading term therefore in $\mathbf{F}$ is

$$\mathbf{F}_0 = -4\pi\rho a^3(\mathbf{U}_0 \cdot \nabla_0)\mathbf{u}_R(\mathrm{o}), \tag{18}$$

where $\mathbf{u}(\mathrm{o})$ indicates the velocity in $\mathbf{x}_0$ and ∇_0 indicates the gradient with respect to $\mathbf{x}_0$. The velocity $\mathbf{u}_R(\mathrm{o})$ is given by

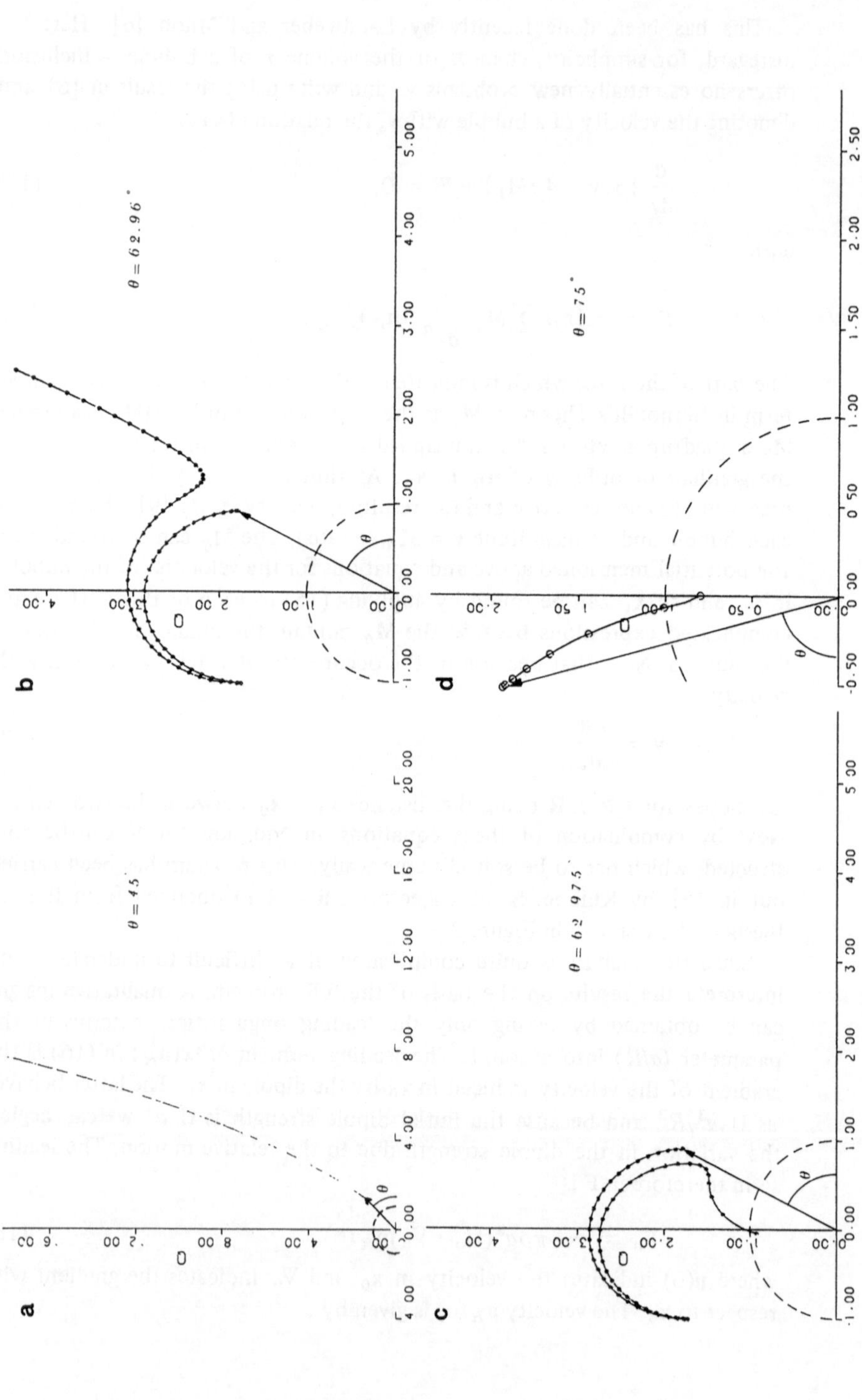
a
θ = 45°
b
θ = 62.96°
c
θ = 62.975
d
θ = 75°

$$\mathbf{u}_R(\mathrm{o}) = -\nabla \frac{\mathbf{U}_0 a^3 \cdot \mathbf{r}_1}{r_1^3} \tag{19}$$

and is in this approximation equal to $\mathbf{u}_R$ (1), the velocity induced in $\mathbf{x}_1$. From subtraction of the equations for $\mathbf{v}_0$ and $\mathbf{v}_1$, we find in this approximation

$$\frac{\mathrm{d}}{\mathrm{d}t}\{-\tfrac{1}{2}\rho\tau(\mathbf{v}_0 - \mathbf{v}_1)\} = 8\pi\rho a^3(\mathbf{U}_0 \cdot \nabla_R)\nabla_R \frac{\mathbf{U}_0 a^3 \cdot \mathbf{R}}{R^3} \tag{20}$$

With $\mathbf{v}_1 - \mathbf{v}_0 = \mathrm{d}\mathbf{R}/\mathrm{d}t$ and working out the right-hand side of (20) we can write this as

$$\frac{\mathrm{d}\dot{\mathbf{R}}}{\mathrm{d}t} + \nabla G = 0,$$

with

$$G = \frac{12a^3 U_0^2}{R^3}(3\cos^2\theta - 1), \tag{21}$$

θ being the angle between $\mathbf{U}_0$ and $\mathbf{R}$ as indicated in Figure 1 and the dot on R indicating the time derivative. The relation (21) means that there is a constant of the motion, the energy, which is in spherical polar coordinates

$$\tfrac{1}{2}(\dot{R}^2 + R^2\dot{\theta}^2) + G = G_0, \text{ say}$$

where account has been taken of the initial conditions on $\dot{R}$ and $\dot{\theta}$.

From analytical mechanics it follows that there is a Lagrangian L,

$$L = \tfrac{1}{2}(\dot{R}^2 + R^2\dot{\theta}^2) - G, \tag{23}$$

the Euler equations of which provide the equations of motion (20). These equations cannot be solved analytically but some important conclusions can nevertheless be drawn. Lack of space prevents to give the analysis here for which we refer to [3]. Here we summarize these conclusions, referring to Figure 3.

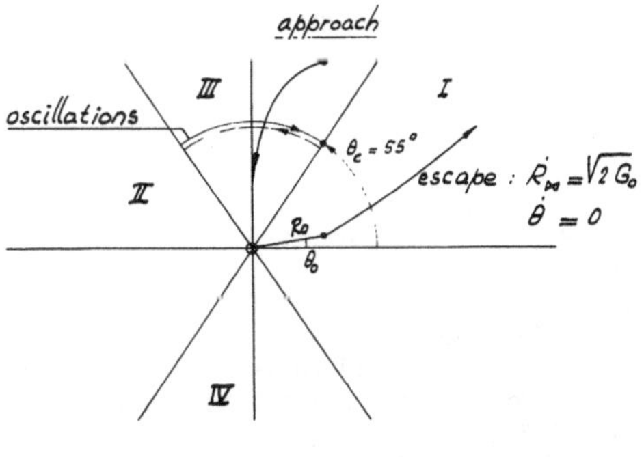

Figure 3. Summary of results for $\mathbf{R}(t)$ according to approximate theory.

(i) When at $t = 0$ the separation vector $\mathbf{R}_0$ ends in regions I and II, where $G > 0$, the bubbles escape from each other. For $R \gg R_0$, $\dot{R} \sim (2G_0)^{1/2}$ and $\dot{\theta} \to 0$. The angle at which G changes sign is $\theta = \theta_c \sim 55°$.

(ii) When initially $G < 0$, which places $\mathbf{R}_0$ in regions III or IV, the bubbles approach each other. The line of centres tends to the vertical in Figure 3.

(iii) When initially $\theta \approx \theta_c$, oscillatory motions are possible. R is approximately constant, which gives, from (21) and (23)

$$\ddot{\theta} - \Omega^2 \sin 2\theta = 0, \qquad \Omega^2 = \frac{36a^3 U_0^2}{R_0^5}$$

or

$$\dot{\theta}^2 = \Omega^2 (\cos 2\theta_0 - \cos 2\theta).$$

However R is only approximately constant. Eventually the separation distance becomes either large or small. These qualitative properties of trajectories agree quite well with the computed trajectories.

Finally we consider the question of the probability distribution. The pair probability distribution $P(\mathbf{x}, \mathbf{x} + \mathbf{R})$ changes, because of the relative motion, according to

$$\frac{\partial P}{\partial t} + \nabla \cdot (\mathbf{V}P) = 0.$$

It can be shown, see [3], that $\nabla \cdot \mathbf{V} = 0$, whence

$$\frac{\mathrm{d}P}{\mathrm{d}t} = 0. \tag{25}$$

This means that if we move along a trajectory in $\mathbf{R}$, t space, P remains constant. If therefore P is random at $t = 0$, e.g. $P = n$, the probability density remains uniform. This is in contrast to what happens in suspensions dominated by viscosity where the probability density is affected by the relative motion, and in some cases cannot even be determined owing to the occurrence of closed trajectories.

References

1. Batchelor GK (1970) The stress system in a suspension of force-free particles. J Fluid Mech 41: 545.
2. Batchelor GK (1974) Transport properties of two-phase materials with random structure. Ann Rev Fluid Mech 6: 227.
3. Biesheuvel A and Van Wijngaarden L (to be published).
4. Jeffrey DJ (1973) Conduction through a random suspension of spheres. Proc R Soc London A 335: 355.
5. Knibbe P (1981) Master's Thesis. Technological University Twente, The Netherlands.

6. Landweber L and Miloh T (1980) Unsteady Lagally theorem for multipoles and deformable bodies. J Fluid Mech 96: 33.
7. Voinov OV and Petrov AG (1977). On the stress tensor in a fluid containing disperse particles. PMM 41: 362.
8. Van Beek P (1981) An O(α)-accurate model for liquid-bubble dispersions. Appl Sc Res.
9. Van Wijngaarden L (1976) Hydrodynamic interaction between gas bubbles in liquid. J Fluid Mech 77: 27.
10. Van Wijngaarden L Jn (1980) On the mathematical modeling of two-phase flows. Proc IVth Int Meeting on water column separation, Cagliari, 1979.

PART SIX

Physico-chemical effects

Surface activity and bubble motion

J.F. HARPER

Mathematics Department, Victoria University of Wellington, Wellington, New Zealand

Abstract. This paper reviews recent progress in the theories of the surface boundary conditions of adsorbed solutes in liquids, and of the effects of those solutes on the steady motion of a bubble or drop in the liquid. Both singular perturbation theory and numerical solutions have useful roles in this problem, and their relationship is explored. In addition, analytical solutions are given to two problems concerning a spherical bubble rising steadily at low Reynolds number in a viscous fluid. One of these is displacement of the internal vortex centre from its position in the absence of surface activity when there is a small stagnant cap of surfactant at the rear. The results agree with experimental data in the direction of that displacement but give only about half its amount. The other problem is the velocity perturbation all round the surface caused by a very dilute solution of a weak surfactant at high Péclet number. This compares quite well with the numerical solution for a Péclet number of 60, having relative errors of order $(60)^{-1/2}$ as would be expected.

Introduction

In recent years two reviews [4, 11] have appeared which describe the motion under gravity of a drop or bubble and the subtle effects on it of surface activity. These reviews make it unnecessary to start again from scratch, and so this paper concentrates on recent work not described in them. It also gives for the first time some theoretical contributions related to that work. One of these new contributions is a pair of integrals giving the first-order displacement of the internal vortex centre in slow (Stokes) flow of a drop moving under gravity in a surfactant solution. The other is the perturbation velocity field near the rear stagnation point of a bubble rising in a very dilute solution of a weak surfactant (in the sense of [13]) at high Péclet number.

This paper begins by reviewing a topic which comes logically before the theory of bubble or drop motion: the surface boundary conditions. It then collects the various theorems for Stokes flow which give properties of the motion of a spherical bubble or drop in terms of integral transforms of the surface pressure distribution, including the new formulae of this kind for the internal vortex position. The next section concerns rear stagnation regions and the difficulties they cause for both numerical and analytical theory. Singular perturbation theory is an essential tool for the latter, but its proper relation to the former still seems to need comment over twenty-five years after its use in fluid mechanics was first elucidated [18].

Surface boundary conditions

It has long been known what the boundary conditions are at a fluid interface: normal stress-difference between the two sides equal to the surface tension

Applied Scientific Research 38: 343–352 (1982) 0003–6994/82/0384–0343 $01.50.

times the sum of the principal curvatures; tangential stress-difference equal to the surface gradient of interfacial tension; and conservation of mass of any surfactants, so that the convective and diffusive fluxes of each adsorbed component along the interface, taken together, must balance the convective and diffusive fluxes towards the interface from the bulk fluids on each side of it. (In problems where the interface changes its shape, this condition requires some care in its mathematical formulation. It is given, for example, in a recent paper [23] on wave propagation on a contaminated surface.)

Most previous proofs of the surface boundary conditions have used phenomenological assumptions about the stress distribution nearby. These can be avoided [2] by considering the intermolecular forces in a system where all properties such as pressure, number densities and chemical potentials are assumed continuous functions of position on a (microscopic) length scale l, but in one special direction (y, say) these properties change so rapidly that when viewed on a coarser (macroscopic) length scale $L \gg l$ they no longer appear to be continuous. Brenner [2] then goes on to point out that the concept of an interface at $y = 0$ is macroscopic, that the interface is in fact diffuse on the microscale, and, the new feature of his work, that singular perturbation theory of kind long used to cope with boundary layers [16, 18] is the natural way to bring surface tension and surface excesses out of the mathematical analysis. The paper does make the common assumption of local chemical equilibrium (at least on the length scale l); it is often a good assumption, but there are cases where it is not, e.g. bipolar solute molecules, which often have adsorption barriers [21].

Integral theorems for stokes flow of spherical bubbles or drops

Suppose that a spherical drop or bubble of radius a moves steadily at speed U because of some external force (usually gravity) in an unbounded fluid. Suppose also that the Reynolds number is much less than one both inside (where the density is ρ_1 and the dynamic viscosity is η_1) and outside (where they are ρ_0, η_0), and that, owing to surface activity, the surface pressure $\Pi(\mu) = \sigma_p - \sigma(\mu)$ varies around the surface. In this equation σ_p and σ are the surface tensions of pure and contaminated fluid, and $\mu = \cos\theta$, where θ is the angular distance from the front stagnation point.

Then the first integral theorem [11] gives the drag coefficient C_D, defined as usual by $C_D = (\text{drag force})/\frac{1}{2}\rho_0 U^2 \pi a^2$, as

$$C_D = \frac{8}{R(\eta_0 + \eta_1)}\left(2\eta_0 + 3\eta_1 - \frac{1}{U}\int_{-1}^{1} \mu\Pi(\mu)\,\mathrm{d}\mu\right), \tag{1}$$

where R is the external Reynolds number equal to $2Ua\rho_0/\eta_0$. The second theorem, first given [13] only for $\eta_1 = 0$, gives the dimensionless rate of strain A at the near stagnation point as

$$A = \frac{1}{2(\eta_0 + \eta_1)}\ \eta_0 + \frac{1}{4\sqrt{2}U}\int_{-1}^{1} \frac{(1-\mu)}{(1+\mu)^{1/2}} \frac{\mathrm{d}\Pi(\mu)}{\mathrm{d}\mu}\,\mathrm{d}\mu\ , \tag{2}$$

where A is the coefficient in the asymptotic expression $u_\theta \sim UA \sin\theta$ for the tangential velocity component u_θ near the rear stagnation point. This theorem is a special case of the following [19], which unfortunately seems to have no simple expression in closed form:

$$\frac{u_\theta}{U \sin\theta} = \frac{\eta_0}{2(\eta_0 + \eta_1)}\left(1 + \sum_{n=1}^{\infty} c_n P'_n(\mu)\right), \tag{3}$$

where

$$c_n = \frac{1}{\eta_0 U}\int_{-1}^{1} \Pi(\mu) P_n(\mu)\,\mathrm{d}\mu, \tag{4}$$

and $P'_n(\mu)$ denotes the derivative of the Legendre polynomial $P_n(\mu)$. All these theorems are proved by expanding the surface pressure in a series of Legendre polynomials in μ, expanding the stream functions inside and out in a series of integrals of Legendre polynomials in μ multiplied by appropriate powers of r, the distance from the centre, and using the boundary conditions and the orthogonality properties of Legendre polynomials.

One can similarly calculate the velocity components (v_r, v_θ) at the point $P : r = a/\sqrt{2},\ \theta = \frac{1}{2}\pi$, which is the centre of Hill's spherical vortex [14], the internal flow when $\Pi =$ constant all round the drop or when the Péclet numbers inside and out are much less than one [11], but not in general otherwise. If the internal vortex centre moves a distance ΔR outwards and ΔZ forwards from P, and if ΔR and ΔZ are both much less than a, the results are

$$v_r = \frac{1}{\eta_0 + \eta_1}\int_{-1}^{1} \Pi(\mu)\, V_r(\mu)\,\mathrm{d}\mu \approx -\frac{\sqrt{2}U\eta_0\, a\,\Delta Z}{4(\eta_0 + \eta_1)}\ , \tag{5}$$

$$v_\theta = \frac{1}{\eta_0 + \eta_1}\int_{-1}^{1} \Pi(\mu)\, V_\theta(\mu)\,\mathrm{d}\mu \approx -\frac{\sqrt{2}U\eta_0\, a\,\Delta R}{4(\eta_0 + \eta_1)}\ ; \tag{6}$$

in these equations the functions V_r and V_θ are given by the following series of Legendre polynomials wich converge fast enough for all relevant μ to be easily computed:

$$V_r(\mu) = \frac{1}{4\sqrt{2}}\sum_{n=2}^{\infty} \frac{n(n+1)P_n(0)P_n(\mu)}{2^{n/2}}\ , \tag{7}$$

$$V_\theta(\mu) = -\frac{1}{8}\sum_{n=2}^{\infty} \frac{n(n+1)P_n(0)P_{n+1}(\mu)}{2^{n/2}}\ . \tag{8}$$

In many practical cases, Π is very small except close to the rear stagnation point $\mu = -1$. Then the integrals in (1) to (6) are insensitive to the values of their integrands except in that neighbourhood. To evaluate (7) and (8) at $\mu = -1$, we use the function $f(x)$ defined for $|x| < \sqrt{2}$ by the series

$$f(x) = \sum_{n=2}^{\infty} n(n+1)P_n(0)\,(x/\sqrt{2})^n$$

$$= -3x/\{2(1+\tfrac{1}{2}x^2)^{5/2}\},$$

and so

$$V_r(-1) = \sqrt{2}\,V_\theta(-1) = f(1)/4\sqrt{2} = -1/6\sqrt{3} \tag{9}$$

Equations (1), (5), (6) and (9) then give, for this case,

$$\Delta Z \approx 4\sqrt{2}\,\Delta R \approx \frac{\sqrt{2}(2\eta_0 + 3\eta_1)\,a\,\Delta C_D}{3\sqrt{3}\,\eta_0\,C_D}, \tag{10}$$

where ΔC_D is the increase in the drag coefficient above its value for uncontaminated fluids. The vortex core thus begins to move, as surface contamination builds up from zero, outwards and forwards, in a direction making an angle initially equal to $\tan^{-1}(4\sqrt{2}) \approx 80°$ with the drop's equatorial plane.

Stagnant caps

Savic [24] first showed that bubbles and drops in surfactant solutions often sweep their surface contamination around to the rear, where it collects in an immobile spherical cap, while the rest of the surface is free to move. Many subsequent authors have studied this 'stagnant-cap' regime [4], experimentally, computationally and analytically in the limit of very small caps. This last calculation can now be simplified slightly and extended to drops as well as bubbles by the use of (1) and (2). One follows the previous work [12] as far as

$$\Pi = \Pi^*(1 - \varphi^2/\varphi^{*2}) \quad \text{for} \quad 0 \leqslant \varphi \leqslant \varphi^* \ll \pi, \tag{11}$$

where $\varphi = \pi - \theta$, φ^* is the cap angle, and Π^* is the surface pressure at the rear stagnation point. Then a stagnant cap obviously requires $A = 0$ in (2), to which a first approximation is

$$\Pi^* = 4\eta_0\,U\varphi^*/\pi, \tag{12}$$

and then (1) yields

$$C_D = \frac{8}{R(\eta_0 + \eta_1)}\,(2\eta_0 + 3\eta_1 + 4\eta_0\varphi^{*3}/3\pi). \tag{13}$$

Because (11) must hold for either a drop or a bubble with a small stagnant cap, so must (12) and (13), unlike their analogues in [12].

To test the theory in (10) and (13) the most accurate experiments appear to be those of Horton reported by Huang and Kintner [15], in which ΔZ and ΔR were reported for vaiious values of φ^*. The results, in Figure 1, show that the simple asymptotic theory gives a good initial direction of vortex-centre displacement but underestimates its magnitude as a function of φ^*. Presumably the asymptotic solution for small φ^* ceases to hold before $\varphi^* = 42.5°$, the smallest cap angle in the experiments [15].

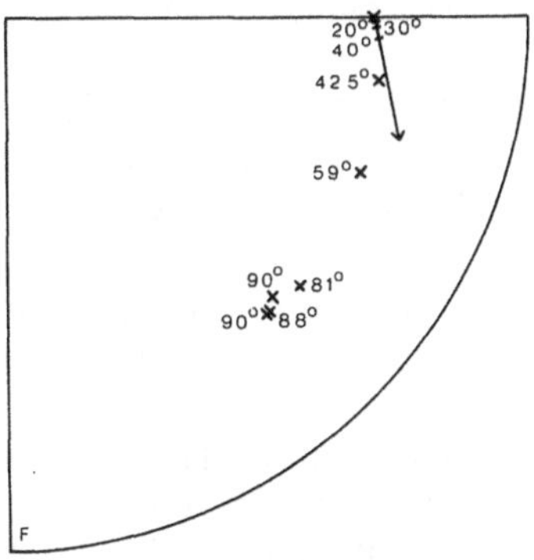

Figure 1. One quadrant of a falling drop, showing the position of the internal vortex core for various stagnant cap angles. Experimental data shown by crosses [15], theoretical points on the arrow making the initial direction (10). F = front stagnation point.

Bubbles and drops with stagnant caps appear to present several as yet unexploited possibilities for both experimenters and theoreticians. They give a problem where non-ideal surfactant solutions can be readily incorporated in the theory, because the fluid mechanics gives Π as a function of θ (or φ) for any given cap angle φ^*, the adsorption isotherm then determines the concentration in the immediately adjacent bulk solutions, and one could proceed to use the theory of convective diffusion in the way already known for ideal solutions [12]. Another problem waiting to be attacked (for any flow regime, including stagnant caps) is a surfactant soluble in the dispersed phase rather than the continuous phase, though there has been some work on the nature of convective diffusion inside a drop [3, 17].

Bubbles in nearly pure liquids with high Péclet numbers

Suppose that a spherical bubble rises at low Reynolds but high Péclet number, and the surfactant is so dilute that the motion is only slightly affected by it, even near the rear stagnation point. Then the theoretician faces in the first instance a Stokes flow problem, with a thin diffusion boundary layer around the surface [7, 8, 9]. An analytical solution is feasible only for very low or very high surface activity, in the sense that k, the ratio of the amount of adsorbed surfactant to the amount in solution in the diffusion layer, has to be either much less or much greater than one [13]. Saville [25] has extended the calculation by computer to intermediate values of k, but one problem remained. The method fails in the rear stagnation region where φ is small, both because the assumptions underlying the approximation of a thin boundary layer fail and because the calculated concentration tends to infinity as $\varphi \to 0$, like $ln\ \varphi$ or $1/\varphi^2$ for $k \ll 1$ and $k \gg 1$ respectively. Saville found behaviour as $ln\ \varphi$ for all finite values of k he investigated: reconciling this with the asymptotic results in an unsolved problem in singular perturbation theory.

Overcoming the infinite values is another singular perturbation problem, whose results for the surfactant distribution have been given [13] in the limiting cases $k \ll 1$ and $k \gg 1$. The boundary-layer solution is treated as an outer approximation and matched asymptotically to a solution of the full diffusion equation valid in the rear stagnation region, where diffusion along streamlines is as important as across them but curvature of the bubble surface can be ignored.

In a new attack on the problem [19], the boundary layer and its singularities have been circumvented by the use of a finite-difference computation. That required particular numerical values to be chosen (a disadvantage of computer solutions when, as here, there are several independent dimensionless parameters): radius $a = 1.095 \times 10^{-5}$ m, viscosity $\eta_0 = 1.140 \times 10^{-3}$ Pa s, diffusivity $D = 1.25 \times 10^{-10}\ \mathrm{m^2\ s^{-1}}$, speed $U = 3.43 \times 10^{-4}\ \mathrm{m\ s^{-1}}$. Reynolds number 6.6×10^{-3}, Péclet number $\mathrm{P} = 2Ua/D = 60$, and surface activity parameter $k = \Gamma \mathrm{P}^{1/2}/2ac = 0.155$, where Γ is the surface excess in equilibrium with a solution of bulk concentration c. In an ideal solution, of course, $\Pi = R_g T\Gamma$ where R_g is the gas constant and T the absolute temperature. The analysis provides a useful test for the applicability of the singular perturbation method, which assumes $\mathrm{P}^{1/2} \gg 1$, $k \ll 1$: two conditions which are barely met in the calculated example.

The results for surfactant concentration are shown in Figure 2. The boundary-layer theory gives a value about 10% lower than the computation near the front stagnation point, rises to be very close to it from about $\theta = 90^\circ$ to $\theta = 150^\circ$, and then tends to infinity as $\theta \to 180^\circ$. The first-order rear stagnation correction removes the singularity and gives an answer only 4% too low at the rear stagnation point. The results are as good as one could expect with the 'small' parameters $\mathrm{P}^{-1/2}$ and k as large as they are here.

Results were also given [19] for the perturbation of surface velocity from that in pure liquid, using equation (3) above and the numerical solution for Γ. That equation can also be used to find the surface velocity from the boundary-layer, theory, except in the rear stagnation region where the number of terms needed increases without bound. It turns out that the

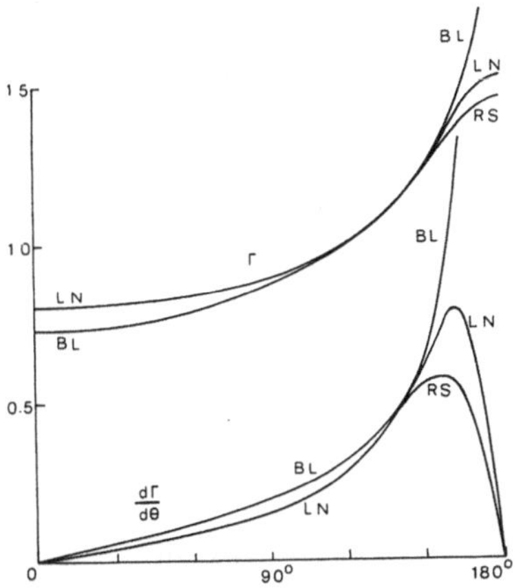

Figure 2. The distribution of adsorbed surfactant on Le Van and Newman's bubble, with θ in degrees from the forward stagnation point, Γ and $d\Gamma/d\theta$ in units of the equilibrium surface excess with no bubble motion. LN indicates LeVan and Newman's solution [19], BL the boundary-layer outer approximation and RS the inner approximation in the rear stagnation region.

velocity perturbation in that region depedns mainly on the surfactant distribution there, that distribution is badly approximated by the boundary-layer solution, but that an analytical inner approximation can be given for the velocity as for Γ (or Π).

That inner approximation is given here (for the first time). Define [13] dimensionless cylindrical polar coordinates (m, s) by

$$(m, s) = (r \sin\theta/\delta, (-r\cos\theta + a)/\delta), \tag{14}$$

which are $0(1)$ in the rear stagnation region. In this equation $\delta = 2\mathrm{P}^{-1/2}a$. The bubble's vertical axis is $m = 0$, and s increases down the wake from (essentially) zero at the surface. The perturbation stream function $\psi(m, s)$ is then zero on $m = 0$ and $s = 0$, and the surface shear stress condition [4] becomes

$$\frac{\partial^2\psi}{\partial s^2} = \frac{16a^2k\Pi_\infty}{\sqrt{\pi}\eta_0\mathrm{P}}(e^{-m^2/2} - 1) = B(e^{-m^2/2} - 1), \text{ say,} \tag{15}$$

on $s = 0$, where Π_∞ is the surface pressure of undisturbed liquid. Put [23] $\psi = m^2 s\psi_3$, where ψ_3 must obey

$$\frac{\partial^2\psi_3}{\partial m^2} + \frac{3}{m}\frac{\partial\psi}{\partial m} + \frac{\partial^2\psi}{\partial s^2} = 0, \tag{16}$$

and then we deduce that

$$\psi = \int_0^\infty ms\, f(\alpha)\, e^{-\alpha s} J_1(\alpha m)\, d\alpha \tag{17}$$

for some function f, where J_1 denotes the usual Bessel function. Equation (15) gives

$$B(e^{-m^2/2} - 1) = \int_0^\infty -2\alpha m f(\alpha) J_1(\alpha m)\, d\alpha,$$

from which [10, p. 717] we find $f(\alpha) = B\, e^{-\alpha^2/2}/2\alpha$. The integrals can now be manipulated [10] to give the perturbation u_m to the surface velocity component for $m = 0(1)$ as

$$u_m = \frac{1}{\delta^2 m} \frac{\partial\psi}{\partial s}\bigg|_{s=0} = -\frac{k\Pi_\infty P^{1/2}}{2\sqrt{2}\eta_0} \sin\theta \cdot e^{-m^2/4} \times$$

$$\{I_0(\tfrac{1}{4}m^2) + I_1(\tfrac{1}{4}m^2)\}, \tag{18}$$

where I_0 and I_1 denote, as usual, Bessel functions of imaginary argument.

The results of this inner approximation for the rear stagnation region are presented in Figure 3, together with the boundary-layer outer approximation and the numerical solution, all three for the same hypothetical bubble mentioned earlier. In this figure, $-\beta(\theta)c_\infty$ is the velocity perturbation due to the surface activity divided by the unperturbed surface velocity, where c_∞ is the surfactant concentration at a great distance from the bubble, i.e.,

$$u_\theta = \tfrac{1}{2} U \sin\theta\, (1 - \beta(\theta)c_\infty). \tag{19}$$

We see that, as was already known [26] the boundary-layer approximation gives a fairly good result over most of the surface by comparison with the numerical solution, with much less effort, though it is not as good for velocity perturbations as it was for the surfactant distribution. Also, as was already known [13, 20], that approximation fails and becomes singular near

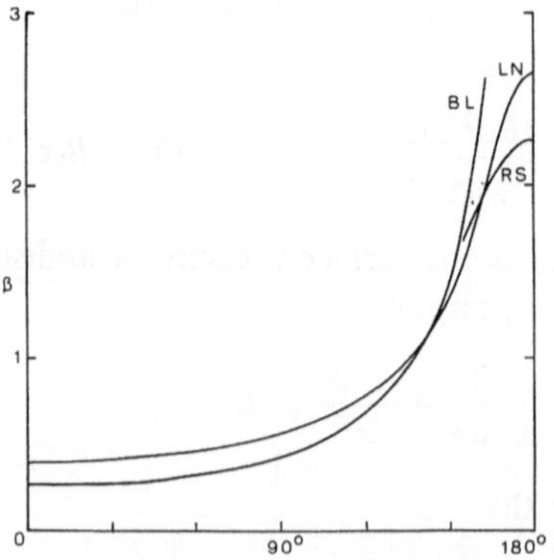

Figure 3. The distribution of surface velocity perturbation on LeVan and Newman's bubble, with β as in (19), in units of 10^3 m^3/mol. Other data as for Figure 2.

the rear stagnation point. What is new in the present work is the appropriate inner approximation, which removes the singularity and reduces the error to about 15% even at the rear stagnation point. As in the theory of the surfactant distribution, the error would become smaller at higher Péclet numbers. This is just the range where numerical solutions are hardest to do, because many terms of the series (3) are needed to convergence and the integrands in (4) for large n are rapidly oscillating.

This theory, which ignores the consequence for convective diffusion of the slowed-down velocity field, even near the rear stagnation point where those consequences are greatest, has little physical importance. Its main practical use seems to be in testing methods of numerical calculation [25]. Like many other singular-perturbation theories, it gives answers for extreme values of the physical parameters at which the solution becomes singular in some way. Numerical solutions, on the other hand, are often better suited to filling in the intermediate values, and a good test for their accuracy is their ability to reproduce the singular-perturbation answers in the appropriate limits. One sometimes has to be depressingly close to those limits, though. The singular perturbation expansion for the drag coefficient of a spherical drop, bubble or solid particle at low R is now known to order $R^4 \ln R$ [1], but it gives better information about the drag than Oseen's order R term only at Reynolds numbers so small that the higher-order corrections are unimportant in practice. This is a penalty that has to be paid for using asymptotic, instead of convergent, series.

References

1. Akiyama T and Yamaguchi K (1975) On the flow past a spherical drop or bubble at low Reynolds number. Can J Chem Eng 53: 695–8.
2. Brenner H (1979) A micromechanical derivation of the differential equation of interfacial statics. J Colloid Interface Sci 68: 422–439.
3. Brignell AS (1975) Solute extraction from an internally circulating spherical liquid drop. Int J Heat Mass Transfer 18: 61–8.
4. Clift R, Grace JR and Weber ME (1978) Bubbles, drops and particles. New York: Academic.
5. Davis RE and Acrivos A (1966) The influence of surfactants on the creeping motion of bubbles. Chem Eng Sci 21: 681–5.
6. Defay R, Prigogine I and Sanfeld A (1977) Surface thermodynamics. J Colloid Interface Sci 58: 498–510.
7. Deryagin BV, Dukhin SS and Lisichenko VA (1959) The kinetics of the attachment of mineral particles to bubbles during flotation. I. The electric field of a moving bubble. Russ J Phys Chem (Engl Transl) 33: 389–393.
8. Deryagin BV, Dukhin SS and Lisichenko VA (1960) The kinetics of the attachment of mineral particles to bubbles during flotation. II. The electric field of a moving bubble when the surface activity of the inorganic substance is high. Russ J Phys Chem (Engl Transl) 34: 248–251.
9. Dukhin SS (1966) An electrical diffusion theory of the Dorn effect and a discussion of the possibility of measuring zeta potentials. Res Surface Forces, Proc 2nd Conf, Moscow 1962, (Engl Transl) 2: 54–74: New York: Consultants Bureau.
10. Gradshteyn IS and Ryzhik IM (1966) Table of integrals, series and products: fourth edition. New York: Academic.

11. Harper JF (1972) The motion of bubbles and drops through liquids. Adv Appl Mech 12: 59–129.
12. Harper JF (1973) On bubbles with small immobile adsorbed films rising in liquids at low Reynolds numbers. J Fluid Mech 58: 539–545.
13. Harper JF (1974) On spherical bubbles rising slowly in dilute surfactant solutions. Q J Mech Appl Math 27: 87–100.
14. Hill MJM (1894) On a spherical vortex. Philos Trans R Soc Lond 185: 213–245.
15. Huang WS and Kintner RC (1969) Effect of surfactants on mass transfer inside drops AIChE J 15: 735–744.
16. Kaplun S (1954) Co-ordinate systems in boundary-layer theory. Z Angew Math Phys 5: 111–135.
17. Kronig R and Brink JC (1950) On the theory of extraction from falling droplets. Appl Sci Res A2: 142–154.
18. Lagerstrom PA and Cole JD (1955) Examples illustrating expansion procedures for Navier-Stokes equations. J Ration Mech Anal 4: 817–882.
19. Le Van MD and Newman J (1976) The effect of surfactant on the terminal and interfacial velocities of a bubble or drop. AIChE J 22: 695–701.
20. Le Van MD and Newman J (1976) Reply (to letter from DA Saville). AIChE J 23: 614.
21. Lucassen J and Giles D (1975) Dynamic surface properties of nonionic surfactant solutions. J Chem Soc Faraday Trans I 71: 217–232.
22. Payne LE and Pell WH (1960) The Stokes flow problem for a class of axially bodies. J Fluid Mech 7: 529–549.
23. Porter D and Dore BD (1974) The effect of a semi-contaminated free surface on mass transport. Proc Camb Philos Soc 75: 283–294.
24. Savic P (1953) Circulation and distortion of liquid drops falling through a viscous medium. Rep MT-22, Div Mech Eng, Nat Res Coun Canada.
25. Saville DA (1973) The effects of interfacial tension gradients on the motion of bubbles or drops. Chem Eng J 5: 251–9.
26. Saville DA (1977) Letter to the Editor. AIChE J 23: 613–4.

Bubble coalescence in pure liquids

A.K. CHESTERS and G. HOFMAN

Laboratory of Aero- and Hydrodynamics, Delft University of Technology,
The Netherlands

Abstract. Solutions of the flow and deformation during the approach of two bubbles along their centre line are presented for the low Weber number case. When viscosity is absent, Weber number and radius ratio disappear from the equations under suitable transformations of the variables and a universal solution is obtained. This solution indicates the formation of a dimple, after which the thinning rate of the film between the bubbles tends to a constant high value. When liquid viscosity is included the Reynolds number, Re, enters the equations. A retardation of the coalescence process is found for Re < 100, while for Re ⩽ 1 dimple formation is suppressed. The influence of gas properties is considered briefly. An extrapolation of the inviscid results to Weber numbers of order unity suggests that the bubbles will bounce apart before coalescence is achieved.

1. Introduction

The evolution of many two-phase flows depends critically on whether bubbles having a relative approach velocity coalesce or not. Situations giving rise to such an approach velocity are, for example: (i) bubbles rising in line, the upper sheltering the lower; (ii) the forced contact of a bubble growing on an orifice with the previous, departing bubble; (iii) the rise of a bubble towards a free surface (which may be seen as a second, infinite bubble).

In liquids of low viscosity, in these and most other situations, both observation and analysis suggest that two processes are in competition:

(a) The liquid between the bubbles is squeezed out, accompanied by a flattening and even dimpling of the bubble surfaces. When the residual film reaches thicknesses of order 100 Å, van der Waals pressures become dominant and a hole is rapidly formed [1]. Surface tension then expands this hole and the bubbles become one.

(b) The deformation of the bubbles increases their surface area, and hence the free energy of the system, at the expense of the kinetic energy associated with the relative motion. The bubbles therefore decelerate and eventually bounce apart if (a) is not yet completed.

Clearly for sufficient slight deformation (b) may be neglected and the problem reduces to that of almost parallel-sided film flow (Figure 1) with initial and outer-boundary conditions corresponding to the constant-velocity approach of the undeformed bubbles. This problem is first solved in the inviscid, gravity-free case. These results are used to deduce: (i) under which conditions the neglect of gravity is justified; (ii) at what critical deformation van der Waals pressures will become dominant; (iii) under which conditions process (b) will arrest the motion before this critical deformation is reached.

Applied Scientific Research 38: 353–361 (1982) 0003–6994/82/0384–0353 $01.35.

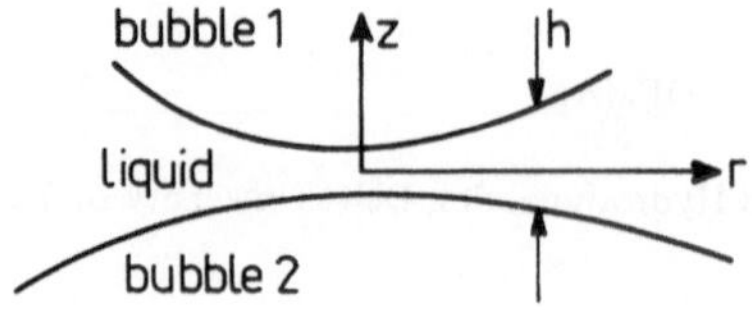

Figure 1. Choice of coordinates in the coalescence problem.

The film-thinning calculations are then extended to include the effects of liquid viscosity, and the influence of the gas properties is briefly considered.

2. Film formation and thinning in the inviscid case

Since the film is almost parallel-sided with freely moving surfaces (a pure, isothermal liquid) the pressure, p, and the r-component of the velocity, u, may be approximated as uniform across the film. For the symmetrical approach of equal bubbles along the line joining their centres the equations governing the film flow are then:

Continuity (volume balance over surfaces at r and $r + \mathrm{d}r$).

$$\frac{\partial h}{\partial t} = -\frac{1}{r}\frac{\partial}{\partial r}(hur) \tag{1}$$

(h local film thickness, t time).

Navier-Stokes (r direction).

$$\frac{\partial u}{\partial t} = -u\frac{\partial u}{\partial r} - \frac{1}{\rho}\frac{\partial p}{\partial r} \tag{2}$$

(ρ liquid density).

The interface condition for the normal stress.

$$p = p_G - \sigma(R_a^{-1} + R_b^{-1}) = 2\sigma/R - \sigma(R_a^{-1} + R_b^{-1}) \tag{3}$$

(R_a, R_b radii of curvature of the surface in the r-z plane, and in the perpendicular plane containing the normal, σ surface tension, R bubble radius, p_G pressure in the gas: the pressure in the undisturbed liquid being taken as zero for convenience).

Now

$$R_a^{-1} = \frac{\partial^2(h/2)}{\partial r^2}\Bigg/\left[1 + \left\{\frac{\partial(h/2)}{\partial r}\right\}^2\right]^{3/2} \simeq \frac{1}{2}\frac{\partial^2 h}{\partial r^2}, \tag{4}$$

since $\{\partial(h/2)/\partial r\}^2 \ll 1$, and (5)

$$R_b^{-1} = \frac{1}{r}\frac{\partial(h/2)}{\partial r}\bigg/\left[1+\left\{\frac{\partial(h/2)}{\partial r}\right\}^2\right]^{1/2} \simeq \frac{1}{2r}\frac{\partial h}{\partial r}. \tag{6}$$

(3) and (2) therefore yield:

$$p = \frac{2\sigma}{R} - \frac{\sigma}{2}\left(\frac{\partial^2 h}{\partial r^2} + \frac{1}{r}\frac{\partial h}{\partial r}\right) \tag{7}$$

and

$$\frac{\partial u}{\partial t} = -u\frac{\partial u}{\partial r} + \frac{\sigma}{2\rho}\left[\frac{\partial^3 h}{\partial r^3} + \frac{1}{r}\frac{\partial^2 h}{\partial r^2} - \frac{1}{r^2}\frac{\partial h}{\partial r}\right]. \tag{8}$$

The initial conditions correspond to undeformed bubbles with relative velocity, V. Making use of (5) the initial variation of h is then

$$h = h_0 + r^2/R, \tag{9}$$

while continuity demands that $\quad\Bigg\} \; t = 0$

$$u = Vr/2h = Vr/2(h_0 + r^2/R). \tag{10}$$

For these conditions to be realistic h_0 must be large enough for deformation to be negligible had the bubbles approached each other from infinity.

The conditions at the r-boundary of the film are:

$$\frac{\partial h}{\partial t} = -V \quad (= \text{const}), \tag{11}$$

$$p = \sigma\left[2/R - \frac{1}{2}\left(\frac{\partial^2 h}{\partial r^2} + \frac{1}{r}\frac{\partial h}{\partial r}\right)\right] = 0. \tag{12}$$

$\Bigg\}$ 'large' r $(= r_{\text{bound}})$

In the case of unequal bubbles the situation is no longer symmetrical and a reference frame with origin in the film centre is in general an accelerating one. This acceleration, which proves equal to the average acceleration of the two bubbles, may be neglected along with g (the acceleration due to gravity) provided it is of similar or smaller order. This condition is satisfied in general. The new film thinning equations then prove to be the same as those for equal bubbles provided R is replaced by an equivalent radius, R_{eq}:

$$R_{\text{eq}}^{-1} = \tfrac{1}{2}(R_1^{-1} + R_2^{-1}) \tag{13}$$

(R_1 and R_2 radii of the two bubbles). The ratio R_{eq}/R_1 is seen to vary from 1 when $R_1 = R_2$ to 2 when $R_2 = \infty$. Coalescence with a free surface is thus equivalent to coalescence of equal bubbles of twice the radius.

The thinning equations (1) and (8), with initial and boundary conditions (9)–(12), contain four parameters: ρ, σ, V and R. By replacing the variables h, u, r, t by their dimensionless equivalents h', u', r' and t' ($h = h'R, u = u'V$,

$r = r'R$, $t = t'R/V$) a new set of equations is obtained containing only one dimensionless parameter, the Weber number, We $(= \rho V^2 R_{eq}/\sigma)$:

$$\frac{\partial h'}{\partial t'} = -\frac{1}{r'}\frac{\partial}{\partial r'}(h'u'r'), \tag{1'}$$

$$\frac{\partial u'}{\partial t'} = -u'\frac{\partial u'}{\partial r'} + \frac{1}{2\mathrm{We}}\left[\frac{\partial^3 h'}{(\partial r')^3} + \frac{1}{r'}\frac{\partial^2 h'}{(\partial r')^2} - \frac{1}{(r')^2}\frac{\partial h'}{\partial r'}\right], \tag{8'}$$

$$\left.\begin{aligned} h' &= h'_0 + (r')^2, & (9')\\ u' &= r'/2(h'_0 + r'^2), & (10') \end{aligned}\right\} \quad t' = 0$$

$$\left.\begin{aligned} \frac{\partial h'}{\partial t'} &= -1, & (11')\\ 2 - \frac{1}{2}\left[\frac{\partial^2 h'}{(\partial r')^2} + \frac{1}{r'}\frac{\partial h'}{\partial r'}\right] &= 0. & (12') \end{aligned}\right\} \quad \text{'large' } r' \; (= r'_{\text{bound}})$$

By a further transformation of variables,

$$h' = h^* \,\mathrm{We}, \qquad u' = u^* \,\mathrm{We}^{-1/2}, \tag{14), (15}$$

$$r' = r^* \,\mathrm{We}^{1/2}, \qquad t' = t^* \,\mathrm{We}, \tag{16), (17}$$

We is eliminated and a single universal set of thinning equations obtained (the same equations as (1′), (8′), (9′)–(12′) except for the absence of We and the replacement of ′ by *).

The solution of these equations, obtained by a modified Lax-Wendroff finite-difference scheme, is shown in Figures 2 and 3. The values of h_0^* and r_{bound}^* proved of little influence provided they were considerably larger than 1. In Figures 2 and 3 these values were 4 and 8, respectively. The transformed time, t^*, is measured from the moment at which the bubbles would have touched had deformation been absent.

The results indicate that the bubbles flatten and then develop dimples, the ring of minimum film thickness ultimately being located at $r^* \simeq 0.4$. In this region very large velocities (and pressure gradients) develop with the result that the thinning rate, $-dh_{\min}^*/dt^*$, no longer diminishes but levels off at an asymptotic value of about 0.1 (i.e. the minimum film thickness decreases at the rate 0.1 V). Even aside from van der Waals pressures therefore, the predicted coalescence time is finite:

$$t_{\text{coal}}^* \simeq 1. \tag{18}$$

Depending on the strength of the van der Waals forces, the effective rupture thickness will vary by the order of 100 Å, producing a variation in coalescence

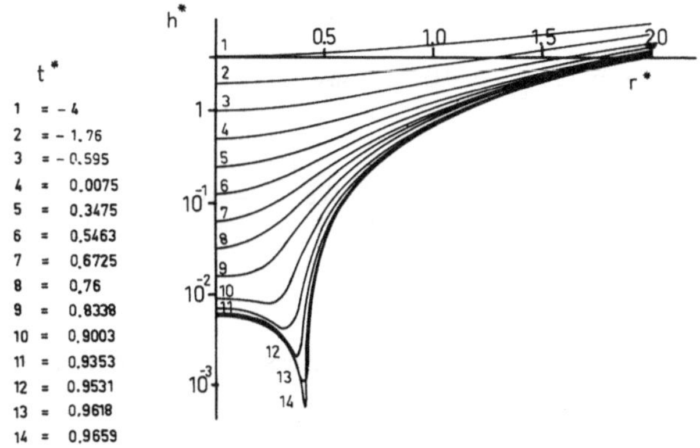

Figure 2. Variation of liquid film thickness, h^*, with r^* and t^* in the inviscid case.

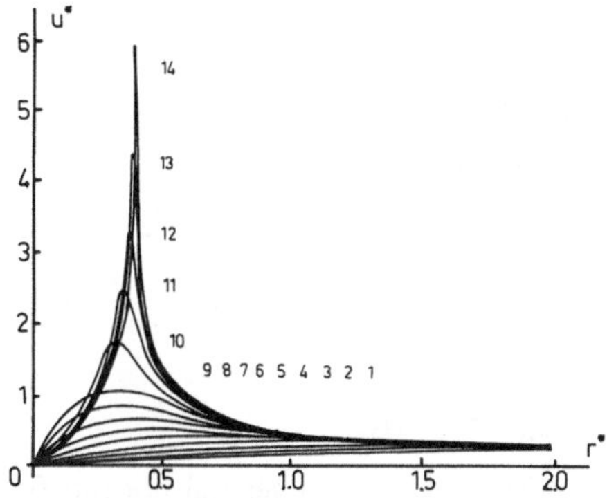

Figure 3. Variation of liquid film velocity, u^*, with r^* and t^* in the inviscid case.

time of

$$\begin{aligned} \delta t &\sim 100\,\text{Å}/(-\mathrm{d}h/\mathrm{d}t) \\ &\lesssim 100\,\text{Å}/(0.1\,V). \end{aligned} \tag{19}$$

For $V \sim 1$ cm/s, (19) yields $\delta t \lesssim 10^{-5}$ s. This is truly negligible: a heartening conclusion from the point of view of two-phase modelling [1].

Assuming the effective rupture thickness to be reached in the final stages of the thinning process, (18) applies. If, further, the deformation is approximated by a film (radius r_{def}) of negligible thickness, outside which the bubbles are undeformed, then it is readily shown that

$$(r'_{\text{def}})^2 = t', \tag{18}$$

i.e.

$$(r^*_{\text{def}})^2 = t^* \simeq 1, \tag{19}$$

at coalescence. At this radius (5) must still hold:

$$\{\partial(h/2)/\partial r\}^2_{r=r_{\text{def}}} \lesssim 10^{-2}, \quad \text{say}. \tag{20}$$

Since

$$\{\partial(h/2)/\partial r\}_{r_{\text{def}}} = r'_{\text{def}} = r^*_{\text{def}}\,\text{We}^{1/2} = \text{We}^{1/2},$$

(20) requires that

$$\text{We} \lesssim 10^{-2}. \tag{21}$$

The restriction to slight deformations thus places an upper limit on the Weber number. For water this limitation yields

$$V^2 R_{\text{eq}} \lesssim 1\ \text{cm}^3\ \text{s}^{-2}. \tag{22}$$

With the help of (19) and (21) the influence of gravity can now be assessed. Gravity (presumed to act in the z-direction) is negligible provided

$$\rho g \Delta z \ll 2\sigma/R_{\text{eq}} \tag{23}$$

(Δz z-variation over the film surface). Δz is greatest initially. Again taking r_{def} as the greatest r-value of interest,

$$(\Delta z)_{\text{max}} = r^2_{\text{def}}/2R_i = (r^*_{\text{def}})^2 R^2_{\text{eq}}\,\text{We}/2R_i = R^2_{\text{eq}}\,\text{We}/2R_i. \tag{24}$$

Combination of (21)–(24) with the fact that $R_{\text{eq}}/R_i \leqslant 2$ leads to

$$R^2_{\text{eq}} \ll 200\sigma/\rho g,$$

$$R_{\text{eq}} \lesssim (2\sigma/\rho g)^{1/2}, \quad \text{say}. \tag{25}$$

Bubbles too large to satisfy (25) will in general not satisfy (21). For water $(2\sigma/\rho g)^{1/2}$ is about 0.4 cm.

3. Bouncing bubbles

For small r'_{def}, the simplified deformation model yields an increase, ΔS, in the surface area of each bubble given by

$$\begin{aligned}\Delta S/4\pi R^2_{\text{eq}} &= (r'_{\text{def}})^4/16 = (r^*_{\text{def}})^4\,\text{We}^2/16 \\ &= \text{We}^2/16,\end{aligned} \tag{26}$$

at coalescence. The free energy associated with this extra surface is $\sigma\Delta S$. In the case of equal bubbles, having velocities $V/2$ and $-V/2$ in the z-direction, the fractional loss of kinetic energy at coalescence, β is therefore given by

$$\sigma \cdot 4\pi\,\text{Re}^2_{\text{eq}} \cdot \text{We}^2/16 = \beta \cdot \tfrac{1}{2} m'_{\text{virt}}\,\pi\rho R^3_{\text{eq}}(V/2)^2,$$

i.e.

$$\beta = 2\,\text{We}/m_{\text{virt}}, \tag{27}$$

where m'_{virt} is the coefficient of virtual mass associated with one of the bubbles.

This result can be extended to unequal bubblés, having unequal velocities. Making use of [4], the value of m'_{virt} for bubbles in close proximity is found to vary from about 1 if $R_1 = R_2$ (compared with 2/3 for an isolated bubble) to about 1/4 if $R_2 = \infty$.

For the low We values at which the preceeding results apply, the assumption of a constant approach velocity is therefore seen to be a good one, β being a few percent or less at coalescence. For

$$\text{We} > m'_{\text{virt}}/2, \tag{28}$$

however, (27) predicts that the motion will be arrested before coalescence can occur, giving rise to the phenomenon of 'bouncing bubbles'. Although only qualitatively reliable in view of the high Weber numbers involved, this conclusion is in accord with observations of bubbles rising to a free surface [3] and of bubbles growing on small orifices [2].

4. The influence of liquid viscosity on film thinning

If liquid viscosity is taken into account an extra term is added to equation (8′):

$$\frac{\partial u'}{\partial t'} = -u'\frac{\partial u'}{\partial r'} + \frac{1}{2\text{We}}\left[\frac{\partial^3 h'}{(\partial r')^3} + \ldots\right]$$
$$+\frac{2}{\text{Re}}\left[2\left\{\frac{\partial^2 u'}{(\partial r')^2} + \frac{1}{r'}\frac{\partial u'}{\partial r'} - \frac{u}{(r')^2}\right\}\right.$$
$$\left. + \frac{1}{h'}\frac{\partial h'}{\partial r'}\left\{\frac{2\partial u'}{\partial r'} + \frac{u'}{r'}\right\}\right] \tag{8a′}$$

(Re Reynolds number, $\rho V R_{\text{eq}}/\mu$; μ liquid dynamic viscosity).

The extra term includes both the viscous forces on z-, r- and θ-faces of an element of film and the effect of viscosity on the pressure equation (7), to which an extra normal-stress term must be added. Forces exerted by the gas are still neglected.

No transformation can be found to remove both We and Re from the new equations and after applying the transformation (14)–(17), the new viscous term remains unaltered except for the replacement of ′ by *. For Re $\geqslant$ 100, the solution is almost identical with the inviscid one (Figure 4), indicating the relevance of the latter to many situations of practical interest. For Re = 10, a noticeable decrease in the thinning rate occurs in the early stages, becoming major in the dimpled stage. For Re = 1 dimple formation itself is suppreśsed.

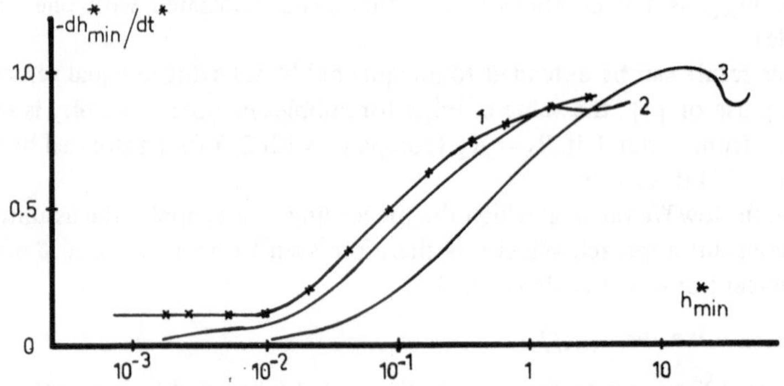

Figure 4. Influence of liquid viscosity on the thinning rate (1: inviscid, x Re = 100, 2 Re = 10, 3 Re = 1).

For still smaller Re values the inertial terms in (8a′) should be negligible. A transformation of the type (14)–(17) is then possible (We/Re playing the role of We), resulting again in a universal solution, valid for all small Re values. This solution will probably closely resemble that for Re = 1, implying that dimpling is not a feature of low-Re bubble coalescence.

5. The influence of gas properties on film thinning

The flow in the bubbles is coupled to that in the film by the surface condition

$$\mathbf{u}_G = \mathbf{u}. \tag{29}$$

At large bubble separations the forces on the film resulting from the gas flow are negligible but when the film becomes very thin even a small component in the r-direction corresponds to an appreciable force per unit mass of liquid.

At first sight the problem is a very complex one, since the forces may depend on the whole time history of the flow in the gas bubbles. An order of magnitude assessment, however, based on the solutions already obtained for the film flow, indicates that: (i) the final film velocities are large in comparison with the initial circulation velocities within the bubbles; (ii) the distance over which vorticity diffuses from the moving surface into the gas during times characteristic of the film flow is small relative to the bubble radii, but large relative to radial distances characteristic of changes in the film flow.

(i) and (ii) lead to the conclusion that the gas flow may be approximated as creeping flow with a prescribed velocity distribution on the $z = 0$ plane at each instant and negligible velocity at large z values. The history effect is thus absent and the extra force per unit mass of film in the radial direction, F_G, is given by

$$F_G = 2\tau_G/\rho h \sim (2/\rho h)\cdot \mu_G u/L \tag{30}$$

(τ_G gas shear stress, L a radial distance characteristic of changes in the film velocity, e.g. $L = |u/\{\partial u/\partial r\}|$).

(30) leads to dimensionless values of F_G comparable with the other terms in (8a′) during the last stages of thinning.

References

1. Chesters AK (1975) The applicability of dynamic similarity criteria to isothermal, liquid-gas two-phase flows without mass transfer. Int J Multiphase Flow 2: 191–212.
2. Chesters AK and Bader P. Bubble Growth on small orifices (as yet unpublished).
3. Farooq SY (1972) Coalescence of bubbles. PhD thesis, Dept Chem Eng, Univ College of Swansea, Univ of Wales.
4. Lamb H (1945) Hydrodynamics (art 98). New York: Dover.

Studies of liquid-vapour phase change by a shock tube

S. FUJIKAWA, T. AKAMATSU, J. YAHARA and H. FUJIOKA

Department of Mechanical Engineering, Kyoto University, Kyoto, Japan

Introduction

Recently, the study of evaporation from a liquid surface or condensation onto it has received much attention in various fields of engineering as well as in physics and chemistry: cavitation, two-phase flow, steam turbines, heat pipes and combustion. A lot of effort has been spent to understand the kinematic behaviour of gas molecules interacting with themselves. The physical properties of the condensed substance including the gas-condensate interaction have not been fully resolved.

The present study aims at clarifying the nonequilibrium condensation and thermal accommodation processes on solid and liquid surfaces by using a shock tube. A stagnant gas of high temperature and high pressure is generated in the region between a reflected shock wave and an end wall of the shock tube. The wall itself being kept at initial room temperature, the gas adjacent to the wall is cooled and becomes supersaturated. There an unsteady thermal boundary layer develops. Then, the vapour begins to condense forming a thin liquid film on the end wall.

The rate of growing of the liquid film on the wall and the temperature discontinuity at the gas-liquid interface have been measured by means of an optical technique, and, mass and thermal accommodation coefficients of vapour molecules are determined. The structure of a thermal boundary layer at a phase-changing interface is theoretically investigated.

Finally, the present results are applied to the problem of vapour bubble collapse. The mechanism of the generation of pressure waves from the bubble is clarified.

Theory

The gas is supposed to behave as an ideal monatomic gas and to condense on the end wall of a shock tube. The basic equations are the normalized one-dimensional unsteady conservation equations [2]:

$$\frac{\partial U_i}{\partial t} + \frac{\partial F_i}{\partial x} = \frac{1}{\mathrm{Re}} \frac{\partial}{\partial x}\left(\mu \frac{\partial S_i}{\partial x}\right), \qquad i = 1, 2, 3 \tag{1}$$

where t is the time, x the coordinate normal to the gas-liquid interface, Re the Reynolds number, μ the normalized shear viscosity, and U_i, F_i and S_i are the local acceleration, convection and diffusion (heat transfer, viscosity) of

Applied Scientific Research 38: 363–372 (1982) 0003–6994/82/0384–0363 $01.50.

mass, momentum and energy. The space coordinate is made dimensionless by the mean free path L_1, the velocity by the initial speed of sound a_1 and the other flow variables by their initial values. The boundary condition for the temperature, taking a thermal accommodation into account, is given by:

$$T(t,0) - T_L(t,0) = g\frac{\partial T}{\partial x}(t,0), \tag{2}$$

where T and T_L are temperatures of gas and liquid at the interface, respectively and g is the temperature jump distance, a function of the thermal accommodation coefficient 'α_T'. The mass flow to the interface is given by:

$$\rho(t,0)u(t,0) = \frac{\alpha_M}{\sqrt{2\pi R}}\left(\frac{p_s}{\sqrt{T_L(t,0)}} - \frac{p}{\sqrt{T(t,0)}}\right), \tag{3}$$

$$= -\rho_L\frac{d\delta}{dt}$$

where 'α_M' is the mass accommodation coefficient, δ the thickness of the liquid film, ρ the density of the gas, ρ_L the density of the liquid film, u the velocity, p_s the saturation vapour pressure, p the actual vapour pressure and R the gas constant.

The liquid film and the solid material in $x<0$ are assumed to have constant thermal properties. Temperatures T_L and T_s of liquid and solid satisfy the diffusion equations,

$$\frac{\partial T_L}{\partial t} = D_L\frac{\partial^2 T_L}{\partial x^2} \qquad (-\delta < x < 0), \tag{4}$$

$$\frac{\partial T_s}{\partial t} = D_s\frac{\partial^2 T_s}{\partial x^2} \qquad (-\infty < x < -\delta), \tag{5}$$

where D_L and D_s are thermal diffusivities of liquid and solid. For equation (4), the following transform variable has been introduced in order to fix the position of the moving boundary [7],

$$\xi = \frac{x}{\delta}. \tag{6}$$

For numerical treatments of equations (1) and (4)–(5), the two-step Lax-Wendroff method and the Crank-Nicholson method have been employed, respectively. Further details of the theory will be described elsewhere [4].

Principle of optical measurements

The present method is based on the dependence of reflection coefficients of light beams on the liquid film thickness and the local refractive index of the

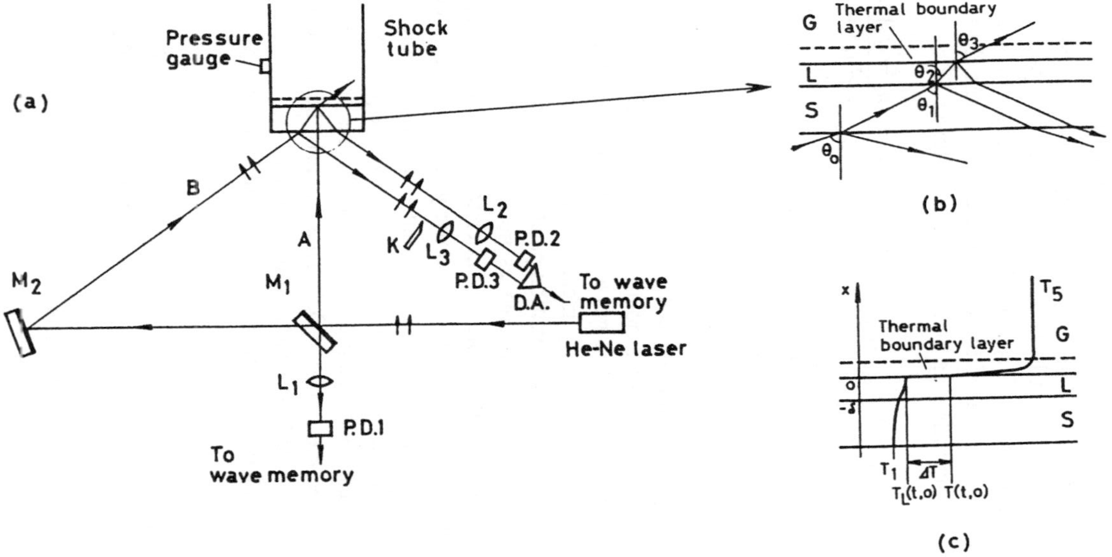

Figure 1. Schematic diagram of experimental setup: (a) optical system, (b) light beam trajectory (*G*: gas, *L*: liquid film, *S*: glass) and (c) temperature profiles.

gas at the gas-liquid interface [9]. The arrangement of the optical apparatus is shown in Figure 1. The shock tube end wall consists of an optically flat BK-7 crown glass: flatness, a tenth of the wavelength of light.

Liquid film thickness

The thickness of the liquid film can be directly obtained from the measurement of the reflection coefficient R_L of a light beam '*A*' at normal incidence on the glass. The reflected light from the liquid film, passing through a half mirror M_1, is brought to focus on a PIN photo-diode P.D.1 by a lens L_1. The electronic response time of the detection circuit is less than 1 μs. Then, the thickness δ of liquid film can be expressed as follows;

$$\delta(t) = \frac{\lambda}{2\pi n_L} \sin^{-1}\left(n_L\left[\frac{(n_s-n_g)^2-(n_s+n_g)^2R_L}{(n_s^2-n_L^2)(n_L^2-n_g^2)(1-R_L)}\right]^{1/2}\right), \quad (7)$$

where λ is the wavelength of light, n the refractive index and the subscripts s for glass, L for liquid, g for gas.

Temperature discontinuity

An interfacial gas temperature can be obtained from the measurement of the reflection coefficient R_T of a polarized light beam '*B*', which is a function of the refractive index n_{gi} of the gas at the interface and the liquid film thickness δ. In order to sensitively measure the refractive index variations, the light beam with plane of polarization parallel to the plane of incidence is used at an angle of incidence as large as possible. Here, an angle of about 80 degrees is chosen. Furthermore, in order to accurately obtain the refractive index, the relative change of the reflectivity is measured. A beam, reflecting from the outer surface of the glass, is used as a reference and is attenuated by means of a knife edge K to obtain an intensity equal to that of the main beam. The intensities of main and reference beams are detected by means of PIN photodiodes P.D.2 and P.D.3. These signals, being led to a differential amplifier, are recorded on a wave form recorder.

Then, the relative variation Y of the reflectivity can be expressed as follows,

$$Y = \frac{R_T - R_{T0}}{R_{T0}} = \frac{1}{R_{T0}}\left(\frac{S_{sL}^2 + S_{Lg}^2 + 2S_{sL}S_{Lg}\cos 2\beta}{1 + S_{sL}^2 S_{Lg}^2 + 2S_{sL}S_{Lg}\cos 2\beta}\right) - 1, \quad (8)$$

where R_{T0}: intensity of reference beam,

$$S_{sL} = \frac{\tan(\theta_s - \theta_L)}{\tan(\theta_s + \theta_L)},$$

$$S_{Lg} = \frac{\tan(\theta_L - \theta_g)}{\tan(\theta_L + \theta_g)},$$

$$\beta = \frac{2\pi n_L \delta \cos \theta_L}{\lambda}.$$

The angle of refraction θ_g varies with the refractive index n_{gi} according to Snell's law:

$$n_{gi} \sin \theta_g = n_L \sin \theta_L = n_s \sin \theta_s = \text{const}, \tag{9}$$

The refractive index n_{gi} is coupled to an interfacial gas density ρ_{gi} by the Lorentz-Lorentz equation:

$$n_{gi} = 1 + K\rho_{gi}, \tag{10}$$

where K is the Gladstone-Dale constant. In consequence, given the pressure of the gas, an interfacial temperature of the gas can be determined using the equation of state. The pressure is measured by means of a piezoelectric pressure transducer, mounted in the side wall of the tube above 30 mm from the inner surface of the glass.

Results and discussion

The optical glass used in experiments has been cleaned in a solution of $K_2Cr_2O_7$ in sulfuric acid, then in boiling ethanol. The glass treated in this way is reasonably free of contaminants. Experiments have been performed at initial downstream conditions of $p_1 = 6.13$ kPa, $T_1 = 290$ K and $M_1 = 1.43$ for a methanol vapour and of $p_1 = 1.47$ kPa, $T_1 = 292$ K and $M_1 = 1.26$ for a water vapour.

Figure 2 shows time histories of a side-wall pressure and reflectivities of light beams at normal and inclined incidences on the glass in a methanol

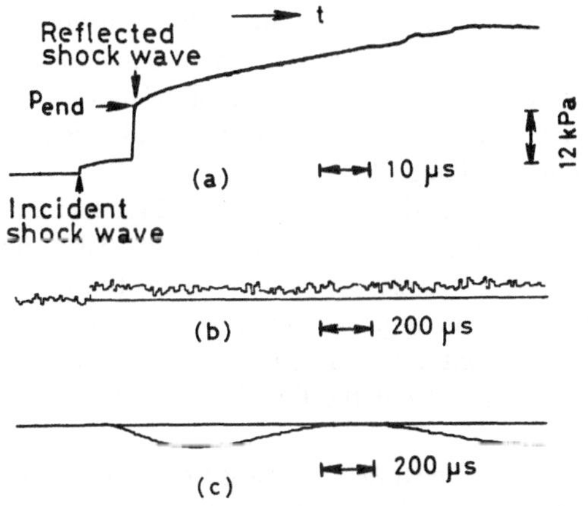

Figure 2. Time histories of side wall pressure and reflectivities of light beams for a methanol vapour: (a) pressure, (b) refractive index and (c) liquid film thickness.

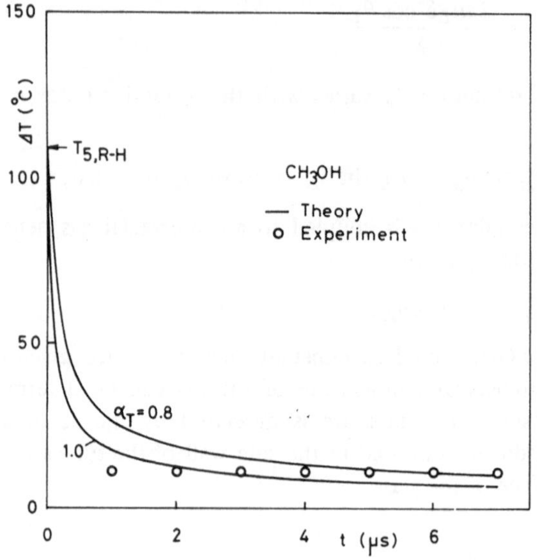

Figure 3. Time history of measured gas temperature at the gas-liquid interface for methanol vapour.

vapour: (a) pressure, (b) refractive index and (c) liquid film thickness. In Figure (a), the rising pressure following arrivals of incidence and reflected shock waves is due to the shock-boundary layer interaction. The pressure on the end wall is indicated by a mark p_{end}. The trace from a transducer mounted in the end wall shows an approximately constant pressure (not shown here). Figure (b) shows that the gas at the interface maintains a constant temperature after a transition of the order of molecular collision time from a reflected shock-heated state. The transition itself cannot be observed here. Figure (c) shows that the vapour condenses in the form of a liquid layer on the end wall after the reflection of shock wave. Judging from the persisting good contrast between maxima and minima, there do not exist irregularities in the liquid film.

Figure 3 shows the time history of a measured gas temperature at the gas-liquid interface for methanol vapour. The interfacial temperature drops, within 1 μs, from 398 K (Rankine-Hugoniot value) to 300 K, and thereafter is kept at this value. From comparison with a numerical result, the thermal accommodation coefficient is found to be nearly equal to unity. This implies that re-emitted molecules from the liquid surface thermally accommodate to the surface.

Figure 4 shows time histories of measured liquid film thickness for methanol and water vapours. They are compared with numerical results. The liquid film initially grows in proportion to the elapsed time after its formation, then to almost the square root of the time [8]. This behaviour of the

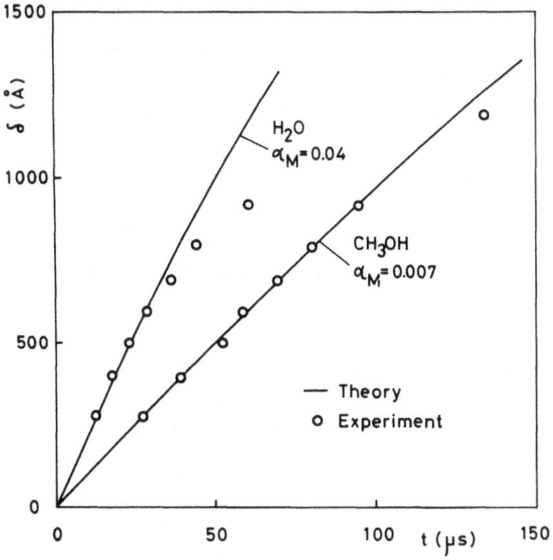

Figure 4. Time histories of measured liquid film thickness for methanol and water vapours.

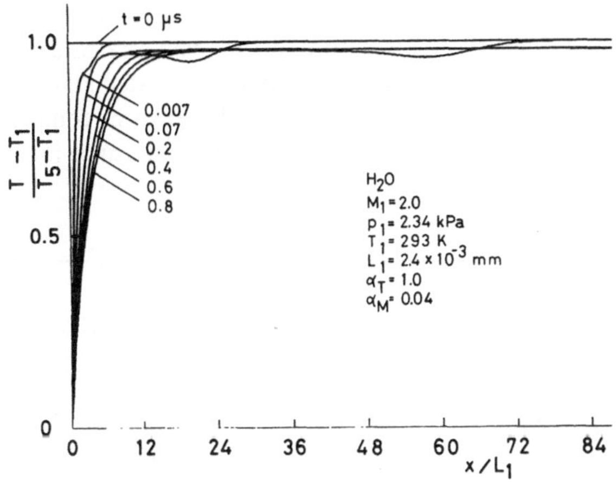

Figure 5. Calculated temperature profiles of water vapour near the gas-liquid interface for mass accommodation coefficient 0.04.

liquid film is due to an increase of the liquid temperature at the gas-liquid interface. The liquid temperature increases with time owing to the release of the latent heat of condensation and, therefore, the saturation vapour pressure does. In consequence, the condensation rate, i.e., the rate of the growing liquid film decreases. From comparison between experiments and numerical results, mass accommodation coefficients have been estimated. They are: $\alpha_M = 0.007$ for methanol vapour and $\alpha_M = 0.04$ for water vapour. The latter agrees with measured values (0.027–0.042) by Delaney et al. [1] and the theoretically predicted one (0.04) by Mortensen and Eyring [6]. On the other hand, Mills and Seban [5] obtain a value of α_M for water vapour close to unity from the measurement of the overall heat-transfer coefficient. In their experiments, however, temperatures which can not be directly measured are introduced in the definition of the heat-transfer coefficient. So, their result does not seem reliable. As to the methanol vapour, the present value is about a sixth of the theoretical result. Further systematic investigations must be made in order to clarify this discrepancy.

Figure 5 shows calculated temperature profiles of the water vapour near the interface: $p_1 = 2.34$ kPa, $T_1 = 293$ K, $M_1 = 2.0$, $\alpha_M = 0.04$ and $\alpha_T = 1.0$. Expansion waves are generated in the gas region because the stagnant high temperature of the gas is instantaneously cooled and contracted by contact with the cold end wall and by the subsequently formed liquid layer on it. After the passage of expansion waves, an unsteady thermal boundary layer develops on the liquid layer. The interfacial temperature of the gas drops, within 0.2 μs, from 657 K to 300 K and is kept thereafter at this value. Temperature profiles for $\alpha_M = 1.0$ are shown in Figure 6 for purposes of comparison with those for 0.04. Expansion waves stronger than those in the case of $\alpha_M = 0.04$ are formed in the gas region owing to strong condensation at the interface, so that the temperature in the main stream decreases to about 80% of the temperature at the reflected shock-heated state. The interfacial temperature of the gas is kept at 360 K and is high in comparison with the case of $\alpha_M = 0.04$, because the temperature jump distance increases owing to

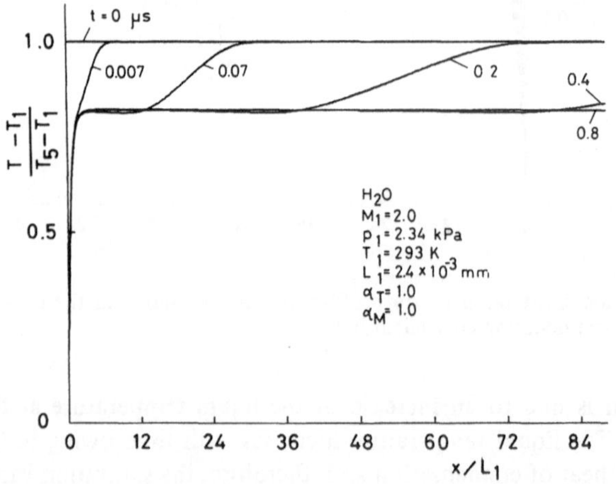

Figure 6. Calculated temperature profiles of water vapour near the gas-liquid interface for mass accommodation coefficient 1.0.

a decrease of the interfacial density. The thermal boundary layer is almost in a steady state.

Application to vapour bubble collapse

The mathematical formulation on the dynamics of a bubble in a liquid has been extensively made by Fujikawa and Akamatsu [3]. Only a numerical result for a vapour bubble collapse will be presented here.

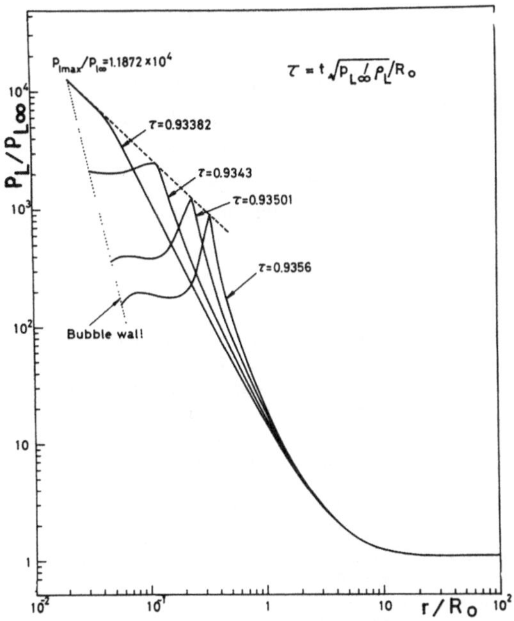

Figure 7. Pressure distributions in water after the collapse of a vapour bubble.

Figure 7 shows the pressure distributions in water after the collapse of a vapour bubble: initial bubble radius $R_0 = 1.0$ mm, water temperature $T_L = 293$ K, vapour pressure $p_v = 2.34$ kPa, ambient pressure $P_{L\infty} = 71.2$ kPa, $\alpha_M = 0.04$ and $\alpha_T = 1.0$. The bubble does rebound and emanate strong pressure waves into the water. This mechanism can be explained as follows. At the initial stages of the collapse the vapour condenses back into the liquid, so that the vapour pressure in the bubble remains equal to the saturation vapour pressure and does not retard the inrushing liquid around the bubble. At the final stages, however, the collapse takes place so rapidly that most of the vapour may then be compressed to a high pressure, which will eventually become large enough to halt the inrushing liquid. Then, the bubble will rebound and radiate pressure waves. At the minimum radius (2.1×10^{-2} mm), 3.2% of the original vapour remains. The maximum impulsive pressure is about 8.45×10^2 MPa and then the temperature is 3.11×10^4 K.

Concluding remarks

A shock tube has been used for studying the nonequilibrium condensation and thermal accommodation processes at a gas-liquid interface. Mass and thermal accommodation coefficients were determined by means of an optical diagnostic technique. The structure of a thermal boundary layer on the liquid film has been theoretically investigated. Effects of the mass accommodation coefficient on it were examined.

The present results have been applied to the collapse of a vapour bubble in water. The vapour bubble was found to produce strong pressure waves at the instant of its rebound owing to nonequilibrium effects of condensation.

Experimental and theoretical investigations will be continued in order to elucidate interfacial condensation phenomena of vapours.

References

1. Delaney LJ, Houston RW and Eagleton LC (1964) The rate of vaporization of water and ice. Chem Engng Sci 19: 105–114.
2. Hänel D (1974) Frontale Stoßwellenreflexion an Wänden mit Adsorption und Akkommodation. Dissertation RWTH Aachen.
3. Fujikawa S and Akamatsu T (1980) J Fluid Mech 97: 481–512.
4. Fujikawa S, Yahara J, Fujioka H and Akamatsu T (1981) Bull of the Japan Soc Mech Engrs (to be presented).
5. Mills AF and Seban RA (1967) The condensation coefficient of water. Int J Heat Mass Transfer 10: 1815–1827.
6. Mortensen EM and Eyring H (1960) Transition coefficients for evaporation and condensation. J Phys Chem 64: 846–849.
7. Ockendon JR and Hodgkins WR (1975) Moving boundary problems in heat flow and diffusion. Oxford: Clarendon.
8. Smith WR (1973) Vapor-liquid condensation in a shock tube. Proc of the 9th int shock tube symp: 785–792.
9. Van Dongen MEH (1978) Thermal diffusion effects in shock tube boundary layers. Dissertation Technische Hogeschool Eindhoven.

Light scattering by bubbles in liquids: Mie theory, physical-optics approximations, and experiments

PHILIP L. MARSTON,* DEAN S. LANGLEY* and DWIGHT L. KINGSBURY**

*Department of Physics, Washington State University, Pullman, WA 99164 USA:
**Department of Electrical Engineering, Washington State University, Pullman, WA 99164 USA

Abstract. Angular structures in the far-field scattering from bubbles are observed and modeled. Mie theory supports a model of diffraction and interference near the critical scattering angle. A new expression for the angular spacing of fine structure is derived. Photographs of scattering show some of the predicted features. Application of these structures to bubble sizing and detection are summarized and the theoretical extinction coefficient in water is plotted.

Mie computations for bubbles in water also reveal backward and forward glory effects. These are partially manifested as cross-polarized scattering. Observed scattering from bubbles in the near backward direction is found to have a strong cross-polarized component.

1. Introduction

Equations for the scattering of plane electromagnetic waves by a dielectric sphere were given by Mie [14] in 1908 and the resulting features of the angular scattering pattern of drops are well known [5, 16, 17]. The Mie solution, though exact, does not give insight into either the scattering process, or changes in the pattern resulting from changes in shape, refractive index, or profile of the incident wave fronts. Reviews of the literature on light scattering [5, 16, 17] reveal a paucity of information about the scattering pattern of bubbles in liquids where the refractive index of the scatterer n_i is less than that of the surroundings n_0. Consequently we have begun [7–10, 12, 13] systematic study of the scattering of light by gas bubbles in liquids. Aspects of the study are: (1) the computation of Mie scattering; (2) the development of simple physical models which give insight; and (3) observations of features in the scattering which differ significantly from both the scattering by drops and the scattering predicted by geometrical optics [3]. In this paper we summarize the main features of the scattering with an emphasis on the critical [7, 10, 13] and back-scatter [12] regions. New experiments and applications will be described. These are the first detailed observations of scattering by bubbles.

Mie's solution [5, 14, 17] to the problem of the scattering efficiency and pattern of a dielectric sphere is a function of the ratio $m = n_i/n_0$. It is usually expressed as a function of $x = ka = 2\pi a/\lambda_0$ where a is the sphere radius,

**$Present address:$* Boeing Commercial Airplane Co, MS 02-15, P.O. Box 3707, Seattle, WA 98124, USA.

Applied Scientific Research 38: 373–383 (1982) 0003–6994/82/0384–0373 $01.65.

k and λ_0 are the wave number and wavelength in the outer dielectric. Optical sources are typically characterized by the wavelength in a vacuum λ_v; we note that $k = 2\pi n_0/\lambda_v$. For an air-filled bubble, $n_i = 1.00029$ and it is usually a good approximation to take $m = 1/n_0$. Most of this paper will deal with scattering by spherical bubbles with plane incident waves. We will gain some insight however, into which features of the scattering pattern should be sensitive to deviations from sphericity.

2. Scattering efficiency

The efficiency factor Q_{sca} is the ratio of the total scattering cross section to the geometric cross section πa^2. When n_i is real valued, Q_{sca} also gives the extinction efficiency [2, 5, 16, 17]. Figure 1 compares Q_{sca} for drops of water in air with $m = 4/3$ with that for a bubble in water with $m = 3/4$. The computations were performed using a slightly modified version [9] of Wiscombe's MIEVO Mie scattering algorithm [18]. A table of Q_{sca} for bubbles exists [19] which is consistent with Figure 1. For a fixed λ_v, and a given value of x, the drop's radius is larger than the bubble's radius by a factor of 4/3. For the case of light from a He-Ne laser, $\lambda_v = 0.6328\ \mu$m; for bubbles in water, x of 10, 100, 1000, and 10 000 give a of 0.75, 7.55, 75.5, and 755 μm, respectively.

The salient feature of Figure 1 is that for drops Q_{sca} exhibits a fine 'ripple' structure [2, 5] but that our calculations of Q_{sca} for bubbles do not reveal a ripple structure. For both drops and bubbles, Q_{sca} has a broad undulation with a quasi-period $\Delta x \simeq \pi/|m-1|$. This approximation, which has been derived from the theory of 'anomalous diffraction' [5, 17], appears to be useful for both bubbles and drops. The ripple structure present for drops is due to optical resonances [2] which are attributed, in part, to internal surface waves. The absence of such structure for bubbles is probably because $m < 1$ does not favor the entrapment of internal surface waves.

3. Critical and Brewster angle scattering

For scattering by drops, diffraction is important for the description of the forward, backward, and rainbow regions [5, 17]. For bubbles, there is no longer a rainbow; however, a new region appears known as the critical scattering region [10]. Diffraction is important in this region because of an abrupt change in the amplitude of the reflected wave as the local angle of incidence θ changes from $\theta < \theta_c$ for small impact parameters to $\theta > \theta_c$ for large ones. Here $\theta_c = \arcsin(m)$ which is the critical angle for a plane surface. Figure 2 illustrates several ray paths which lead to a scattering angle ϕ (the deviation from the forward direction) of 50°. The number on the left specifies the number of internal chords p; θ_p and ρ_p denote the angle of incidence and refraction of the pth ray. The reflected ray (which has

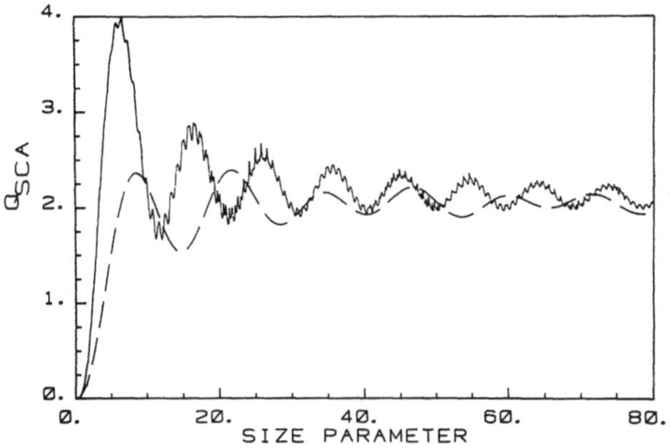

Figure 1. Scattering efficiency from Mie theory as a function of the size parameter x. The solid curve is for drops and the dashed curve is for bubbles.

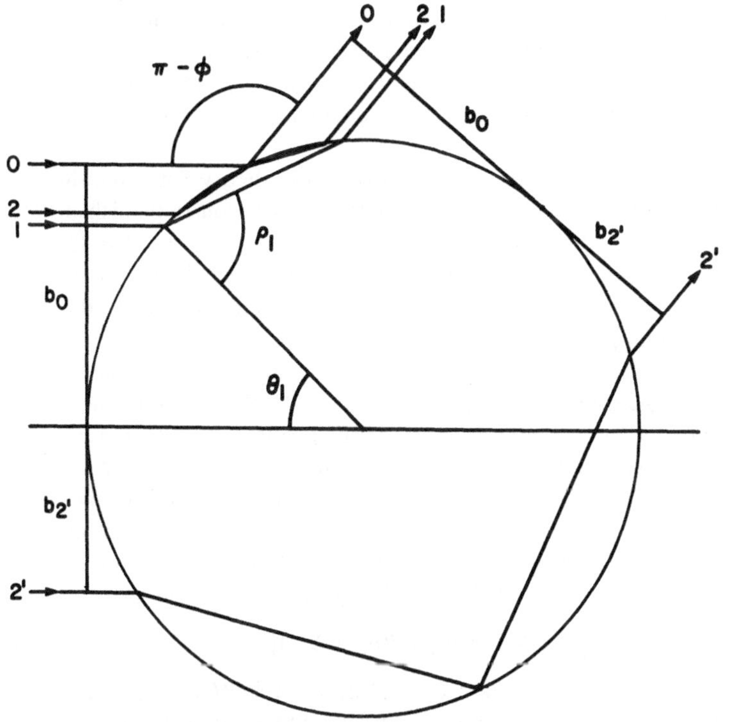

Figure 2. Ray paths for a bubble with $m = 3/4$. The number of internal chords p is given a prime if the ray enters the bubble below the center-line.

$p = 0$) has a scattering angle $\phi = \pi - 2\theta_0$. For an air bubble in water with $m = 3/4$, the critical scattering angle is $\phi_c = \pi - 2\theta_c = 82.82°$. Geometric optics [3, 10, 13] predicts that $|dI_j/d\phi| \to \infty$ as ϕ approaches ϕ_c from above ϕ_c. Here I_j is the normalized scattered intensity defined as follows: the actual j-polarized intensity at a distance $R \gg a$ from the bubble's center is the incident j-polarized intensity multiplied by $I_j(a/R)^2/4$. For the electric vector perpendicular to the scattering plane, $j = 1$; for the parallel case, $j = 2$. This normalization is appropriate for bubbles since geometric optics predicts that if the intensity of the $p = 0$ reflection could be considered by itself I_j $(\phi \leqslant \phi_c) = 1$ due to total reflection.

Our observations [10], model [10, 13], and Mie computations [7–9] demonstrate that instead of a divergence of $|dI_j/d\phi|$, the rise in I_j is spread out over the region $|\phi - \phi_c| \simeq \Omega$ where $\Omega = \arcsin[0.8(1 - m^2)^{-1/4}(\lambda_0/a)^{1/2}] \simeq 144x^{-1/2}$ deg for water. The model makes use of a physical optics approximation which is to (a) use ray optics along with the reflection coefficients of a plane surface to compute the complex amplitudes of virtual waves which simulate the scattering, and (b) use Fraunhofer's approximation to compute the diffraction of the virtual wave to the far field where $R \gg xa$. Step (b) yields the improvements over the geometrical predictions [3]. This procedure is analogous to Airy's model of diffraction near the rainbow [17]. Additional approximations used in the model are: only the two most intense virtual waves for $\phi \simeq \phi_c$ are included ($p = 0$ and 1); and the divergence of the derivative of the reflectivity is simulated by truncating the $p = 0$ reflection when $\theta_0 < \theta_c$.

To compare the model results with Mie theory,* we consider examples of bubbles in water which complement those previously published [7]. Figure 3 shows both theories for $x = 100$ and $j = 1$. It shows that there is a coarse structure in the Mie scattering for $\phi < \phi_c$ which has a quasi-period $\lesssim \Omega \simeq 14.4°$. This structure is described by the model except in the near-forward direction ($\phi \lesssim 20°$).

The Mie result also has a superposed fine structure where the magnitude of the quasi-period $< \lambda_0/a$ rad. $= 360° x^{-1}$. Near ϕ_c, this fine structure arises primarily from the interference of the $p = 0$ wave with the wave due to the $p = 2'$ ray in Figure 2. Its quasi-period Ω_f may be estimated from the lateral separation $b_0 + b_{2'}$ of virtual point sources which would simulate the scattering at a given ϕ. As shown in Figure 2, this separation is the sum of the impact parameters b_p for these rays. Standard relations for the far-field interference applied to these sources gives:

$$\Omega_f(\phi) \simeq \arcsin\,[\lambda_0/(b_0 + b_{2'})] \tag{1}$$

where $b_p/a = \sin\theta_p$, $\theta_0 = (\pi - \phi)/2$, $\phi = \pi + 2(\theta_{2'} - 2\rho_{2'})$ and from Snell's law, $m \sin\rho_p = \sin\theta_p$. Equation (1) is only an approximation because it fails

*The conversion from the Mie amplitudes S_j [5, 17, 18] to the I_j is $I_j = (2|S_j|/x)^2$.

to include other (such as $p = 1$) virtual waves and it does not completely account for the longitudinal spacing of virtual sources and for how those sources vary with ϕ. With $\phi = \phi_c$ and $m = 3/4$, $b_0 + b_{2'} \simeq a/0.825$ and $\Omega_f \simeq \arcsin(5.18x^{-1})$. This gives $2.97°$ at $x = 100$ which is in reasonable agreement with Figure 3. Equation (1) also omits effects of a phase shift in reflection present when $\theta_0 > \theta_c$.

Figure 4 shows the Mie and physical-optics results for $x = 10\,000$ and $j = 1$. The fine structure quasi-period is greatly reduced; equation (1) gives $\Omega_f(\phi_c) \simeq 0.0297°$ which agrees with the Mie result of $0.03°$. The amplitudes of the coarse undulations ($\Omega \simeq 1.4°$) decrease slightly with decreasing ϕ until $\phi \simeq 75°$. A graph of I_1 for $\phi < 75°$ shows coarse undulations increasing in amplitude with decreasing ϕ. The explanation is that for $x = 10\,000$, with

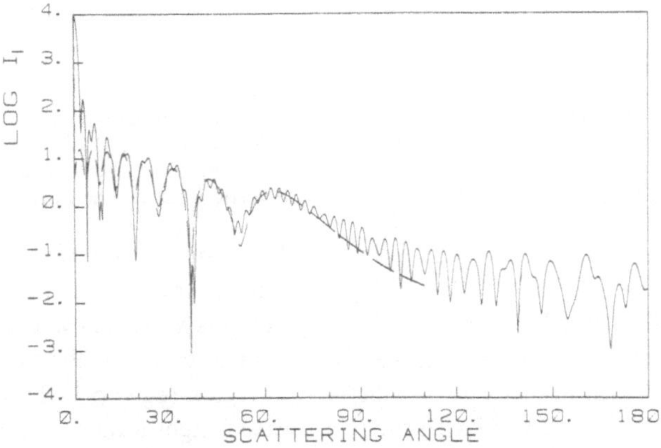

Figure 3. Logarithm (base 10) of the normalized scattered intensity predicted by Mie theory (solid curve) for $x = 100$. The dashed curve is the physical-optics approximation [13] of the coarse structure.

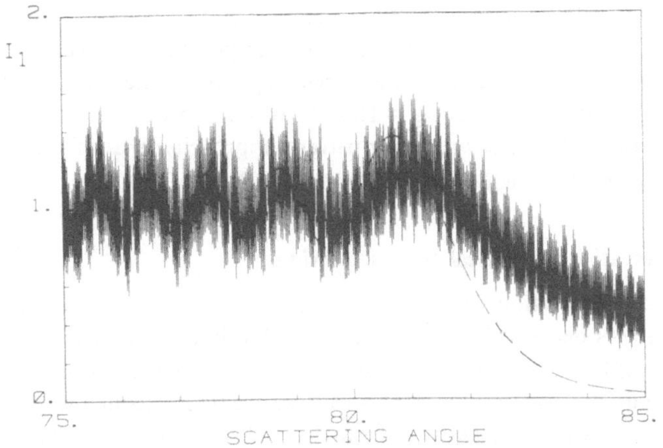

Figure 4. Like Figure 3 but with a linear scale and $x = 10\,000$.

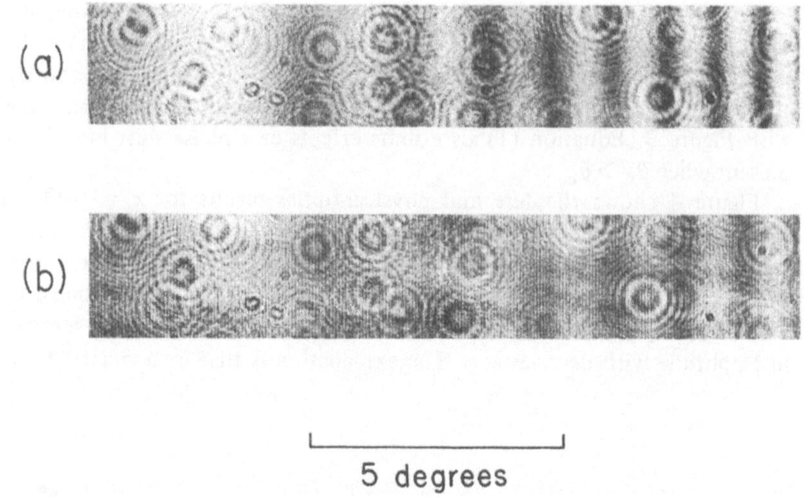

Figure 5. Photographs of far-field scattering for (a) the bubble radius $a \simeq 480\,\mu m$ ($x \simeq 6360$, $\Omega \simeq 1.8°$) and $j = 2$; (b) $a \simeq 330\,\mu m$ ($x \simeq 4370$, $\Omega \simeq 2.2°$) and $j = 1$. The critical scattering angle ϕ_c is near the left edge of the photograph; ϕ decreases from left to right. Coarse structure is manifest as broad vertical bands. The rings are artifacts.

$\phi \gtrsim 75°$ the coarse structure is primarily due to diffraction of the $p = 0$ wave [10] but for $\phi \lesssim 75°$ it is primarily from the interference of the $p = 0$ wave with $p > 0$ waves [13]. Plots [7] of I_2 for this x also show this transition and that fine structure is significantly weaker when $j = 2$. Figure 3 does not show the transition because the diffraction region ($\phi_c - \phi \lesssim \Omega$) overlaps the region where interference with the $p = 1$ wave is significant.

As θ_0 approaches the Brewster angle, $\theta_B = \arctan(m)$, the reflectivity of the $p = 0$, $j = 2$ polarized ray vanishes [17]. Consequently, according to geometric optics, there is no contribution to I_2 from this ray at the Brewster scattering angle $\phi_B = \pi - 2\theta_B = 2\arctan(m^{-1})$. Because the reflectivity varies slowly near ϕ_B, diffraction is less important than near ϕ_c. A coarse minimum is evident in the Mie I_2 near ϕ_B for bubbles in glass [8] and for bubbles in water where $\phi_B \simeq 106.3°$. When x drops below 5, its location shifts toward 90°; the scattering pattern approaches that of a dipole radiator predicted by Rayleigh scattering theory [5, 8, 17].

4. Observations of critical angle scattering

Previous observations [10] were limited to the ϕ region where the coarse undulations decreased with decreasing ϕ. We have a new apparatus similar to that in [10] except that it permits observations with ϕ down to 70°. As in [10], bubbles were attached to a vertical needle in distilled water. They were illuminated with a plane wave from a He-Ne laser with $\lambda_v = 632.8$ nm. They were photographed via a window with a camera focused on infinity to place the film in the far field. Because of a technical problem, we have not obtained precise direct measurements of a in the (horizontal) scattering plane along with the photographs, however, we have demonstrated that $\Omega_f(\phi_c)$ from equation (1) gives an a which is consistent with the observed and modeled coarse structure. Thus the a values quoted in Figure 5 were determined from the coarse and fine structures and not from direct observations. Figure 5 serves more to illustrate phenomena described in Section 3 than to rigorously test the models with real bubbles. The angle scale is for a ϕ change in water. The horizontal displacement is not exactly linear in ϕ due to refractive corrections at the water-window-air interfaces. In Figure 5(a), the broad bright region on the left is the first coarse maximum with $\phi < \phi_c$. On the right, the coarse undulations increase in amplitude as ϕ decreases, which is the modeled behavior. Figure 5(b) is for a smaller bubble (verified with direct observations). The coarse undulations are spread out and the fine structure is clearly visible with $\Omega_f \simeq 0.07°$.

5. Backward and forward glory

Backscattering from drops is known to be enhanced by the axial focusing of certain rays [5, 17]. A complete description of this 'glory' for water drops is complicated due to the necessity of including surface waves [6]. We have observed and modeled backscattering from air bubbles [12] and find that some aspects can be explained with a physical-optics approximation that does not include surface waves. For the following discussion, it is convenient to define $\gamma = \pi - \phi$; $\gamma = 0$ corresponds to exact backscattering.

Compound Mie intensities show that backscattering from bubbles may be significantly larger than that expected from reflection of the $p = 0$ ray. For $\gamma = 0$, the plane of scattering is no longer defined and $I_1 = I_2 = I$. Figure 6 shows several broad peaks in I which are significantly larger than the normalized intensity of the $p = 0$ reflection which, by itself, is $(m-1)^2/(m+1)^2 = 0.02$. Other axial (e.g. $p = 2$) rays are too weak to explain the magnitude of I.

For polarized incident light, scattering near $\gamma = 0$ may be described in part with (non-cross) polarized $I^{(1)}$ and cross-polarized $I^{(2)}$ normalized intensities. For $I^{(2)}$, the plane of polarization is roated by 90° from incident polarization; symmetry gives $I^{(2)}(\gamma = 0) = 0$ while $I^{(1)}(\gamma = 0) = I_j(\gamma = 0)$. For $\gamma \neq 0$, the $I^{(\ell)}$ depend on both γ and the angle ξ which the scattering plane makes with the incident electric vector.* Figure 7(a) illustrates the computed scattering. Geometric models of $I^{(1)}(\gamma = 0)$ predict that the contributions of off-axis ($p = 3$ and 4) rays diverge as $\gamma \to 0$ because of a factor which accounts for focusing. This divergence is present in previous ray models [3],

*The conversion from the complex Mie amplitudes S_j to the $I^{(\ell)}(\gamma \ll 10°, \xi)$ are $I^{(1)} = |S^- - S^+ \cos 2\xi|^2/x^2$ and $I^{(2)} = |S^+ \sin 2\xi|^2/x^2$ where $S^\pm = S_1 \pm S_2$. Evidently the ξ dependence of equation (2) is of the correct type.

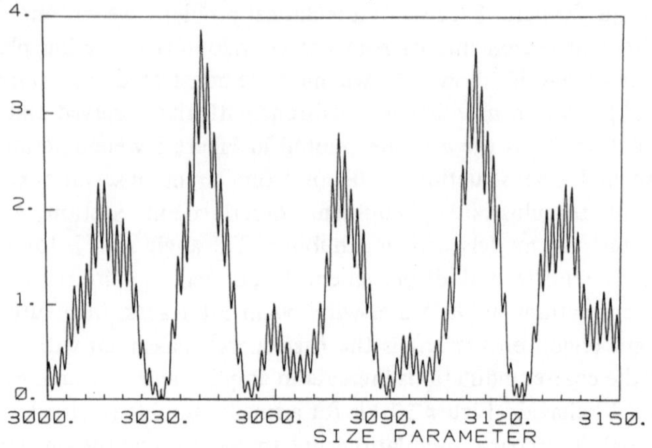

Figure 6. Normalized Mie back scattering from a spherical air bubble in water ($m = 3/4$) plotted as a function of the size parameter x; for green light, $a \simeq 200\ \mu m$.

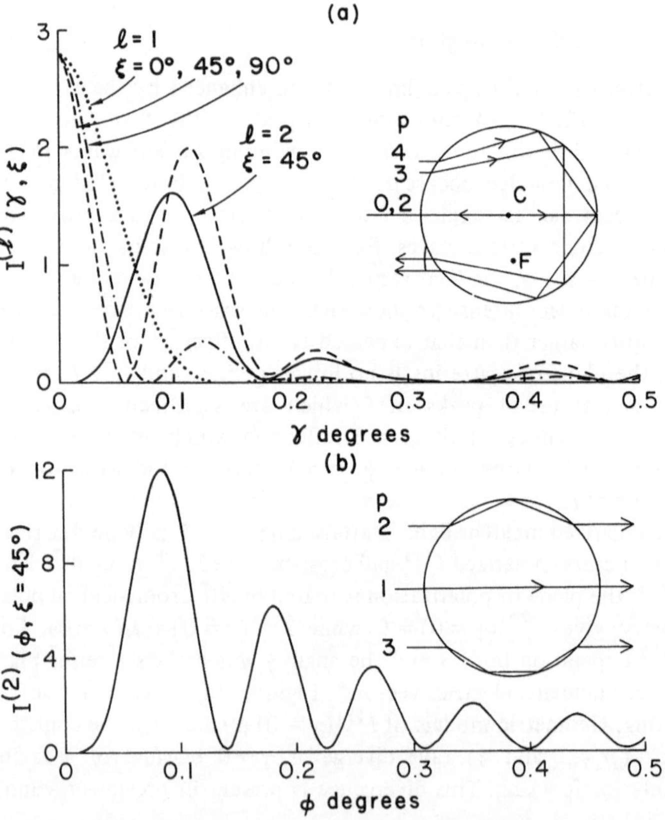

Figure 7. Normalized Mie near-back scattering (a) the near-forward scattering (b) for $m = 3/4$ and $x = 3040$. For clarity, the computed $I^{(2)}$ was multiplied by two in (a) before plotting. The inserts illustrate some of the significant rays.

but was not discussed; it is also present for drops [17] for certain m. Because of this divergence, diffraction provides an essential correction to ray optics.

The scattering due to the $p = 3$ ray in Figure 7(a) (insert) is a back-facing toroidal wave front which appears to originate at a virtual ring-like source known as the focal circle in the analogous [17] $p = 2$ scattering from drops. This source at point F is ring-like because the figure may be rotated around the optic axis through the center C. For large x, the stationary-phase approximation of far-field diffraction integrals gives the following proportionalities for off-axis ($p > 2$) contributions to the $I^{(\ell)}$ taken separately:

$$I^{(1)}_{(p)} \propto x\,[(c_1 + c_2)J_0(u) + (c_1 - c_2)J_2(u)\cos 2\xi]^2\,, \tag{2a}$$

$$I^{(2)}_{(p)} \propto x\,[(c_1 - c_2)J_2(u)\sin 2\xi]^2\,, \tag{2b}$$

where the c_j are combined Fresnel reflectivities and transmissivities of the j-polarized fields, $u = kb'_p \sin\gamma$, and b'_p is the impact parameter of the exactly backscattered pth ray. Because $I^{(\ell)}_{(p)} \propto x$, the unnormalized intensity is proportional to ka^3 while that for the $p = 0$ ray is proportional to a^2. Consequently, off-axis waves are dominant when a is sufficiently large. It is evident from Figures 6 and 7(a) that the modeling of $I^{(\ell)}$ may require summation of electric fields from several virtual sources. (In physical optics, intensities do not add but fields do.) Interference of various ring-like and axial virtual sources depends on x and could lead to the broad structures in Figure 6. Detailed computation of $I^{(1)}_{(p)}$ ($\gamma = 0$) give magnitudes sufficient to explain the Mie scattering. Due to the Bessel functions in equation (2), the actual scattering should be peaked at or near $\gamma = 0$.

We have observed backscattering from single air bubbles in a dimethylsiloxane-polymer liquid which were nearly immobilized by viscosity. In these experiments, $m^{-1} = 1.403$, $\lambda_v = 632.8$ nm, $a \simeq 0.3$–0.8 mm and $x \simeq 4000$–$11\,000$. The far-field cross-polarized intensity had a dependence on γ and ξ similar to that predicted by equation (2b) with $p = 3$. Equation (2b) predicts that when both $\sin\gamma \simeq \gamma$ and $u \gg 1$, the minima in $I^{(2)}_{(3)}$ should be spaced by $\Delta\gamma \simeq \pi(kb'_3)^{-1}$ radians where $b'_3 = 0.447a$. The prediction correctly described the observed $\ell = 2$ scattering which is apparently dominated here by the $p = 3$ virtual source. The focal circle was viewed by focusing the camera on the bubble. The $\ell = 1$ scattering was not dominated by a single class of rays. Cross-polarized scattering from a polydispersion of bubbles in water has also been seen.

Figure 7(b) shows an enhancement of near-forward cross-polarized scattering due to axial focusing. Focal circles due to the $p = 2$ and 3 rays largely contribute to the $\ell = 2$ scattering. The forward $\ell = 1$ scattering is dominated by ordinary diffraction [17] when x is large. Mie theory gives $I^{(1)} \simeq 10^7$ when $\phi = 0$ for the x and m of Figure 7. Forward optical glory has also been displayed in Mie scattering by water drops [15]. Its description for drops is complicated by optical surface waves.

5. Discussion

Figures 3–7 demonstrate that for several angular regions, the intensity exceeds that of geometric scattering from a perfectly reflecting sphere of the same a (for which $I_j = 1$). This information should be useful in optical devices which size or detect bubbles [1, 4, 7]. Due to diffraction near ϕ_c, it is preferable to detect scattering with $\phi \lesssim \phi_c - \Omega$ than to observe it with $\phi \simeq 90^\circ$ which is the usual practice [4] for bubbles in water. For an unpolarized source, the normalized total scattering is $(I_1 + I_2)/2$.

Though this paper has emphasized the far-field ϕ dependence, the results are also applicable to the imaging of bubbles. For example, if it desired to resolve the virtual sources of the $p = 0$ and 1 rays in Figure 2, the aperture of the imaging system should have an angular width $\gtrsim \Omega$. It is easier to resolve the $p = 0$ and $2'$ virtual sources since the angular width requirements are reduced to Ω_f. Resolution of identifiable virtual sources reveals the size of a bubble.

Fine structure in the far-field scattering, such as that shown in Figure 5(b), arises from the interference of widely spaced rays. Consequently the positions of the maxima will be quite sensitive to changes in the bubble's shape. That part of the coarse structure due to the interference of $p = 0$ and 1 rays will also depend on the shape but more weakly. As is the case for drops [11], these shape dependences may be useful for detecting mechanical resonances.

Scattered intensities for bubbles in water purported to be from Mie theory have been used in a study of cavitation nuclei [4]. Comparison of our Mie results and model of the coarse structure (which are consistent for the a in question) with those in [4], show that the latter err significantly. Plots in [4] predict extra coarse maxima at $\phi = 100^\circ$ for $a = 5\,\mu$m and $\phi = 120^\circ$ for $a = 7.5\,\mu$m which are not present in our Mie computations. We cannot find any physical justification for coarse maxima at these ϕ for x near 100.

Acknowledgments

We are grateful to W.J. Wiscombe for providing the initial computer program from which the program used here was derived. This research was supported in part by the Washington State University Research and Arts Committee and by the Office of Naval Research. P.L. Marston is an Alfred P. Sloan Research Fellow.

References

1. Ben-Yosef N, Ginio O, Mahlab D and Weitz A (1975) Bubble size distribution measurement by Doppler velocimeter. J Appl Phys 46: 738–740.
2. Chýlek P, Kiehl JT, Ko MK and Ashkin A (1980) Surface waves in light scattering by spherical and non-spherical particles. In Scheuerman D (ed) Light scattering by irregularly shaped particles, p 153. New York: Plenum.

3. Davis GE (1955) Scattering by an air bubble in water. J Opt Soc Am 45: 572–581.
4. Keller A (1972) The influence of the cavitation nucleus spectrum on cavitation inception. J Basic Eng 94: 917–925.
5. Kerker M (1969) The scattering of light. New York: Academic.
6. Khare V and Nussenzveig HM (1977) Theory of the glory. Phys Rev Lett 38: 1279–1282.
7. Kingsbury DL and Marston PL (1981) Mie scattering near the critical angle of bubbles in water. J Opt Soc Am 71: 358–361.
8. Kingsbury DL and Marston PL (1981) Scattering by bubbles in glass: Mie theory and physical optics approximation. Appl Opt 20: 2348–2350.
9. Kingsbury DL (1981) MS thesis, Washington State University, Pullman.
10. Marston PL (1979) Critical angle scattering by a bubble: physical-optics approximation and observations. J Opt Soc Am 69: 1205–1211; (1980) Erratum 70: 353.
11. Marston PL (1980) Rainbow phenomena and the detection of nonsphericity in drops. Appl Opt 19: 680–684.
12. Marston PL and Langley DS (1980) Glory and depolarization in backscattering from air bubbles. J Opt Soc Am 70: 1607; Langley DS and Marston PL (1981) Glory in optical backscattering from air bubbles. Phys Rev Lett 47: 913–916.
13. Marston PL and Kingsbury DL (1981) Scattering by an air bubble in water near the critical angle: interference effects. J Opt Soc Am 71: 192–196; (1981) Erratum 71: 917.
14. Mie G (1908) Beiträge zur optik truber medien, speziell kolloidaler metallösungen. Ann Phys (Leipzig) 25: 377–445.
15. Nussenzveig HM and Wiscombe WJ (1980) Forward optical glory. Opt Lett 5: 455–457.
16. Shipley ST and Weinman JA (1978) A numerical study of scattering by large dielectric spheres. J Opt Soc Am 68: 130–134.
17. Van de Hulst HC (1957) Light scattering by small particles. New York: Wiley.
18. Wiscombe WJ (1980) Improved Mie scattering algorithms. Appl Opt 19: 1505–1509.
19. Zelmanovich IL and Shifrin KS (1968) Tables of light scattering part III. Leningrad: Hydrometeorological Publishing House.

ANNOUNCEMENTS

Institute of Environmental Sciences
to present
Three Tutorial Lecture Series

On Tuesday, April 20, 1982 at the Marriott Hotel (downtown), Atlanta, GA, the Institute of Environmental Sciences will present three concurrent Tutorial Lectures Series. The tutorials are dedicated to the concept of keeping the IES members and their colleagues aware of continually changing technological developments.

The tutorials will cover the following subjects:

- Contamination control technology
  Dr. Irving Stowers, Lawrence Livermore National Laboratories – instructor

- Fundamentals of aerosol science and engineering
  Prof. Benjamin Liu, University of Minnesota – instructor

- Basics of acoustic emission
  Satirious J. Vahaviolos, James Mitchell, Physical Acoustics Corp. – instructors

The enrollment for these lecture series is limited. Pre-registration is required. For further information and program please contact:

Betty L. Peterson
Executive Director
Institute of Environmental Sciences
940 E. Northwest Highway
Mt. Prospect, IL 60056, USA
Tel. 312/255/1561

12th Space Simulation Conference co-sponsored by Institute of Environmental Sciences, Jet Propulsion Laboratory, National Aeronautics and Space Administration, AIAA and ASTM

The 12th Space Simulation Conference will be hosted by the Institute of Environmental Sciences (IES) and supported by the American Institute of Aeronautics and Astronautics (AIAA), American Society for Testing and Materials (ASTM), the National Aeronautics and Space Administration, and Jet Propulsion Laboratory.

The theme of the conference is *'Shuttle plus one, a new view of space'*. The purpose of the conference is to provide a forum for the review and exchange of information and ideas on current space simulation technology and closely related disciplines as well as projections for testing requirements and technology development for the 1980s.

Subject areas are:

- Space simulation facilities
- Spacecraft testing
- Thermal protection
- Space program trends
- Unique facilities
- Remote sensing
- Facility management issues
- Life sciences
- Space physics
- Vacuum/cryogenics
- Contamination experiments
- Contamination equipment
- Thermal simulation
- Shuttle flight results

George F. Wright, Jr., Sandia National Laboratories, is General Chairman; *John W. Harrell*, Jet Propulsion Laboratory is the Technical Program Chairman; *Russell T. Hollingsworth*, NASA Goddard Space Flight Center, is the Publications Chairman and *Robert Daniel*, Jet Propulsion Laboratory, is Facilities Chairman. *John D. Campbell*, Perkin-Elmer Corp. will serve as the IES Meeting Manager.

The meeting will be held at the Pasadena Hilton Hotel, Pasadena, CA, May 17–19, 1982. For further information contact:

Betty L. Peterson
Executive Director
Institute of Environmental Sciences
940 E. Northwest Highway
Mt. Prospect, IL 60056, USA
Tel. 312/255/1561

Enhancement of Quality through Environmental Technology
'The Application of Environmental Technology Through the Production Cycle'
28th Annual Technical Meeting and Equipment Exposition

The Institute of Environmental Sciences will hold its 28th Annual Technical Meeting and Equipment Exposition at the Marriott Hotel (downtown), Atlanta, GA, April 20–23, 1982.

Dr. Harry Plumblee, Chief Scientist and Director of Lockheed International Research Institute, will give the Keynote Address on Wednesday, April 21, 1982. Dr. Plumblee is internationally recognized in the field of acoustics. He received his doctorate from South Hampton in England. He travels the world discussing aircraft performance with other countries and will present some of these findings in his address.

Dr. Noah Langdale, President of Georgia State University, will give the address on Thursday, April 22, 1982 at the IES Annual Awards Banquet.

The seminar format of this meeting will include four seminars conducted concurrently for the full three days:

Seminar I – Engineering Methods

'Getting our act together'
'Cost effective testing'
'Successful test tailoring'
'Combined environment testing'
'Current aspects of dynamic testing'
'Impact of computer technology on environment test operations'
'Climatic environment simulation'
'Simulation of low frequency vibration environments'
'Environmental standards'
'IES recommended practices'

Seminar II – Environmental Stress Impact

'Getting our act together'
'Electrostatic discharge'
'Successful test tailoring'
'Combined environment testing'
'The ultimate test'
'Reliability growth'
'Reliability modelling'
'Reliability in production cycle'
'Environmental stress screening of electronic hardware'
'IES recommended practices'

Seminar III – Contamination Control

'Getting our act together'
'Education and training in contamination control'
'Clean room equipment and design'
'Effects of contaminants – case histories'
'Biological hazards'
'Filtration of liquids'
'Aerosol filtration'
'Particle contamination identification and analysis'
'Dealing and measuring cleanliness'
'Standards and practices – reports from committees'
'Clean room garments, gloves and wipes'
'Precision cleaning of surfaces'

Seminar IV – Energy and the Environment

'Getting our act together'
'Alternate energy sources'
'Air quality impact'
'Precipitation chemistry'
'Water quality impact'
'Recycling and resource recovery'
'Effects assessment'
'Hazardous waste management'

Joseph J. Popolo, Grumman Aerospace Corp., is General Chairman, with *Harvey Golinger,* Arkwin Industries, serving as Vice Chairman. *Edward A. Szymkowiak,* Westinghouse Electric Corp., is the Technical Program Chairman; *Dr. Irving Stowers,* Lawrence Livermore National Laboratory, is Technical Program Co-Chairman; *Ludwig Pulaski,* IBM Corp., and *Charles F. Conrad* are Exhibits Co-Chairmen; *John Breen,* RCA Corp., will serve as Facilities Chairman; *Joseph L. Stecher,* NASA/GSFC, is the Registration Chairman.

For further information and a copy of the complete program please contact:

Betty L. Peterson
Executive Director
Institute of Environmental Sciences
940 E. Northwest Highway
Mt. Prospect, IL 60056, USA
Tel. 312/255/1561

Designing Electronic Equipment for Random Vibration Environments Seminar

The Institute of Environmental Sciences will hold a seminar on '*Designing Electronic Equipment for Random Vibration Environments*' on March 25–26, 1982 at the Marriott Hotel, Los Angeles, California.

David S. Steinberg, Litton Guidance and Control Systems, will be the General Chairman and Technical Program Chairman.

The subjects to be covered are:

- Speaking in a random fashion
- Random vibration basics for electronics packaging design
- Response spectrum analysis for random vibration
- Design guides for random vibration
- Non-design for random vibration
- Guidelines for random vibration design of electronic equipment
- Electronic housing design for random vibration environment
- Design for long fatigue life in random vibration equipment
- Vibration-fatigue reliability analysis
- Electronics packaging design for random vibration
- Analysis and test of ceramic substrates for packaging of leadless chip carriers
- Computer-aided interactive structural optimization of printed circuit-board design
- A survey of F. E. computer programs for random vibration analysis
- Random vibration effects on piecepart application
- Hinge damage in a random vibration environment
- Vibration isolation system development for the F-111 tail POD electronics
- Use of cathode ray tubes in a random vibration environment
- Power supplies designed for random vibration
- Wear of connector contact exposed to relative motion
- Test data on leadless chip carriers with ceramics substrates in severe random vibration environment
- Component lead wire strain relief for random vibration environment
- Computing the fatigue life of component lead wires in severe random vibration environments
- A random vibration test for the evaluation of stiff sensitive component parts

Registration fees are: IES members – $ 100.00 – includes copy of preprinted *Proceedings* available at the seminar. Non-members – $ 140.00. For further information contact:

Betty L. Peterson, Executive Director,
Institute of Environmental Sciences
940 E. Northwest Highway, Mt. Prospect, IL 60056, USA
Tel. 312/255/1561

Index of authors and papers